Advances in Boundary Element Methods for Fracture Mechanics

International Series on Computational Engineering

Aims:
Computational Engineering has grown in power and diversity in recent years, and for the engineering community the advances are matched by their wider accessibility through modern workstations.

The aim of this series is to provide a clear account of computational methods in engineering analysis and design, dealing with both established methods as well as those currently in a state of rapid development.

The series will cover books on the state-of-the-art development in computational engineering and as such will comprise several volumes every year covering the latest developments in the application of the methods to different engineering topics. Each volume will consist of authored work or edited volumes of several chapters written by the leading researchers in the field. The aim will be to provide the fundamental concepts of advances in computational methods as well as outlining the algorithms required to implement the techniques in practical engineering analysis.

The scope of the series covers almost the entire spectrum of engineering analysis. As such, it will cover Stress Analysis, Inelastic Problems, Contact Problems, Fracture Mechanics, Optimization and Design Sensitivity Analysis, Plate and Shell Analysis, Composite Materials, Probabilistic Mechanics, Fluid Mechanics, Groundwater Flow, Hydraulics, Heat Transfer, Geomechanics, Soil Mechanics, Wave Propagation, Acoustics, Electromagnetics, Electrical Problems, Bioengineering, Knowledge Based Systems and Environmental Modelling.

Professor J. Dominguez
Escuela Superior de Ingenieros Industriales
Av. Reina Mercedes
41012 Sevilla
Spain

Professor G.S. Gipson
School of Civil Engineering
Engineering South 207
Oklahoma State University
Stillwater, OK 74078-0327
USA

Professor S. Grilli
The University of Rhode Island
Department of Ocean Engineering
Kingston, RI 02881-0814
USA

Professor D.B. Ingham
Department of Applied Mathematical Studies
School of Mathematics
The University of Leeds
Leeds LS2 9JT
UK

Professor P. Molinaro
Ente Nazionale per l'Energia Elettrica
Direzione Degli Studi e Ricerche
Centro di Ricerca Idraulica e Strutturale
Via Ornato 90/14
20162 Milano
Italy

Professor Dr. K. Onishi
Department of Mathematics II
Science University of Tokyo
Wakamiya-cho 26
Shinjuku-ku
Tokyo 162
Japan

Professor H. Pina
Instituto Superior Tecnico
Av. Rovisco Pais
1096 Lisboa Codex
Portugal

Dr. A.P.S. Selvadurai
Department of Civil Engineering
Room 277, C.J. Mackenzie Building
Carleton University
Ottawa
Canada K1S 5B6

Professor A. Giorgini
Purdue University
School of Civil Engineering
West Lafayette, IN 47907
USA

Professor W.G. Gray
Department of Civil Engineering and
Geological Sciences
University of Notre Dame
Notre Dame, IN 46556
USA

Dr. S. Hernandez
Department of Mechanical Engineering
University of Zaragoza
Maria de Luna
50015 Zaragoza
Spain

Professor G.D. Manolis
Aristotle University of Thessaloniki
School of Engineering
Department of Civil Engineering
GR-54006, Thessaloniki
Greece

Dr. A.J. Nowak
Silesian Technical University
Institute of Thermal Technology
44-101 Gliwice
Konarskiego 22
Poland

Professor P. Parreira
Departamento de Engenharia Civil
Avenida Rovisco Pais
1096 Lisboa Codex
Portugal

Professor D.P. Rooke
DRA (Aerospace Division)
Materials and Structures Department
R50 Building
RAE Farnborough
Hampshire GU14 GTD
UK

Professor R.P. Shaw
S.U.N.Y. at Buffalo
Department of Civil Engineering
School of Engineering and Applied Sciences
212 Ketter Hall
Buffalo, New York 14260
USA

Professor P. Skerget
University of Maribor
Faculty of Technical Sciences
YU-62000 Maribor
Smetanova 17
P.O. Box 224
Slovenia

Professor M.D. Trifunac
Department of Civil Engineering, KAP 216D
University of Southern California
Los Angeles, CA 90089-2531
USA

Dr P.P. Strona
Centro Ricerche Fiat S.C.p.A.
Strada Torino, 50
10043 Orbassano (TO)
Italy

Professor N.G. Zamani
University of Windsor
Department of Mathematics and Statistics
401 Sunset
Windsor
Ontario
Canada N9B 3P4

Acknowledgement is made to A. Portela *et al.* for the use of Figure 18 on page 39, which appears on the front cover of this book.

Advances in Boundary Element Methods for Fracture Mechanics

Editors:
M.H. Aliabadi and C.A. Brebbia

Computational Mechanics Publications
Southampton Boston

Co-published with

Elsevier Applied Science
London New York

M.H. Aliabadi
Wessex Institute of Technology
University of Portsmouth
Ashurst Lodge, Ashurst
Southampton SO4 2AA
UK

C.A. Brebbia
Wessex Institute of Technology
University of Portsmouth
Ashurst Lodge, Ashurst
Southampton SO4 2AA
UK

Co-published by

Computational Mechanics Publications
Ashurst Lodge, Ashurst, Southampton, UK

Computational Mechanics Publications Ltd
Sole Distributor in the USA and Canada:

Computational Mechanics Inc
25 Bridge Street, Billerica, MA 01821, USA

and

Elsevier Science Publishers Ltd
Crown House, Linton Road, Barking, Essex IG11 8JU, UK

British Library Cataloguing-in-Publication Data

A Catalogue record for this book is available
from the British Library

ISBN 1-85166-860-8 Elsevier Applied Science, London, New York
ISBN 1-85312-102-9 Computational Mechanics Publications, Southampton
ISBN 1-945824-85-8 Computational Mechanics Publications, Boston

Library of Congress Catalog Card Number 92-70559

Printed and bound in Great Britain by Bookcraft Ltd

CONTENTS

PREFACE

The boundary element method (BEM) has emerged over the past few years as the most powerful numerical technique for the solution of linear elastic crack problems in fracture mechanics. While much progress has been made, there are still many new frontiers to be explored with this method.

The aim of this book is to present the state-of-the-art in applications of the boundary element method to crack problems. As such, it includes chapters written by some of the leading researchers in the field describing the recent advances of the method, as well as presenting new ideas for further development.

It is well-known that the straightforward application of BEM to crack problems leads to a mathematical degeneration in the numerical formulation, if the two crack surfaces are co-planar. The first two chapters of the book deal with methods of removing this difficulty. In Chapter 1, a dual boundary element formulation is presented where the displacement integral equation and the traction integral equation are used independently on crack surfaces. The application of the dual boundary element method to crack growth analysis is also presented utilizing an incremental crack extension technique. The stress intensity factors are evaluated using a mixed mode J-integral method. In Chapter 2, a boundary element displacement discontinuity method is presented for elastodynamic problems. Formulations for bodies containing interior, surface breaking and near-surface cracks are presented. Also discussed are problems associated with interface cracks for which multi-domain analysis is used.

In Chapter 3 the application of BEM to analyse delamination of composite laminates is described. The numerical treatment of delaminated structures is carried out by using special crack-tip elements and the energy method. In Chapter 4, an indirect boundary element formulation known as the body force method is described. Applications of the body force method to crack problems in two-dimensional, axisymmetric and three-dimensional geometries are presented; a number of examples are given to demonstrate the accuracy of the method. Finally, the extension of the body force method to elastoplastic and elastodynamic problems is discussed.

The remaining chapters of the book concentrate on the application of BEM to three-dimensional crack problems. Chapter 5 describes different crack-front elements, including the quarter-point element which can be used to model the displacement and stress fields in the vicinity of the crack front. A detailed discussion on the numerical treatment of these crack-front elements is given with particular emphasis on special integration techniques required to accurately implement them in BEM. Chapter 6 presents an application of BEM to the analysis of three-dimensional stress intensity factor weight functions. A weight function formulation based on the notion of fundamental fields is derived. The fundamental displacements and stress fields for rectilinear and penny-shaped crack fronts are given. A regularization procedure called subtraction of fundamental fields is developed to remove highly singular stress and displacement fields along the crack front, reducing the singularity of the problem to the usual order as in the case of traction free cracks. Special crack-front elements are subsequently used to model the remaining fields. Stress intensity factors for a number of three-dimensional crack geometries including circular, elliptical, quarter-circular and quarter-elliptical cracks in finite geometries are presented.

The final chapter describes the application of the boundary element method to anisotropic crack problems. A detailed discussion on the numerical implementation of the anisotropic fundamental solution is described. A number of test examples, including centre crack in anisotropic plate, semi-circular surface crack in transversely isotropic square rod and circular crack in anisotropic cylindrical bar are presented.

We are grateful to all the authors for their contribution to this book. Special thanks are due to Mrs Juliet Booker for her outstanding work in preparation of the final manuscript.

M.H. Aliabadi
C.A. Brebbia

Chapter 1

Dual boundary element analysis of fatigue crack growth

A. Portela,[a] M.H. Aliabadi[a] & D.P. Rooke[b]

[a] *Wessex Institute of Technology/University of Portsmouth, Ashurst Lodge, Ashurst, Southampton SO4 2AA, UK*

[b] *Defence Research Agency, Aerospace Division, RAE Farnborough, Hants GU14 7TU, UK*

Abstract

The dual boundary element method is formulated for the two-dimensional linear elastic analysis of fatigue crack-growth problems. Crack growth is simulated with an incremental crack-extension analysis based on the maximum principal stress criterion. For each increment of the crack extension, the dual boundary element method is used to perform a stress analysis of the structure. The dual equations considered are the displacement and the traction boundary integral equations. When the dual boundary integral equations are simultaneously applied along the crack boundaries, general crack problems can be solved in a single-region formulation, with both crack surfaces discretized with discontinuous boundary elements. If the crack extension is modelled with new boundary elements, remeshing is not required because of the single-region analysis, an intrinsic feature of the dual boundary element method. The technique used for the evaluation of the stress intensity factors was selected between the singularity subtraction and the J-integral techniques. The singularity subtraction technique is a regularization procedure that uses a singular particular solution of the crack problem to introduce the stress intensity factors as additional problem unknowns. Its application to general crack problems requires the use of artificial boundaries. On the other hand, the J-integral is an efficient post-processing technique based on a path-independent contour integral that can be used to solve general crack problems. The decomposition technique, used to decouple the stress intensity factors in the analysis of mixed-mode problems with the J-integral, is implemented by considering a circular contour path around each crack tip. Several examples of cracked plates are analyzed with the singularity subtraction and J-integral techniques. The efficiency of the J-integral technique make it ideal for the study of crack growth problems under mixed-mode conditions with the dual boundary element method. Some examples of fatigue crack growth are presented, in the context of damage tolerance analysis.

1 Introduction

Catastrophic fracture failure of engineering structures is caused by cracks that extend beyond a safe size. Cracks, present to some extent in all structures, either as a result of manufacturing fabrication defects or localized damage in service, may grow by mechanisms such as fatigue, stress-corrosion or creep. The crack growth leads to a decrease in the structural strength. Thus, when the service loading cannot be sustained by the current residual strength, fracture occurs leading to the failure of the structure. Fracture, the final catastrophic event that takes place very rapidly, is preceded by crack growth which develops slowly during normal service conditions, mainly by fatigue due to cyclic loading.

Damage tolerance assessment is a procedure that defines whether a crack can be sustained safely during the projected service life of the structure. Damage tolerance assessment is, therefore, required as a basis for any fracture control plan, generating the following information, upon which fracture control decisions can be made:

- The effect of cracks on the structural residual strength, leading to the evaluation of their maximum permissible size.
- The crack growth as a function of time, leading to the evaluation of the life of the cracks to reach their maximum permissible size, from which the safe operational life of the structure is defined.

Linear elastic fracture mechanics can be used in damage tolerance analyses to describe the behaviour of cracks. The fundamental postulate of linear elastic fracture mechanics is that the crack behaviour is determined solely by the values of the stress intensity factors which are a function of the applied load and the geometry of the cracked structure. The stress intensity factors, thus play a fundamental role in linear elastic fracture mechanics applications.

Crack-growth processes are simulated with an incremental crack-extension analysis. For each increment of the crack extension, a stress analysis is carried out and the stress intensity factors are evaluated. The crack path, predicted on an incremental basis, is computed by a criterion defined in terms of the stress intensity factors.

In general, numerical methods must be used for the evaluation of the stress intensity factors in engineering structures, because of their complex shape and the continuously changing path of the growing cracks. The finite element method has a long and well documented history in fracture mechanics applications, see Gallagher[1] and Rice & Tracey.[2] More recently it has been applied to study crack-growth processes, see Refs 3-I7. An intrinsic feature of the finite element method, common to all these formulations, is the need for continuous remeshing to follow the crack extension; this is a practical disadvantage of the method.

The boundary element method is a well established numerical technique in fracture mechanics, see Aliabadi & Rooke,[8] which has also been applied in incremental analysis of crack extension problems by Ingraffea, Blandford & Ligget.[9] The boundary element method which is based on the displacement boundary integral equation, has been successfully applied to linear elastic problems in domains containing no degenerated geometries, see Brebbia & Dominguez.[10] These degeneracies, either internal or edge surfaces that include no area or volume and across which the displacement field is discontinuous, are defined as mathematical cracks. For symmetric crack problems only one side of the crack need be modelled and a single-region boundary element analysis may be used. However, the solution of general crack problems cannot be achieved in a single-region analysis with the direct application of the boundary element method, because the coincidence of the crack boundaries gives rise to a singular system of algebraic equations. The equations for a point located at one of the boundaries of the crack are identical to those equations for the point with the same coordinates but on the opposite surface, because the same integral equation is collocated with the same integration path, at both coincident points.

Some special techniques have been devised to overcome this difficulty. Among these, the most important are: the crack Green's function method, Snyder & Cruse;[11] the displacement discontinuity method, Crouch;[12] the subregions method, Blandford, Ingraffea & Liggett;13 and the dual boundary element method, Portela, Aliabadi & Rooke.[14] The crack Green's function method which eliminates the need for discretization of the crack, is limited to problems with a single straight traction-free crack. In the displacement discontinuity method, where the cracks are modelled by a single line of elements, new variables are introduced into the boundary integrals, see Sladek, Sladek & Balas[15] and Cruse.[16] The subregions method introduces artificial boundaries into the body which connect the crack-tips to the boundary in such a way that the domain is divided into subregions without cracks. In an incremental crack-extension analysis, these artificial boundaries must be repeatedly re-introduced for each increment of the crack extension. The main drawback of this method is that the introduction of artificial boundaries is not unique and thus, cannot be easily implemented as an automatic procedure in incremental analyses of crack extension. In addition, the method generates a larger system of algebraic equations than it is strictly required.

The use of dual integral equations in crack problems was first reported by Bueckner.[17] In the boundary element method, dual integral equations were first presented by Watson,[18] in a formulation based on the displacement equation and its normal derivative. The theoretical basis of the dual boundary element method were presented by Hong & Chen,[19] in a general formulation that incorporates the displacement and the traction boundary integral equations. General crack problems can be solved with the dual boundary element method in a single-region formulation when both the displacement and the traction boundary integral equations are simultaneously applied along the crack boundaries. Although the integration path is still the same for coincident points on the crack boundaries, the respective boundary integral equations are now different. Dual integral equations have been applied to solve problems in three-dimensional potential theory by Gray,[20] and in three-dimensional elastostatics by Gray, Martha & Ingraffea.[21] An essential ingredient of both these formulations is the analytic evaluation of the hypersingular integrals over flat elements. This feature, which limits their analysis to the use of flat elements, requires non-standard boundary element interpolation functions for the integrations around the collocation node. Furthermore, the extension to edge crack problems was not dealt with in their analysis. Dual integral equations have also been applied to two-dimensional problems. In elastostatics Watson[18] presented a formulation limited to embedded cracks. In potential theory Rudolphi, Krishnasamy, Schmerr & Rizzo[22] presented results exhibiting unexplained oscillations. An alternative crack modelling strategy that caters for the analysis of general embedded and edge crack problems with the dual boundary element method was recently presented by Portela, Aliabadi & Rooke.[23] In this formulation, both crack boundaries are discretized with discontinuous quadratic boundary elements. This results in a practical advantage because the problem of collocation at crack tips, crack kinks and crack-edge corners is automatically circumvented. The dual boundary element method has also been applied in the incremental analysis of crack-extension problems, as reported by Portela, Aliabadi & Rooke,[23] who have shown that remeshing is not necessary in the dual boundary element method, when the crack extension is discretized with new boundary elements.

The techniques used in the dual boundary element method for the evaluation of stress intensity factors are described by Portela.[24] The singularity subtraction technique is a regularization procedure that uses a singular particular solution of the crack problem to introduce the stress intensity factors as additional unknowns in the problem. It was employed by Aliabadi, Rooke & Cartwright[25] to solve symmetrical crack problems using the boundary element method. The extension of the singularity subtraction technique to pure opening mode analysis of sharp notches was reported by Portela, Aliabadi & Rooke.[26] Analysis of symmetrical problems with the singularity subtraction technique is straightforward, because the singular tractions are among the problem unknowns, when only half of the problem is considered with the proper boundary conditions along the symmetry line. In the case of non-symmetrical problems, the singular tractions are not among the boundary element unknowns and consequently, the additional conditions of the singularity subtraction technique cannot be applied. The partition of the problem with an artificial boundary, passing through the crack-tip, is an obvious way to circumvent this difficulty, as shown by Aliabadi.[27] However, artificial boundaries are not strictly necessary in the analysis; a new strategy was devised by Portela, Aliabadi and Rooke,[28] who introduced the stress of an internal point, approaching the notch-tip, as primary unknowns in the boundary element formulation. The accuracy and the efficiency of the internal point procedure in the singularity subtraction technique was demonstrated with the mixed-mode analysis of notched plates, see Portela, Aliabadi & Rooke.[28] The single-region analysis of the dual boundary element method introduces some difficulties in the singularity subtraction technique, regarding the availability of singular particular solutions to use in a global regularization of the domain. These difficulties can be overcome with a modelling strategy based on an automatic partitioning procedure, such that the singularity subtraction technique is applied locally in a region around each crack tip, as suggested by Portela, Aliabadi & Rooke.[29]

The J-integral technique is based on a path-independent integral, see Rice.[30] A simple procedure, based on the decomposition of the elastic field into its respective symmetric and antisymmetric mode components, is used to decouple the stress intensity factors of a mixed-mode problem.

The present work is concerned with the formulation and numerical implementation of the dual boundary element method to the two-dimensional analysis of crack-growth problems in linear elastic fracture mechanics. The dual boundary integral equations are presented and the crack modelling strategy defined. The effective treatment of the improper integrals of the dual equations, a matter of crucial importance in the dual boundary element method, is then dealt with. For curved boundary elements, the natural definition of ordinary finite-part integrals can be applied to regularize the improper integrals. For flat boundary elements, analytic integration is carried out. The singularity subtraction and the J-integral techniques are introduced for the stress intensity factors evaluation. Several examples of crack problems are analysed to assess the performance of these techniques, in order to select the most efficient one to be used in the analysis of crack extension.

An incremental crack-extension analysis is performed to determine the crack path. For each increment of the analysis, in which the crack extension is modelled with new boundary elements, the dual boundary element method is applied for the stress analysis and the J-integral is used for the stress intensity factors evaluation. The incremental analysis is based on a prediction-correction technique to define the direction of the crack-extension increment. The maximum principal stress criterion is applied to predict the tangent direction of the continuous crack path and then a correction is introduced to determine the actual direction of the increment of crack-extension. Results of incremental crack-extension analyses are presented for several geometries. Fatigue crack-growth is caused by cyclic loading: in the simplest case, the loading cycle has a constant amplitude and is described by a static load level with a stress amplitude ratio. Results of fatigue crack-growth analysis are presented for a simple problem.

The dual boundary integral equations

The dual boundary element method allows the analysis of general crack problems to be performed in a single-region boundary element formulation. The dual equations, on which the dual boundary element method is based, are the displacement and the traction boundary integral equations. Throughout this work the cartesian tensor notation is used and Einstein's summation convention is assumed. In the absence of body forces, the boundary integral representation of the displacement components u_i, at an internal source point $\mathbf{x}'$, is given by

$$u_i(\mathbf{x}') + \int_\Gamma T_{ij}(\mathbf{x}',\mathbf{x})\, u_j(\mathbf{x})\, \mathrm{d}\Gamma(\mathbf{x}) = \int_\Gamma U_{ij}(\mathbf{x}',\mathbf{x})\, t_j(\mathbf{x})\, \mathrm{d}\Gamma(\mathbf{x}) \tag{1}$$

where $T_{ij}(\mathbf{x}',\mathbf{x})$ and $U_{ij}(\mathbf{x}',\mathbf{x})$ represent the Kelvin traction and displacement fundamental solutions, respectively, at a field point $\mathbf{x}$ on the boundary Γ. The distance between the source point and the current field point is denoted by r and given by $|\mathbf{x} - \mathbf{x}'|$. Since for an internal source point $r \neq 0$, the integrals in eqn (1) contain regular integrands. Consider now the limit transition of eqn (1) when the source point tends to the boundary, such that $r \to 0$. This operation can be implemented by taking the source point to the boundary and augmenting the problem domain by a semicircular region, with boundary Γ_ϵ^* and radius ϵ, centered at the source point, as shown in Fig. 1. With this configuration, the whole boundary is divided into two parts, $\Gamma = (\Gamma - \Gamma_\epsilon) + \Gamma_\epsilon^*$, and the limit transition of eqn (1) is now carried out with the parameter ϵ, in the form

$$u_i(\mathbf{x}') + \lim_{\epsilon\to 0}\int_{\Gamma-\Gamma_\epsilon+\Gamma_\epsilon^*} T_{ij}(\mathbf{x}',\mathbf{x})\, u_j(\mathbf{x})\, \mathrm{d}\Gamma(\mathbf{x}) = \lim_{\epsilon\to 0}\int_{\Gamma-\Gamma_\epsilon+\Gamma_\epsilon^*} U_{ij}(\mathbf{x}',\mathbf{x})\, t_j(\mathbf{x})\, \mathrm{d}\Gamma(\mathbf{x}) \tag{2}$$

In eqn (2), the right-hand side integral contains a weakly singular integrand of order

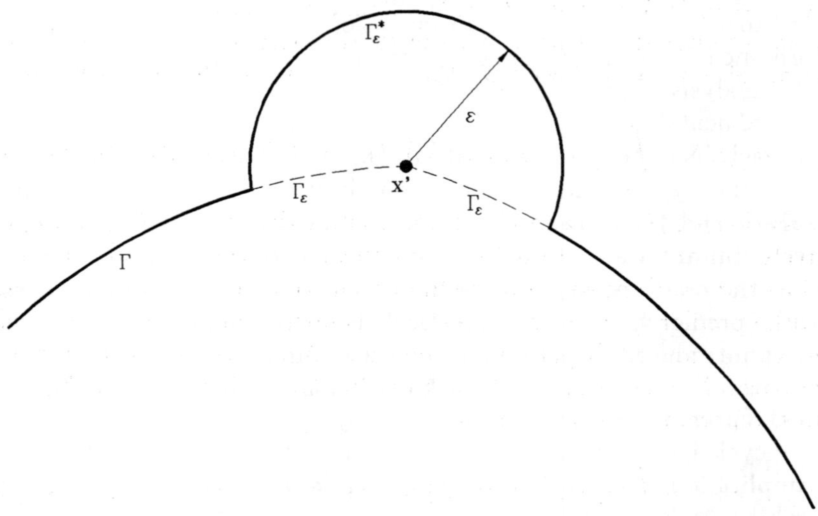

Figure 1: Source point located on the boundary, surrounded by a semicircular region

$\ln \frac{1}{r}$ and is integrable as an improper integral when the limit and the integration order are interchanged. The left-hand side integral contains a strongly singular integrand of order $\frac{1}{r}$ and can be regularized with the first term of a Taylor's expansion of the displacements, about the source point, to give:

$$\begin{aligned}\lim_{\epsilon\to 0}\int_{\Gamma-\Gamma_\epsilon+\Gamma_\epsilon^*} T_{ij}(\mathbf{x}',\mathbf{x})\, u_j(\mathbf{x})\, \mathrm{d}\Gamma(\mathbf{x}) &= \lim_{\epsilon\to 0}\int_{\Gamma_\epsilon^*} T_{ij}(\mathbf{x}',\mathbf{x})\, [u_j(\mathbf{x}) - u_j(\mathbf{x}')]\, \mathrm{d}\Gamma(\mathbf{x}) + \\ &\quad + u_j(\mathbf{x}') \lim_{\epsilon\to 0}\int_{\Gamma_\epsilon^*} T_{ij}(\mathbf{x}',\mathbf{x})\, \mathrm{d}\Gamma(\mathbf{x}) + \\ &\quad + \lim_{\epsilon\to 0}\int_{\Gamma-\Gamma_\epsilon} T_{ij}(\mathbf{x}',\mathbf{x})\, u_j(\mathbf{x})\, \mathrm{d}\Gamma(\mathbf{x})\end{aligned} \tag{3}$$

Assuming that the displacements are Hölder continuous, that is there are constants $|C| < \infty$ and $0 < \alpha \leq 1$, such that the inequality

$$|u_j(\mathbf{x}) - u_j(\mathbf{x}')| \leq C r^\alpha$$

holds, the first term of the right-hand side of eqn (3) is integrable and vanishes in the limiting process. The second limit expression of the right-hand side of eqn (3) leads to a jump in displacements, $A_{ij}(\mathbf{x}')u_j(\mathbf{x}')$, in which $A_{ij}(\mathbf{x}')$ is a constant that depends on the local geometry and elastic constants. Finally, the third limit expression results in an improper integral that is taken in a Cauchy principal-value sense. Therefore, as $\epsilon \to 0$ the source point tends to the boundary and, in the limit, eqn (2) can be written in the form

$$c_{ij}(\mathbf{x}')u_j(\mathbf{x}') + \fint_{\Gamma} T_{ij}(\mathbf{x}',\mathbf{x})\, u_j(\mathbf{x})\, d\Gamma(\mathbf{x}) = \int_{\Gamma} U_{ij}(\mathbf{x}',\mathbf{x})\, t_j(\mathbf{x})\, d\Gamma(\mathbf{x}) \tag{4}$$

where $\fint$ stands for the Cauchy principal-value integral and the coefficient $c_{ij}(\mathbf{x}')$ is given by $\delta_{ij} + A_{ij}(\mathbf{x}')$, in which δ_{ij} is the Kronecker delta.

At an internal source point $\mathbf{x}'$, the stress components σ_{ij} are obtained by differentiation of eqn (1) followed by the application of Hooke's law; they are given by

$$\sigma_{ij}(\mathbf{x}') + \int_{\Gamma} S_{ijk}(\mathbf{x}',\mathbf{x})\, u_k(\mathbf{x})\, d\Gamma(\mathbf{x}) = \int_{\Gamma} D_{ijk}(\mathbf{x}',\mathbf{x})\, t_k(\mathbf{x})\, d\Gamma(\mathbf{x}) \tag{5}$$

In this equation, $S_{ijk}(\mathbf{x}',\mathbf{x})$ and $D_{ijk}(\mathbf{x}',\mathbf{x})$ contain derivatives of $T_{ij}(\mathbf{x}',\mathbf{x})$ and $U_{ij}(\mathbf{x}',\mathbf{x})$, respectively. Since for an internal source point $r \neq 0$, the integrals in eqn (5) contain regular integrands. Consider now the limit transition of eqn (5) when the source point tends to the boundary, such that $r \to 0$. This operation can be performed exactly in the same way as for the displacement equation. Again, the whole boundary is divided into two parts, $\Gamma = (\Gamma - \Gamma_\epsilon) + \Gamma_\epsilon^*$, and the limit transition of eqn (5) is now carried out with the parameter ϵ in the form

$$\sigma_{ij}(\mathbf{x}') + \lim_{\epsilon\to 0} \int_{\Gamma-\Gamma_\epsilon+\Gamma_\epsilon^*} S_{ijk}(\mathbf{x}',\mathbf{x})\, u_k(\mathbf{x})\, d\Gamma(\mathbf{x}) = \lim_{\epsilon\to 0} \int_{\Gamma-\Gamma_\epsilon+\Gamma_\epsilon^*} D_{ijk}(\mathbf{x}',\mathbf{x})\, t_k(\mathbf{x})\, d\Gamma(\mathbf{x}) \tag{6}$$

In this equation, the right-hand side integral contains a strongly singular integrand of order $\frac{1}{r}$ and can be regularized with the first term of a Taylor's expansion of the tractions, about the source point, to give:

$$\begin{aligned}\lim_{\epsilon\to 0} \int_{\Gamma-\Gamma_\epsilon+\Gamma_\epsilon^*} D_{ijk}(\mathbf{x}',\mathbf{x})\, t_k(\mathbf{x})\, d\Gamma(\mathbf{x}) &= \lim_{\epsilon\to 0} \int_{\Gamma_\epsilon^*} D_{ijk}(\mathbf{x}',\mathbf{x})\, [t_k(\mathbf{x}) - t_k(\mathbf{x}')]\, d\Gamma(\mathbf{x}) + \\ &\quad + t_k(\mathbf{x}') \lim_{\epsilon\to 0} \int_{\Gamma_\epsilon^*} D_{ijk}(\mathbf{x}',\mathbf{x})\, d\Gamma(\mathbf{x}) + \\ &\quad + \lim_{\epsilon\to 0} \int_{\Gamma-\Gamma_\epsilon} D_{ijk}(\mathbf{x}',\mathbf{x}) t_k(\mathbf{x})\, d\Gamma(\mathbf{x}) \end{aligned} \tag{7}$$

Assuming Hölder-continuous tractions, the first term of the right hand side of eqn (7) is integrable and vanishes in the limiting process. The second limit expression of the right-hand side of eqn (7) leads to a jump in tractions, given by $A_{ijk}(\mathbf{x}')t_k(\mathbf{x}')$, in which $A_{ijk}(\mathbf{x}')$ is a constant that depends on elastic constants and coordinate transformations. The third limit expression in the right-hand side of eqn (7) results in an improper integral that is taken in a Cauchy principal-value sense. Consider now the left-hand side integral of eqn (6). It contains a hypersingular integrand of order $\frac{1}{r^2}$ that can be regularized with the first two terms of a Taylor's expansion of the displacements, about the source point, to give:

$$\lim_{\epsilon\to 0}\int_{\Gamma-\Gamma_\epsilon+\Gamma_\epsilon^*} S_{ijk}(\mathbf{x}',\mathbf{x})\, u_k(\mathbf{x})\, \mathrm{d}\Gamma(\mathbf{x}) =$$

$$= \lim_{\epsilon\to 0}\int_{\Gamma_\epsilon^*} S_{ijk}(\mathbf{x}',\mathbf{x})\, [u_k(\mathbf{x}) - u_k(\mathbf{x}') - u_{k,m}(\mathbf{x}')(x_m - x_m')]\, \mathrm{d}\Gamma(\mathbf{x}) +$$

$$+ u_k(\mathbf{x}')\lim_{\epsilon\to 0}\int_{\Gamma_\epsilon^*} S_{ijk}(\mathbf{x}',\mathbf{x})\, \mathrm{d}\Gamma(\mathbf{x}) +$$

$$+ u_{k,m}(\mathbf{x}')\lim_{\epsilon\to 0}\int_{\Gamma_\epsilon^*} S_{ijk}(\mathbf{x}',\mathbf{x})(x_m - x_m')\, \mathrm{d}\Gamma(\mathbf{x}) +$$

$$+ \lim_{\epsilon\to 0}\int_{\Gamma-\Gamma_\epsilon} S_{ijk}(\mathbf{x}',\mathbf{x}) u_k(\mathbf{x})\, \mathrm{d}\Gamma(\mathbf{x}) \tag{8}$$

where the comma between two indices denotes differentiation. Assuming that the displacement derivatives are Hölder continuous, that is there are constants $|C| < \infty$ and $0 < \alpha \leq 1$, such that the inequality

$$|u_k(\mathbf{x}) - u_k(\mathbf{x}') - u_{k,m}(\mathbf{x}')(x_m - x_m')| \leq C|x_m - x_m'|^{\alpha+1}$$

holds, the first term of the right-hand side of eqn (8) is integrable and vanishes in the limiting process. The second limit expression in the right-hand side of the same equation is unbounded and will be considered later. The third limit expression leads to a jump in displacement derivatives, given by $B_{ijkm}(\mathbf{x}')u_{k,m}(\mathbf{x}')$, in which $B_{ijkm}(\mathbf{x}')$ is a constant that depends on elastic constants and coordinate transformations. Finally, it can be shown that the fourth limit expression of the right-hand side of eqn (8), together with the second limit expression, are always bounded. Thus, the fourth limit expression is considered alone, in a Hadamard principal-value sense.[31] Collecting the results of eqns (7) and (8), it can be shown that for a source point in a smooth boundary the jump terms are equivalent to boundary stresses, that is

$$A_{ijk}(\mathbf{x}')t_k(\mathbf{x}') - B_{ijkm}(\mathbf{x}')u_{k,m}(\mathbf{x}') = \frac{1}{2}\sigma_{ij}(\mathbf{x}') \tag{9}$$

Therefore, as $\epsilon \to 0$, the source point tends to the boundary and, in the limit, on a smooth boundary, eqn (6) can be written in the form

$$\frac{1}{2}\sigma_{ij}(\mathbf{x}') + \mathop{=\!\!\!\!\!\!\int}_{\Gamma} S_{ijk}(\mathbf{x}',\mathbf{x})\, u_k(\mathbf{x})\, \mathrm{d}\Gamma(\mathbf{x}) = \mathop{-\!\!\!\!\!\!\int}_{\Gamma} D_{ijk}(\mathbf{x}',\mathbf{x})\, t_k(\mathbf{x})\, \mathrm{d}\Gamma(\mathbf{x}) \tag{10}$$

where $=\!\!\!\!\!\!\int$ stands for the Hadamard principal value integral. Notice that eqn (10) could be obtained by direct differentiation of eqn (4), followed by the application of Hooke's law, when the relationship between Cauchy and Hadamard principal-value integrals is applied, as shown by Portela.[24]

The traction components t_j are given by

$$\frac{1}{2}t_j(\mathbf{x}') + n_i(\mathbf{x}')\mathop{=\!\!\!\!\!\!\int}_{\Gamma} S_{ijk}(\mathbf{x}',\mathbf{x})u_k(\mathbf{x})\mathrm{d}\Gamma(\mathbf{x}) = n_i(\mathbf{x}')\mathop{-\!\!\!\!\!\!\int}_{\Gamma} D_{ijk}(\mathbf{x}',\mathbf{x})t_k(\mathbf{x})\mathrm{d}\Gamma(\mathbf{x}) \tag{11}$$

where n_i denotes the i-th component of the unit outward normal to the boundary, at the source point. Equations (4) and (11) constitute the basis of the dual boundary element method. A more complete description of the dual boundary element method is given by Portela.[24]

3 Crack Modelling

The necessary conditions for the existence of the principal-value integrals obtained in the derivation of the dual boundary integral equations impose restrictions on the discretization. Consider that both the geometry and boundary field variables are described by a piece-wise continuously differentiable approximation and that collocation is always performed with the source point at the boundary element nodes. Under these considerations, the continuity requirements of the Cauchy principal-value integral in the displacement equation can be satisfied by any Lagrangian continuous or discontinuous boundary element. However, in the traction equation the continuity requirements of the Hadamard principal-value integral are satisfied only by discontinuous elements, since all the nodes are internal points of the element where a continuously differentiable approximation is defined. Moreover, the requirement of the smoothness of the geometry at a collocation point in the traction equation is implicitly satisfied by the discontinuous element. Furthermore, discontinuous traction fields can be modelled by discontinuous elements, since the tractions must satisfy Hölder continuity only at the nodes. It is important to realize that if the element approximation does not satisfy these necessary continuity requirements then the finite-part integrals do not exist.

For the sake of efficiency and to keep the simplicity of the standard boundary elements, the present formulation uses discontinuous quadratic elements for the crack modelling, as shown in Fig. 2. The general modelling strategy, developed by Portela, Aliabadi & Rooke,[14] can be summarized as follows:

- The traction eqn (11) is applied for collocation on one of the crack boundaries.
- The displacement eqn (4) is applied for collocation on the opposite crack boundary and remaining boundaries.
- The crack boundaries are discretized with discontinuous quadratic elements, as shown in Fig. 2.
- Continuous quadratic boundary elements are used along the remaining boundaries of the problem domain, except at the intersection between a crack and an edge, where discontinuous or semi-discontinuous elements are required, in order to avoid nodes at the intersection.

This simple strategy is robust and allows the dual boundary element method to effectively model general edge or embedded crack problems; crack tips, crack-edge corners and crack kinks do not require special treatment, since they are not located at nodal points where the collocation is carried out. Other alternatives to discontinuous elements can be used to model the crack, but they are not as efficient. For instance, Hermitian boundary elements, Watson,[18] and Overhouser boundary elements, Walters, Ortiz, Gibson & Bewer,[32] also suit the continuity requirements. However, they

are computationally expensive and are not readily suitable to deal with edge crack problems.
The numerical implementation of Cauchy and Hadamard principal-value integrals is performed in terms of double-sided finite-part integrals of first and second order, respectively. A description of the use of finite-part integrals, in the context of the dual boundary element method, is given by Portela.[24]

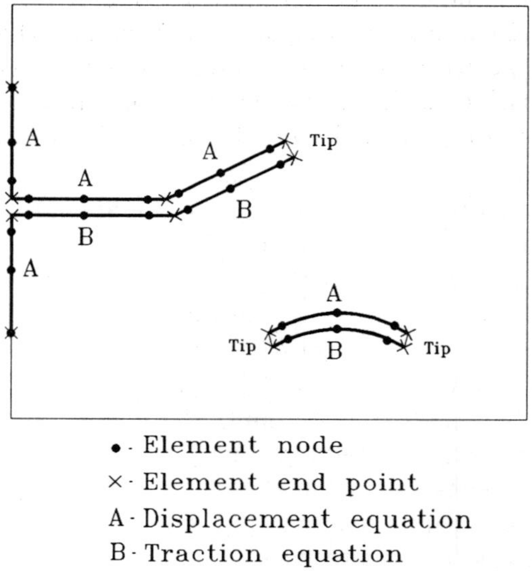

Figure 2: Crack modelling with discontinuous quadratic boundary elements

4 Algebraic equations

By means of nodal collocation and integration over the boundary elements, the boundary integral eqns (4) and (11) are transformed into a system of linear algebraic equations. The unknown boundary tractions and displacements can then be obtained by solving this system of equations which includes the boundary conditions. As the collocation point passes through all the nodal points, the following system of linear algebraic equations is obtained:

$$\mathbf{Hu} = \mathbf{Gt} \tag{12}$$

where the matrices $\mathbf{H}$ and $\mathbf{G}$ contain integrals of the fundamental solutions T_{ij} and U_{ij} respectively, in the case of eqn (4), or integrals of S_{ijk} and D_{ijk} respectively, in the case of eqn (11). The vectors $\mathbf{t}$ and $\mathbf{u}$ contain boundary traction components and boundary displacement components respectively. Equations (12) are then rearranged in the form:

$$\mathbf{Ax} = \mathbf{By} = \mathbf{f} \tag{13}$$

where the vector $\mathbf{x}$ contains the boundary unknowns t_i, u_i and the vector $\mathbf{y}$ contains the boundary conditions $\overline{t}_i$ and $\overline{u}_i$. The matrices $\mathbf{A}$ and $\mathbf{B}$ result from the rearrangement of **H** and **G**. From the system of eqns (13) a unique solution can then be obtained.

When the dual boundary element method is applied in an incremental crack-extension analysis, such that each new crack-extension increment is modelled with new discontinuous boundary elements, it becomes apparent that remeshing is not required. The new discontinuous boundary elements of the crack-extension increment will generate new equations and up-date the ones already existing with new unknowns. In other words, new boundary elements will generate new rows and new columns in the coefficient matrix of the system of equations. Assuming that the crack extension is traction-free, the right-hand side of the system of equations is only extended for the positions corresponding to the new unknowns introduced. This procedure is illustrated schematically in Fig. 3. If the *LU* decomposition method is adopted for the

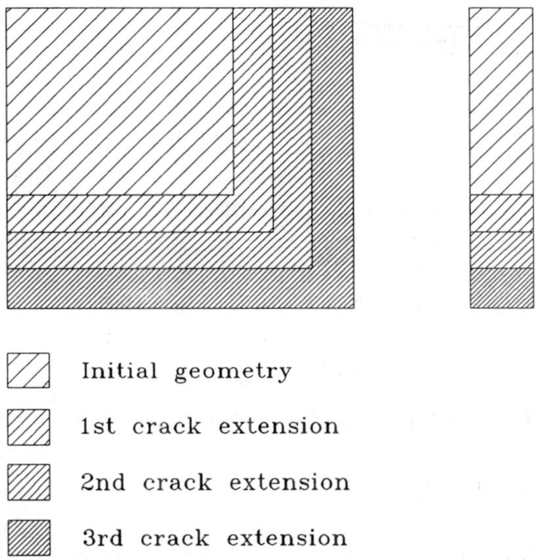

Figure 3: Schematic representation of the generation of the system of equations

solution of the system of equations, a very efficient incremental analysis can be carried out. For each increment of the analysis, only the new rows and new columns need to be *LU*-decomposed. The existing rows and columns, already decomposed, are carried over from the previous increment into the current one.

5 Computation of principal-value integrals

The numerical implementation of the principal-value integrals that arise in the dual integral equations is easily carried out by the classical method of singularity subtraction, Davis & Rabinowitz,[43] which leads to the natural definition of ordinary finite-part integrals. In the vicinity of a collocation node the regular part of the integrand is expressed as a Taylor's expansion. If a sufficient number of terms of the expansion are subtracted from the regular part of the integrand and then separately added back, the singularity can be isolated. The original improper integral is thus transformed into the sum of a regular integral and an integral of the singular function. This latter integral is then evaluated analytically, while standard Gaussian quadrature is used for numerical integration of the regular integral. The procedure is general and applicable to any type of boundary element. See Portela[24] for basic definitions of finite-part integrals, in the context of the dual boundary element method.

5.1 General-shape discontinuous elements

On a traction-free crack, consider a discontinuous quadratic boundary element of general shape Γ_e that contains the collocation node. On this element, eqns (4) and (11) become

$$c_{ij}(\mathbf{x}')u_j(\mathbf{x}') + ⨍_{\Gamma_e} T_{ij}(\mathbf{x}',\mathbf{x})\, u_j(\mathbf{x})\, \mathrm{d}\Gamma(\mathbf{x}) = 0 \tag{14}$$

and

$$n_i(\mathbf{x}') ⨎_{\Gamma_e} S_{ijk}(\mathbf{x}',\mathbf{x})u_k(\mathbf{x})\mathrm{d}\Gamma(\mathbf{x}) = 0 \tag{15}$$

respectively. Coincident with this element, consider also a continuous quadratic boundary element which will be used for approximation of the local geometry. For both elements the local parametric coordinate ξ is defined, as usual, in the range $-1 \leq \xi \leq +1$. The collocation node ξ' is mapped onto $\mathbf{x}'$, via the continuous element shape functions, while the displacement components u_j are approximated in the local coordinate system by means of the nodal values u_j^n and the discontinuous element shape functions. The Cauchy principal value integral of eqn (14) can be expressed in the local coordinate as

$$⨍_{\Gamma_e} T_{ij}(\mathbf{x}',\mathbf{x})\, u_j(\mathbf{x})\, \mathrm{d}\Gamma(\mathbf{x}) = u_j^n ⨍_{-1}^{+1} \frac{f_{ij}^n(\xi)}{\xi - \xi'}\, \mathrm{d}\xi \tag{16}$$

where $f_{ij}^n(\xi)$ is a regular function, given by the product of the fundamental solution, a shape function and the Jacobian of the coordinate transformation, multiplied by the term $\xi - \xi'$. The integral of the right-hand side of eqn (16) can be transformed with the aid of the first term of a Taylor's expansion of the function f_{ij}^n, around the collocation node, to give

$$⨍_{-1}^{+1} \frac{f_{ij}^n(\xi)}{\xi - \xi'}\, \mathrm{d}\xi = \int_{-1}^{+1} \frac{f_{ij}^n(\xi) - f_{ij}^n(\xi')}{\xi - \xi'}\, \mathrm{d}\xi + f_{ij}^n(\xi') ⨍_{-1}^{+1} \frac{\mathrm{d}\xi}{\xi - \xi'} \tag{17}$$

Now, the first integral of the right-hand side is regular and the second one can be integrated analytically to give

$$⨍_{-1}^{+1} \frac{\mathrm{d}\xi}{\xi - \xi'} = \ln\left|\frac{1-\xi'}{1+\xi'}\right| \tag{18}$$

In eqn (17), the existence of the first-order finite-part integral requires the continuity of the first derivative of f_{ij}^n, or at least the Hölder continuity of f_{ij}^n, at the collocation node. For a discontinuous element, this requirement is automatically satisfied, because the nodes are internal points of the element, where f_{ij}^n is continuously differentiable.

The Hadamard principal value integral of eqn (15) can be expressed in the local parametric coordinate as

$$\fint_{\Gamma_e} S_{ijk}(\mathbf{x}', \mathbf{x})\, u_k(\mathbf{x})\, \mathrm{d}\Gamma(\mathbf{x}) = u_k^n \fint_{-1}^{-1} \frac{g_{ijk}^n(\xi)}{(\xi - \xi')^2}\, \mathrm{d}\xi \tag{19}$$

where $g_{ijk}^n(\xi)$ is a regular function, given by the product of the fundamental solution, a shape function and the Jacobian of the coordinate transformation, multiplied by the term $(\xi - \xi')^2$. The integral of the right-hand side of eqn (19) can be transformed with the aid of the first and second terms of a Taylor's expansion of the function g_{ijk}^n, in the neighbourhood of the collocation node, to give

$$\begin{aligned}
\fint_{-1}^{+1} \frac{g_{ijk}^n(\xi)}{(\xi - \xi')^2}\, \mathrm{d}\xi \;=\; & \int_{-1}^{+1} \frac{g_{ijk}^n(\xi) - g_{ijk}^n(\xi') - g_{ijk}^{n(1)}(\xi')\,(\xi - \xi')}{(\xi - \xi')^2}\, \mathrm{d}\xi \;+ \\
& +\; g_{ijk}^n(\xi') \fint_{-1}^{+1} \frac{\mathrm{d}\xi}{(\xi - \xi')^2} \;+ \\
& +\; g_{ijk}^{n(1)}(\xi') \fint_{-1}^{+1} \frac{\mathrm{d}\xi}{\xi - \xi'}
\end{aligned} \tag{20}$$

where $g_{ijk}^{n(1)}$ denotes the first derivative. At the collocation node, the function g_{ijk}^n is required to have continuity of its second derivative, or at least a Hölder-continuous first derivative, for the finite-part integral to exist. This requirement is automatically satisfied by a discontinuous element, since the nodes are internal points of the element. Now, in eqn (20), the first integral of the right-hand side is regular and the third integral is identical with the one given in eqn (18). The second integral of the right hand side of eqn (20) can be integrated analytically to give

$$\fint_{-1}^{+1} \frac{\mathrm{d}\xi}{(\xi - \xi')^2} = -\frac{1}{1 + \xi'} - \frac{1}{1 - \xi'} \tag{21}$$

Equations (17) and (20) are just the ordinary double-sided first and second order finite-part integrals, respectively, as defined by Kutt.[33]

Numerical integration of singular integrands is also possible; quadrature rules for finite-part integrals were published by Ossicini,[34] Kutt,[35] Ioakimidis[36] and Ladopoulos.[37]

5.2 Straight discontinuous elements

In general, cracking processes can extend the cracks along curved paths. However, in practice, curved paths are usually modelled with flat increments which lead to piece-wise flat cracks. For piece-wise flat cracks, the integrals in eqns (16) and (19) are carried out most effectively by direct analytic integration which is presented in the following. Consider a flat discontinuous quadratic boundary element, with the nodes positioned arbitrarily at the points $\xi = -\frac{2}{3}$, $\xi = 0$ and $\xi = +\frac{2}{3}$. The shape functions of this element are given by:

$$N_1 \;=\; \xi \left(\frac{9}{8}\xi - \frac{3}{4} \right)$$

$$N_2 = \left(1-\frac{3}{2}\xi\right)\left(1+\frac{3}{2}\xi\right)$$

$$N_3 = \xi\left(\frac{9}{8}\xi+\frac{3}{4}\right)$$

For this element, the integral of eqn (16) is represented by:

$$\int_{\Gamma_e} T_{ij}(\mathbf{x}',\mathbf{x})\, u_j(\mathbf{x})\, \mathrm{d}\Gamma(\mathbf{x}) = u_j^n \int_{-1}^{+1} T_{ij}(\xi',\mathbf{x}(\xi))\, N_n(\xi)\, J(\xi)\, \mathrm{d}\xi = \mathbf{h}_i^n \mathbf{u}^n \tag{22}$$

where $\mathbf{u}^n$ denotes the nodal displacement components and $J(\xi)$ is the Jacobian of the coordinate transformation. Because of the assumed flatness of the element, $J = \frac{l}{2}$, where l represents the element length and the matrix $\mathbf{h}^n$ is given by

$$\mathbf{h}^n = \frac{1-2\nu}{4\pi(1-\nu)} \begin{bmatrix} 0 & -1 \\ +1 & 0 \end{bmatrix} \int_{-1}^{+1} \frac{N_n}{\xi-\xi'}\, \mathrm{d}\xi \tag{23}$$

The first order finite-part integrals are integrated analytically to give:

$$\int_{-1}^{+1} \frac{N_1}{\xi-\xi'}\, \mathrm{d}\xi = \frac{3}{4}\left(\frac{\xi'(3\xi'-2)}{2}\ln\left|\frac{1-\xi'}{1+\xi'}\right| + 3\xi' - 2\right) \tag{24}$$

$$\int_{-1}^{+1} \frac{N_2}{\xi-\xi'}\, \mathrm{d}\xi = \frac{1}{2}\left(\frac{(3\xi'-2)(3\xi'+2)}{2}\ln\left|\frac{1+\xi'}{1-\xi'}\right| - 9\xi'\right) \tag{25}$$

and

$$\int_{-1}^{+1} \frac{N_3}{\xi-\xi'}\, \mathrm{d}\xi = \frac{3}{4}\left(\frac{\xi'(3\xi'+2)}{2}\ln\left|\frac{1-\xi'}{1+\xi'}\right| + 3\xi' + 2\right) \tag{26}$$

The integral of eqn (19) is represented by

$$\int_{\Gamma_e} S_{ijk}(\mathbf{x}',\mathbf{x})\, u_k(\mathbf{x})\, \mathrm{d}\Gamma(\mathbf{x}) = u_k^n \int_{-1}^{+1} S_{ijk}(\xi',\mathbf{x}(\xi))\, N_n(\xi)\, J(\xi)\, \mathrm{d}\xi = \overline{\mathbf{h}}_{ij}^n \mathbf{u}^n \tag{27}$$

where the matrix $\overline{\mathbf{h}}^n$ is given by:

$$\overline{\mathbf{h}}^n = \frac{E}{4\pi(1-\nu^2)}\frac{2}{l}\mathbf{S}' \int_{-1}^{+1} \frac{N_n}{(\xi-\xi')^2}\, \mathrm{d}\xi \tag{28}$$

The matrix $\mathbf{S}'$ is given by

$$\mathbf{S}' = \begin{bmatrix} +n_1(2n_2^2+1) & -n_2(-2n_2^2+1) \\ +n_1(2n_1^2-1) & -n_2(-2n_1^2-1) \\ -n_2(2n_1^2-1) & +n_1(-2n_2^2+1) \end{bmatrix} \tag{29}$$

where n_1 and n_2 are the components of the unit outward normal to the element. The second-order finite-part integrals of eqn (28) are integrated analytically to give:

$$\int_{-1}^{+1} \frac{N_1}{(\xi-\xi')^2}\, \mathrm{d}\xi = \frac{3}{4}\left((3\xi'-1)\ln\left|\frac{1-\xi'}{1+\xi'}\right| + \frac{6\xi'^2-2\xi'-3}{\xi'^2-1}\right) \tag{30}$$

$$\int_{-1}^{+1} \frac{N_2}{(\xi-\xi')^2}\, \mathrm{d}\xi = \frac{1}{2}\left(9\xi'\ln\left|\frac{1+\xi'}{1-\xi'}\right| - \frac{18\xi'^2-13}{\xi'^2-1}\right) \tag{31}$$

and

$$\oint_{-1}^{+1} \frac{N_3}{(\xi-\xi')^2}\,\mathrm{d}\xi \;=\; \frac{3}{4}\left((3\xi'+1)\ln\left|\frac{1-\xi'}{1+\xi'}\right| + \frac{6\xi'^2+2\xi'-3}{\xi'^2-1}\right) \qquad (32)$$

Equation (28) shows that the terms of the matrix $\overline{\mathbf{h}}^n$ are inversely proportional to the element length l. Notice that this property does not occur with the matrix $\mathbf{h}^n$ of eqn (23).

5.3 The Rigid Body Condition

When collocation is performed at a crack node there are always two elements on opposite faces that contain the collocation point, because both crack boundaries are discretized. This means that, along the crack, the finite-part integrals in eqns (14) and (15) are required twice: once on the element that contains the collocation node, that is the self-point element and again, on the opposite element, which is also a self-point element, since it contains the node which is coincident with the collocation node. This feature of the dual boundary element method puts restrictions on the use of the rigid body condition at crack nodes, as explained in the following. Consider a constant displacement field, with components $u_i(\mathbf{x}) = C$, defined throughout the body. In this circumstance the traction components are zero and eqn (4) gives

$$c_{ij}(\mathbf{x}') + \oint_{\Gamma} T_{ij}(\mathbf{x}',\mathbf{x})\,\mathrm{d}\Gamma(\mathbf{x}) = 0 \qquad (33)$$

while eqn (11) gives:

$$n_i(\mathbf{x}') \oint_{\Gamma} S_{ijk}(\mathbf{x}',\mathbf{x})\,\mathrm{d}\Gamma(\mathbf{x}) = 0 \qquad (34)$$

Equations (33) and (34) express the usual rigid body condition that must be satisfied by the dual boundary integral equations at every collocation point. According to eqn (33), the coefficient $c_{ij}(\mathbf{x}')$ need not be dealt with directly. When the equation is discretized, this coefficient, together with the finite-part integral, is determined by the *row sum* technique. However, for collocation at a crack node, the *row sum* technique can no longer be used if the integration along opposite self-point elements has any symmetric terms, since a pair of symmetric values always cancel each other in the sum. It can be shown, through eqns (22) to (26), that the off-diagonal terms of the first and last nodes of the elements used here are symmetric, with the finite-part integrals given by $(\ln 5 - 3)$ and $(-\ln 5 + 3)$, respectively. Bearing in mind that opposite self-point elements have their first and last nodes interchanged, it is evident that two symmetric off-diagonal terms are obtained from the collocation at either one of these nodes and integration along opposite self-point elements. According to eqn (34), it appears that the *row sum* technique can be used to evaluate indirectly the diagonal terms of the discretized form of the traction equation. Again, it can be shown, through eqns (27) to (32), that the diagonal terms of any crack node are symmetric with respect to its opposite node. These features invalidate the use of the standard rigid body condition, to evaluate indirectly the diagonal terms at crack nodes.

6 Singularity subtraction technique

In order to avoid the numerical difficulties that arise from the presence of a singularity in the stress field, it is convenient to modify the original problem by introducing a particular solution before it is solved by the dual boundary element method.

6.1 Formulation

Under the linear elasticity assumptions, where the principle of superposition is valid, the tractions and the displacements of the original elastic field can be written as:

$$t_i = (t_i - t_i^S) + t_i^S = t_i^R + t_i^S \tag{35}$$

and

$$u_i = (u_i - u_i^S) + u_i^S = u_i^R + u_i^S \tag{36}$$

where t_i^S and u_i^S are the components of a particular solution of the original problem. When the appropriate particular solution is chosen to represent the singular field, eqns (35) and (36) regularize the original problem, in the sense that the tractions t_i^R are non-singular. As a consequence of this regularization, the numerical analysis can now be carried out on the regular fields t_i^R and u_i^R only; the fields t_i^S and u_i^S automatically satisfy the field equations, because they are defined as a particular solution of the original problem. The boundary conditions for the regularized problem are obtained from the boundary conditions of the original problem, $t_i = \bar{t}_i$ and $u_i = \bar{u}_i$, transformed with eqns (35) and (36), respectively. They become:

$$t_i^R = \bar{t}_i - t_i^S \tag{37}$$

on the boundary Γ_t and

$$u_i^R = \bar{u}_i - u_i^S \tag{38}$$

on the boundary Γ_u, where $\Gamma = \Gamma_u + \Gamma_t$ is the total boundary. These modified boundary conditions introduce additional unknowns into the problem, which are the constants of the particular solution used to represent the singular field. As the stress intensity factors, defined in terms of these constants, are not known at this stage of a general problem, they become additional primary unknowns. This feature of the singularity subtraction technique can be advantageous, since no further post-processing analysis of the results is required to obtain the stress intensity factors. In order to get a unique solution of the regularized problem, it is necessary to specify additional constraints, as many as the number of additional unknowns, that is one for each deformation mode included in the analysis. These extra conditions can be applied by constraining the regularized traction field directly at the crack-tip, that is

$$t_i^R = 0 \tag{39}$$

which actually implies the presence of a singularity in the original field at the crack-tip. A complete discussion of the extra conditions of the singularity subtraction technique was presented by Portela.[24]

6.2 Algebraic equations

As the analysis is performed in terms of the regularized variables, eqns (12) and (13) are now written as

$$\mathbf{H}\mathbf{u}^R = \mathbf{G}\mathbf{t}^R \tag{40}$$

and

$$\mathbf{A}\mathbf{x}^R = \mathbf{B}\mathbf{y}^R \tag{41}$$

When the boundary conditions (37) and (38) are introduced, eqns (41) can be written as

$$\mathbf{A}\mathbf{x}^R = \mathbf{B}(\mathbf{y} - \mathbf{E}\mathbf{z}) = \mathbf{f} - \mathbf{C}\mathbf{z} \tag{42}$$

in which the product $\mathbf{Ez}$ represents the components of the singular field, evaluated at the boundary, with the additional unknowns contained in $\mathbf{z}$ and known closed-form functions contained in $\mathbf{E}$. A rearrangement of eqns (42) leads to the following system of equations

$$\left[\, \mathbf{A} \;\vdots\; \mathbf{C} \,\right] \left\{ \begin{array}{c} \mathbf{x}^R \\ \cdots \\ \mathbf{z} \end{array} \right\} = \{\mathbf{f}\} \tag{43}$$

which is underdetermined, because of the additional unknowns contained in $\mathbf{z}$.

6.3 The internal point approach

The additional eqns (39) rely on the assumption that the unbounded tip tractions are boundary element unknowns. In general, this will not be the case, since the crack faces are traction-free and consequently, the crack-tip boundary element unknowns are displacements. In such a case, the unbounded tip tractions must somehow be introduced in the formulation, before attempting the regularization. The general procedure, defined by Portela, Aliabadi & Rooke[28] for the boundary element singularity subtraction technique, consists of the following:

- introduction of the unbounded tip tractions as boundary element unknowns, by means of the stress of an internal point approaching the tip along the crack axis;
- subtraction of the crack-tip singularity, which regularizes the elastic field and introduces the stress intensity factors among the primary unknowns of the problem, using the internal point equations for the additional extra conditions.

For an internal point p, the discrete form of eqns (11), with the boundary conditions (37) and (38), can be written as

$$\mathbf{t}_p^R + \mathbf{H}_p\, \mathbf{u}^R = \mathbf{G}_p\, \mathbf{t}^R \tag{44}$$

and

$$\mathbf{t}_p^R + \mathbf{A}_p\, \mathbf{x}^R = \mathbf{B}_p(\mathbf{y} - \mathbf{E}\,\mathbf{z}) = \mathbf{f}_p - \mathbf{C}_p\, \mathbf{z} \tag{45}$$

in which matrices $\mathbf{H}_p$ and $\mathbf{G}_p$ contain integrals of S_{ijk} and D_{ijk}, respectively. Matrices $\mathbf{A}_p$ and $\mathbf{B}_p$ result from rearrangement of $\mathbf{H}_p$ and $\mathbf{G}_p$ in system (44), in order to keep all the boundary unknowns u_j^R and t_j^R in the vector $\mathbf{x}$ and the original boundary conditions in the vector $\mathbf{y}$. If the point p approaches the tip, along the crack axis, the additional conditions (39) can now be imposed at that point. Hence, after the introduction of the condition (39), eqns (45) can be written as:

$$\left[\ \mathbf{A}_p\ \vdots\ \mathbf{C}_p\ \right]\left\{\begin{array}{c}\mathbf{x}^R\\ \cdots\\ \mathbf{z}\end{array}\right\} = \{\mathbf{f}_p\} \tag{46}$$

These additional equations are now coupled with the system of eqns (43), to give the final system as:

$$\left[\begin{array}{ccc}\mathbf{A} & \vdots & \mathbf{C}\\ \cdots & & \cdots\\ \mathbf{A}_p & \vdots & \mathbf{C}_p\end{array}\right]\left\{\begin{array}{c}\mathbf{x}^R\\ \cdots\\ \mathbf{z}\end{array}\right\} = \left\{\begin{array}{c}\mathbf{f}\\ \cdots\\ \mathbf{f}_p\end{array}\right\} \tag{47}$$

From this system of equations a unique solution can now be obtained.

In eqns (44) the collocation at the point p, exactly located at the crack-tip, is replaced with a quadratic extrapolation procedure. For this purpose, three internal points, p_1, p_2 and p_3, are located on the crack axis, at distances ε, 2ε and 3ε respectively, from the tip. At each one of these points, eqns (44) are generated by collocation and then quadratically extrapolated to the point p, at the crack-tip, to give

$$\mathbf{H}_p = 3(\mathbf{H}_{p_1} - \mathbf{H}_{p_2}) + \mathbf{H}_{p_3} \tag{48}$$

and

$$\mathbf{G}_p = 3(\mathbf{G}_{p_1} - \mathbf{G}_{p_2}) + \mathbf{G}_{p_3} \tag{49}$$

The minimum value for the distance ε is usually limited to about one half of the length of the boundary elements that meet at the crack-tip.

6.4 General modelling strategy

The reliability of the singularity subtraction technique depends on the availability of a closed-form, singular, particular solution to be used globally in the regularization of the problem. Consider, for instance, the problem of a flat edge crack in a finite plate: the singular term of the Williams' eigenexpansion[38] can be used as a particular solution valid in the global domain of the problem. In the case of a flat centre crack in a finite plate, the Williams' field is not a particular solution valid in the global domain of the problem. In this case, the Westergaard solution[39] can be used globally in the regularization of the problem. For an arbitrary piece-wise flat crack, neither the Williams' nor the Westergaard elastic fields are particular solutions of the problem, valid in the global domain. In other words, there is not a general closed-form field solution available to be used globally in the regularization of the problem. This apparent difficulty of the singularity subtraction technique can be readily overcome with a modelling strategy, based on a partitioning procedure, such that the Williams' field can always be used locally, in the neighbourhood of each crack tip, as a particular solution in the regularization. This modelling strategy considers the problem domain to be partitioned into complementary regions, by artificial interface boundaries, as shown in Fig. 4. The first region, which surrounds each crack tip, is called the

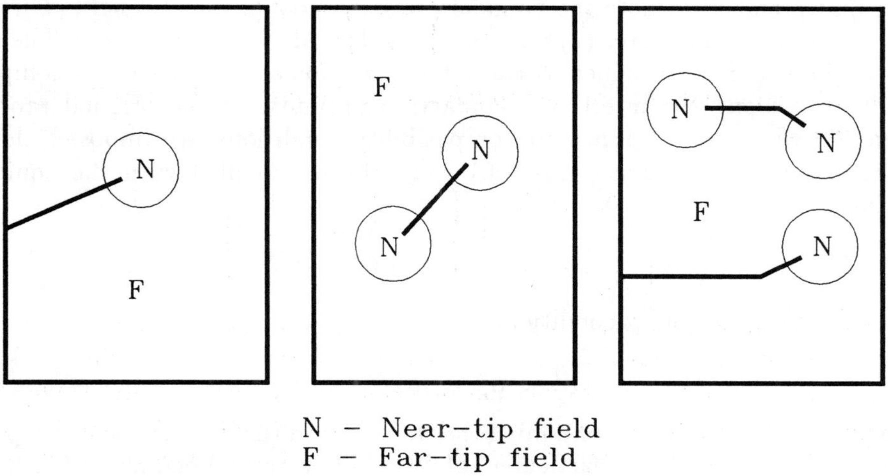

Figure 4: Complementary regions of the problem domain

near-tip field region (N), while the second region, which is that between the interface boundaries and the boundary of the global domain of the problem, is called the far-tip field region (F). The singularity subtraction technique is now applied in the near-tip field region, only. Hence, in this region, eqns (47) can be written as:

$$\begin{bmatrix} -\mathbf{G}_I^N & \vdots & \mathbf{H}_I^N & \vdots & \mathbf{H}_C^N \\ \cdots & & \cdots & & \cdots \\ -\mathbf{G}_{pI}^N & \vdots & \mathbf{H}_{pI}^N & \vdots & \mathbf{H}_{pC}^N \end{bmatrix} \left\{ \begin{array}{c} \mathbf{t}_I^{NR} \\ \cdots \\ \mathbf{u}_I^{NR} \\ \cdots \\ \mathbf{u}_C^{NR} \end{array} \right\} = \left\{ \begin{array}{c} \mathbf{o} \\ \cdots \\ \mathbf{o} \end{array} \right\} \tag{50}$$

where $\mathbf{t}_I^{NR}$, $\mathbf{u}_I^{NR}$ and $\mathbf{u}_C^{NR}$ replaced $\mathbf{x}^R$ in eqns (47) and represent the regularized (R) boundary unknowns of the near-tip field region (N), defined along the interface boundary (I) and the traction-free crack (C), respectively. Matrices $\mathbf{G}_I^N$, $\mathbf{H}_I^N$ and $\mathbf{H}_C^N$ are partition matrices that replaced matrix $\mathbf{A}$ in eqns (47) and satisfy eqns (40), written for the near-field region. Matrices $\mathbf{G}_{pI}^N$, $\mathbf{H}_{pI}^N$ and $\mathbf{H}_{pC}^N$ are partition matrices that replaced matrix $\mathbf{A}_p$ in eqns (47). Notice that matrices $\mathbf{C}$ and $\mathbf{C}_p$, in eqns (47), are now identically null because they are derived exclusively from the traction-free crack surfaces that belong to the boundary of the near-tip field region. Because of this particular feature, the vector $\mathbf{z}$ does not appear in eqns (50). The system of eqns (50) is underdetermined, because both traction and displacement components are unknown along the interface boundary. In the far-tip field region, the standard dual boundary element method is applied in terms of the original variables of the problem. Hence, in this region, eqns (13) can now be written as:

$$\left[-\mathbf{G}_I^F \;\vdots\; \mathbf{H}_I^F \;\vdots\; \mathbf{A}^F \right] \left\{ \begin{array}{c} \mathbf{t}_I^F \\ \cdots \\ \mathbf{u}_I^F \\ \cdots \\ \mathbf{x}^F \end{array} \right\} = \{ \mathbf{f} \} \tag{51}$$

where $\mathbf{t}_I^F$, $\mathbf{u}_I^F$ and $\mathbf{x}^F$ denote the boundary unknowns of the far-tip field (F), defined along the interface boundary (I) and remaining boundaries, respectively. The system of eqns (51) is underdetermined, because both traction and displacement components are unknown along the interface boundary. Continuity of the original problem is maintained when equilibrium and compatibility conditions are imposed along the interface boundaries. Thus, eqns (50) and (51) are coupled with the equilibrium conditions:

$$\mathbf{t}_I^F = -\mathbf{t}_I^N = -(\mathbf{t}_I^{NR} + \mathbf{t}^S) \tag{52}$$

and with the compatibility conditions:

$$\mathbf{u}_I^F = \mathbf{u}_I^N = (\mathbf{u}_I^{NR} + \mathbf{u}^S) \tag{53}$$

where $\mathbf{t}^S$ and $\mathbf{u}^S$ are vectors containing the traction and the displacement components, respectively, of the singular particular solution. They are defined along the interface boundary as:

$$\mathbf{t}^S = \mathbf{E}^t \mathbf{z} \tag{54}$$

and

$$\mathbf{u}^S = \mathbf{E}^u \mathbf{z} \tag{55}$$

where $\mathbf{E}^t$ and $\mathbf{E}^u$ contain the closed-form functions of the tractions and displacements, respectively, of the singular particular solution. From eqns (50) to (55), a unique solution of the problem can now be found, in which the stress intensity factors are among the primary unknowns, as a result of the near-tip field local regularization. The performance of this strategy depends on an efficient definition of the near-tip field regions.

In order to use the Williams' field as a particular solution, valid for the local regularization, each near-tip field can be arbitrarily extended throughout the problem domain, provided it encloses a single crack tip of a flat crack. For instance, the flat edge crack problem can be analyzed with the near-tip field extended throughout the whole problem domain.

In order to deal with several cracks simultaneously, the near-tip fields can be automatically generated in a small circular region surrounding each crack tip, as shown in Fig. 4. This arrangement has an implicit guarantee that, in each near-tip field region, the Williams' field is a particular solution valid for the local regularization of the region. Furthermore, it allows the method to be efficiently used for the study of general crack growth problems, because the matrix equations of each near-tip field region can be generated once and for all, provided the relative position of the interface boundary elements, to the crack tip elements, is kept constant during the crack growth process.

The smallness of the near-tip field is not expected to affect the solution accuracy, provided the usual half-length ratio is adopted between neighbour elements. Also, the closeness of the interface boundary to the crack tip should not affect the problem accuracy because, in the near-tip field, the problem is regularized with the cancellation of the tractions right at the tip.

7 The J–integral method

Consider a cartesian reference system, defined at the tip of a traction-free crack, as shown in Fig. 5. The J-integral is defined as

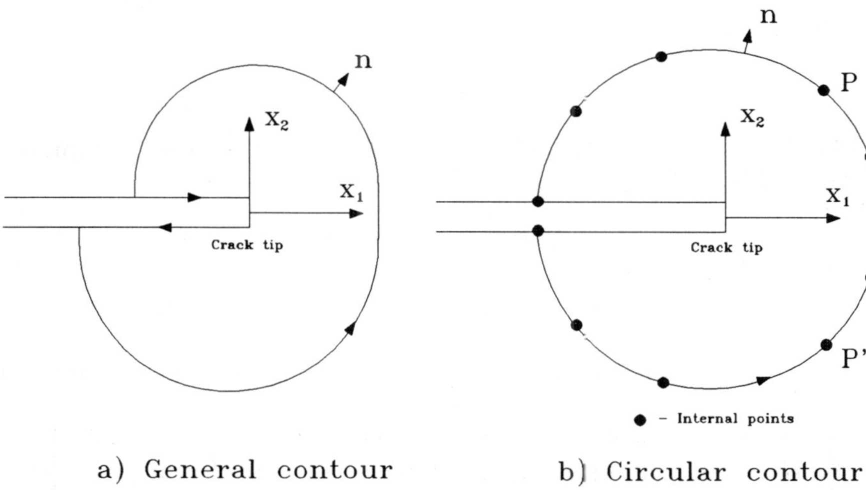

Figure 5: Coordinate reference system and contour path for J-integral

$$J = \int_S (W n_1 - t_j u_{j,1}) \, \mathrm{d}S \tag{56}$$

where S is an arbitrary contour surrounding the crack tip; W is the strain energy density, given by $\frac{1}{2}\sigma_{ij}\varepsilon_{ij}$, where σ_{ij} and ε_{ij} are the stress and strain tensors, respectively; t_j are traction components, given by $\sigma_{ij}n_i$, where n_i are the components of the unit outward normal to the contour path. The relationship between the J-integral and the stress intensity factors is given by

$$J = \frac{K_I^2 + K_{II}^2}{E'} \tag{57}$$

where E' is the elasticity modulus E for plane stress conditions and $E' = E/(1-\nu^2)$ for plane strain conditions.

In order to decouple the stress intensity factors in eqn (57), the integral J is represented by the sum of two integrals as follows:

$$J = J^I + J^{II} \tag{58}$$

where the superscripts indicate the deformation mode. For this representation to be possible, it is sufficient to decompose the displacement and stress fields into their symmetric and antisymmetric components, as described below. Consider two points, $P(x_1, x_2)$ and $P'(x_1, -x_2)$, that are symmetric relative to the crack axis, as shown in Fig. 5. At these points the displacement field can be expressed as a combination of symmetric and antisymmetric components in the form:

$$\left\{ \begin{array}{c} u_1 \\ u_2 \end{array} \right\} = \left\{ \begin{array}{c} +u_1^I + u_1^{II} \\ +u_2^I + u_2^{II} \end{array} \right\}, \quad \left\{ \begin{array}{c} u_1' \\ u_2' \end{array} \right\} = \left\{ \begin{array}{c} +u_1^I - u_1^{II} \\ -u_2^I + u_2^{II} \end{array} \right\} \tag{59}$$

From eqns (59), the symmetric and antisymmetric displacement components are given by:

$$\left\{ \begin{array}{c} u_1^I \\ u_2^I \end{array} \right\} = \frac{1}{2} \left\{ \begin{array}{c} u_1 + u_1' \\ u_2 - u_2' \end{array} \right\}, \quad \left\{ \begin{array}{c} u_1^{II} \\ u_2^{II} \end{array} \right\} = \frac{1}{2} \left\{ \begin{array}{c} u_1 - u_1' \\ u_2 + u_2' \end{array} \right\} \tag{60}$$

At the same points $P(x_1, x_2)$ and $P'(x_1, -x_2)$, the stress field can be expressed as a combination of symmetric and antisymmetric components in the form:

$$\left\{ \begin{array}{c} \sigma_{11} \\ \sigma_{22} \\ \sigma_{12} \end{array} \right\} = \left\{ \begin{array}{c} +\sigma_{11}^I + \sigma_{11}^{II} \\ +\sigma_{22}^I + \sigma_{22}^{II} \\ +\sigma_{12}^I + \sigma_{12}^{II} \end{array} \right\}, \quad \left\{ \begin{array}{c} \sigma_{11}' \\ \sigma_{22}' \\ \sigma_{12}' \end{array} \right\} = \left\{ \begin{array}{c} +\sigma_{11}^I - \sigma_{11}^{II} \\ +\sigma_{22}^I - \sigma_{22}^{II} \\ -\sigma_{12}^I + \sigma_{12}^{II} \end{array} \right\} \tag{61}$$

From eqns (61), the symmetric and antisymmetric stress components are given by:

$$\left\{ \begin{array}{c} \sigma_{11}^I \\ \sigma_{22}^I \\ \sigma_{12}^I \end{array} \right\} = \frac{1}{2} \left\{ \begin{array}{c} \sigma_{11} + \sigma_{11}' \\ \sigma_{22} + \sigma_{22}' \\ \sigma_{12} - \sigma_{12}' \end{array} \right\}, \quad \left\{ \begin{array}{c} \sigma_{11}^{II} \\ \sigma_{22}^{II} \\ \sigma_{12}^{II} \end{array} \right\} = \frac{1}{2} \left\{ \begin{array}{c} \sigma_{11} - \sigma_{11}' \\ \sigma_{22} - \sigma_{22}' \\ \sigma_{12} + \sigma_{12}' \end{array} \right\} \tag{62}$$

When eqns (59) and (61) are introduced in eqn (56), eqn (58) is obtained, with the J-integral components given by

$$J^m = \int_S \left(W^m \, n_1 - t_j^m \, u_{j,1}^m \right) \mathrm{d}S \tag{63}$$

for $m = I$ or $m = II$. Finally, the following relationships hold:

$$J^I = \frac{K_I^2}{E'}, \quad J^{II} = \frac{K_{II}^2}{E'} \tag{64}$$

The implementation of this procedure into the boundary element method is straightforward. A circular contour path, around the crack tip, is defined with a set of internal points, located at symmetrical positions, relative to the crack axis, as shown in Fig. 5. The two contour points on the crack faces are the first and the last points of the path respectively. In a circular path, at these points it is always verified that $n_1 = -1$

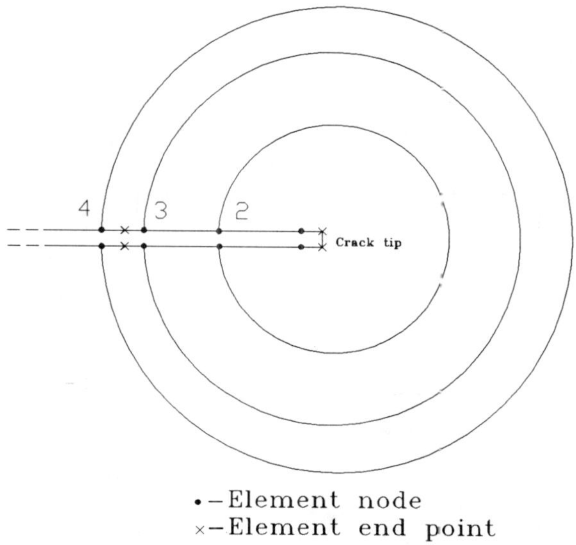

Figure 6: Circular path numbering system for J-integral contours

and $n_2 = 0$ and thus, for a traction-free crack, $t_2 = 0$. The integration along the contour path can be accomplished with the trapezoidal or Simpson's rule or by Gaussian quadrature. For the sake of simplicity only circular paths, centered at the crack-tip and containing a pair of crack nodes, were considered; each path is referred to by a path number which increases as the radius of the contour increases, see Fig. 6.
The J-integral is an effective method for the determination of stress intensity factors, because the interior elastic field can be accurately determined along the contour path with the boundary element method, since the exact variation of the interior elastic field is built into the fundamental solution.

8 Numerical results

The edge crack problem

Consider a square plate with a single edge crack, as represented in Fig. 7. The crack length is denoted by a and the width of the plate is denoted by w. The plate is subjected to the action of a uniform traction $\bar{t}$, applied symmetrically at the ends in the direction perpendicular to the crack.

Results have been obtained from the singularity subtraction and the J-integral techniques and compared with those published by Civelek & Erdogan.[40] Five cases were considered, $a/w = 0.2, 0.3, 0.4, 0.5$ and 0.6. A convergence study was carried out with three different meshes of 32, 40 and 64 quadratic boundary elements, in which the crack was discretized with 4, 5 and 8 quadratic discontinuous elements on each surface. Convergence was achieved with these meshes and therefore, the mesh of 32 elements with a crack discretization graded towards the tip with the ratios 0.4, 0.3, 0.2 and 0.1, was selected for the analysis. The results obtained, presented in Table 1, show a high level of accuracy when compared with those of Ref. 40.

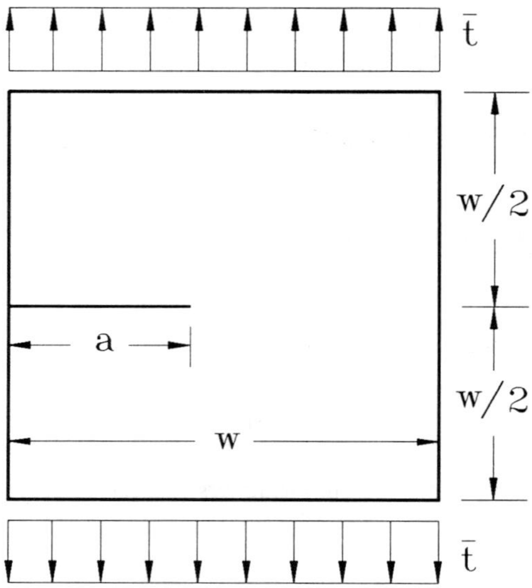

Figure 7: Square plate with a single edge crack

a/w	$K_I/(\bar{t}\sqrt{\pi a})$						
	SST	J-Integral Contour Path					Ref.
		2	3	4	5	8	40
0.2	1.483	1.496	1.495	1.495	1.494	1.495	1.488
0.3	1.841	1.860	1.859	1.858	1.857	1.858	1.848
0.4	2.309	2.340	2.338	2.338	2.336	2.335	2.324
0.5	2.980	3.032	3.029	3.028	3.025	3.021	3.010
0.6	4.092	4.188	4.185	4.184	4.179	4.168	4.152

Table 1: Stress intensity factors for a single edge crack in a square plate

For the analysis of this problem with the singularity subtraction technique (SST), the near-tip field region was defined on the whole problem domain. A particular solution, representing the singular field around the crack-tip of a semi-infinite crack and valid in the global domain of the problem, was defined by the first term of the Williams' eigenexpansion:[38] the stresses are given by

$$\sigma_{xx} = \frac{K_I}{\sqrt{2\pi r}} \cos\frac{\theta}{2}(1 - \sin\frac{\theta}{2}\sin\frac{3\theta}{2}) + \frac{K_{II}}{\sqrt{2\pi r}} \sin\frac{\theta}{2}(2 + \cos\frac{\theta}{2}\cos\frac{3\theta}{2}) \tag{65}$$

$$\sigma_{yy} = \frac{K_I}{\sqrt{2\pi r}} \cos\frac{\theta}{2}(1 + \sin\frac{\theta}{2}\sin\frac{3\theta}{2}) - \frac{K_{II}}{\sqrt{2\pi r}} \sin\frac{\theta}{2}\cos\frac{\theta}{2}\cos\frac{3\theta}{2} \tag{66}$$

and

$$\sigma_{xy} = \frac{K_I}{\sqrt{2\pi r}} \cos\frac{\theta}{2}\sin\frac{\theta}{2}\cos\frac{3\theta}{2} + \frac{K_{II}}{\sqrt{2\pi r}} \cos\frac{\theta}{2}(1 - \sin\frac{\theta}{2}\sin\frac{3\theta}{2}) \tag{67}$$

and the associated displacements are given by

$$u = \frac{K_I}{4\mu}\sqrt{\frac{r}{2\pi}}[(2\kappa - 1)\cos\frac{\theta}{2} - \cos\frac{3\theta}{2}] + \frac{K_{II}}{4\mu}\sqrt{\frac{r}{2\pi}}[(2\kappa + 3)\sin\frac{\theta}{2} + \sin\frac{3\theta}{2}] \tag{68}$$

and

$$v = \frac{K_I}{4\mu}\sqrt{\frac{r}{2\pi}}[(2\kappa + 1)\sin\frac{\theta}{2} - \sin\frac{3\theta}{2}] + \frac{K_{II}}{4\mu}\sqrt{\frac{r}{2\pi}} + [(2\kappa - 3)\cos\frac{\theta}{2} + \cos\frac{3\theta}{2}] \tag{69}$$

where K_I and K_{II} are the stress intensity factors of the opening and sliding modes, respectively. The constant μ is the shear modulus and the constant κ is given by $3 - 4\nu$ for plain strain and $(3 - \nu)/(1 + \nu)$ for plain stress, where ν is Poisson's ratio. The polar coordinate reference system (r, θ) is centred at the crack-tip, such that $\theta = 0$ is the crack axis, coincident with the cartesian x axis.

For the J-integral computations, 10 internal points and the trapezoidal rule were considered. The stability of the J-integral results, for any contour path, is noticeable. Fig. 8 shows the initial and deformed boundary element meshes for the case $a/w = 0.6$. Notice that, although discontinuous elements are used to model the crack faces, no oscillations are present in the results. This is an important feature of the present formulation that clearly differs from that reported in the work of Rudolphi, Krishnasamy, Schmerr & Rizzo[22] and Krishnasamy, Rudolphi, Schmerr & Rizzo.[41]

8.2 The centre crack problem

Consider, now, the analysis of a central slant crack in a rectangular plate, represented in Fig. 9. The ratio between the height and the width of the plate is given by $h/w = 2$. The crack has a length $2a$ and makes an angle of $\varphi = 45°$ with the direction perpendicular to the applied traction. The plate is loaded with a uniform traction $\bar{t}$, applied symmetrically at the ends.

To solve this problem, a mesh of 36 quadratic boundary elements was set up, in which 6 discontinuous elements were used on each face of the crack, graded from the centre towards the tips, with the ratios 0.25, 0.15 and 0.1. Accurate results for this

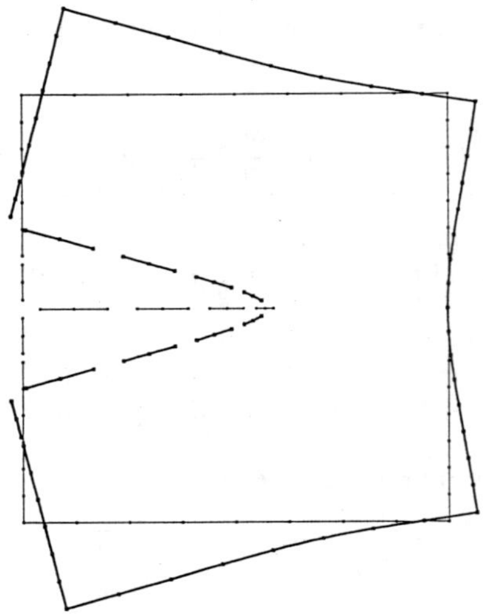

Figure 8: Initial and deformed boundary element meshes for the edge crack problem ($a/w = 0.6$)

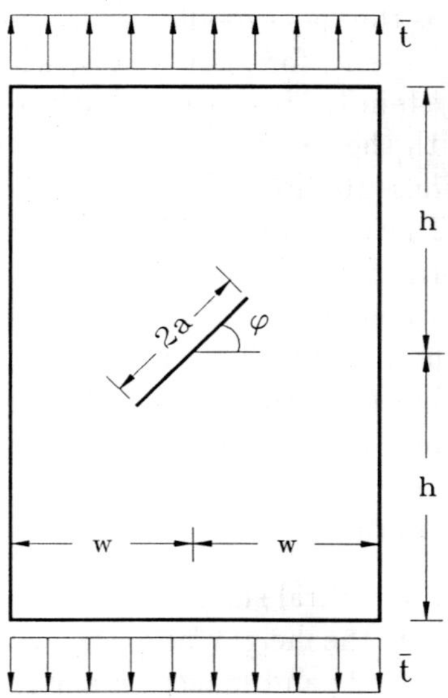

Figure 9: Rectangular plate with a central slant crack ($h/w = 2$, $\varphi = 45°$)

problem were published by Murakami.[42] The results obtained, presented in Tables 2 and 3, show an excellent accuracy when compared with the results of the Ref. 42.

a/w	$K_I/(t\sqrt{\pi a})$						
	SST	J-Integral Contour Path					Ref.
		2	3	4	5	8	42
0.2	0.518	0.521	0.519	0.521	0.521	0.521	0.518
0.3	0.540	0.544	0.542	0.544	0.544	0.544	0.541
0.4	0.571	0.575	0.574	0.576	0.576	0.576	0.572
0.5	0.610	0.616	0.614	0.617	0.617	0.616	0.612
0.6	0.657	0.666	0.665	0.667	0.667	0.666	0.661

Table 2: Mode I stress intensity factors for a central slant crack in a rectangular plate ($h/w = 2$, $\varphi = 45°$)

a/w	$K_{II}/(t\sqrt{\pi a})$						
	SST	J-Integral Contour Path					Ref.
		2	3	4	5	8	42
0.2	0.507	0.499	0.499	0.501	0.503	0.508	0.507
0.3	0.516	0.508	0.508	0.511	0.512	0.517	0.516
0.4	0.529	0.521	0.521	0.523	0.525	0.529	0.529
0.5	0.546	0.538	0.538	0.541	0.542	0.547	0.546
0.6	0.567	0.560	0.560	0.562	0.564	0.569	0.567

Table 3: Mode II stress intensity factors for a central slant crack in a rectangular plate ($h/w = 2$, $\varphi = 45°$)

For the SST analysis of this problem, the near-tip field region can still be defined on the whole problem domain, provided a particular solution representing the elastic field around the crack-tips of a finite flat crack in an infinite plate and valid in the global problem domain, is available. Such a particular solution was defined in terms of the Westergaard[39] stress functions. For an infinite plate, containing a crack of length $2a$, subjected to a remote loading, the stress field components are given by

$$\sigma_{xx} = \frac{1}{\sqrt{\pi a}}[\frac{K_I r}{\sqrt{r_1 r_2}}\cos(\theta - \frac{\theta_1+\theta_2}{2}) + \frac{K_I a^2 r}{\sqrt{(r_1 r_2)^3}}\sin\theta\sin\frac{-3(\theta_1+\theta_2)}{2} + +2\frac{K_{II} r}{\sqrt{r_1 r_2}}\sin(\theta - \frac{\theta_1+\theta_2}{2}) - \frac{K_{II} a^2 r}{\sqrt{(r_1 r_2)^3}}\sin\theta\cos\frac{-3(\theta_1+\theta_2)}{2}] \quad (70)$$

$$\sigma_{yy} = \frac{1}{\sqrt{\pi a}}[\frac{K_I r}{\sqrt{r_1 r_2}}\cos(\theta - \frac{\theta_1+\theta_2}{2}) - \frac{K_I a^2 r}{\sqrt{(r_1 r_2)^3}}\sin\theta\sin\frac{-3(\theta_1+\theta_2)}{2} + +\frac{K_{II} a^2 r}{\sqrt{(r_1 r_2)^3}}\sin\theta\cos\frac{-3(\theta_1+\theta_2)}{2}] \quad (71)$$

and

$$\sigma_{xy} = \frac{1}{\sqrt{\pi a}}[\frac{K_I a^2 r}{\sqrt{(r_1 r_2)^3}}\sin\theta\cos\frac{-3(\theta_1+\theta_2)}{2} + +\frac{K_{II} r}{\sqrt{r_1 r_2}}\cos(\theta - \frac{\theta_1+\theta_2}{2}) + \frac{K_{II} a^2 r}{\sqrt{(r_1 r_2)^3}}\sin\theta\sin\frac{-3(\theta_1+\theta_2)}{2}] \quad (72)$$

where (r_1, θ_1), (r_2, θ_2) and (r, θ) are polar coordinate reference systems centred at each crack tip and at the centre of the crack, as shown in Fig. 10. For plain strain, the

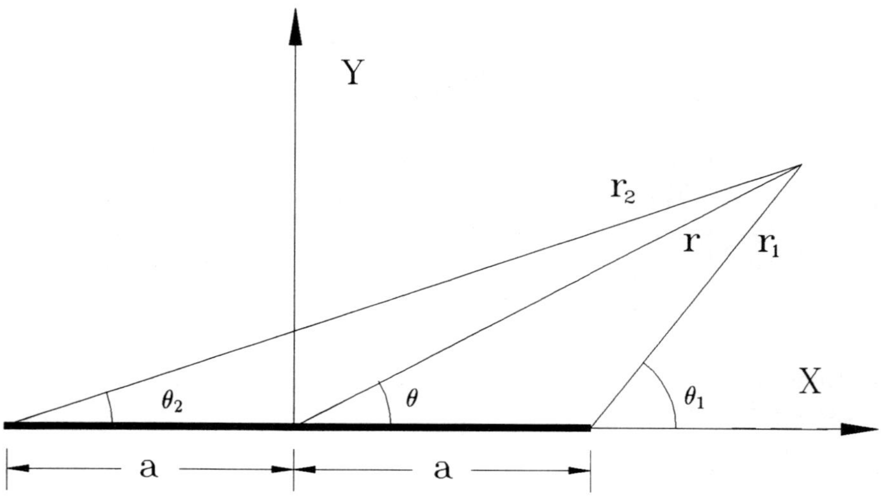

Figure 10: Coordinate reference systems for the Westergaard elastic field

associated displacements are written as follows:

$$u = \frac{1}{2\mu\sqrt{\pi a}}[(1-2\nu)K_I\sqrt{r_1 r_2}\cos\frac{\theta_1+\theta_2}{2} - \frac{K_I r^2}{\sqrt{r_1 r_2}}\sin\theta\sin(\theta - \frac{\theta_1+\theta_2}{2}) + +2(1-\nu)K_{II}\sqrt{r_1 r_2}\sin\frac{\theta_1+\theta_2}{2} + \frac{K_{II} r^2}{\sqrt{r_1 r_2}}\sin\theta\cos(\theta - \frac{\theta_1+\theta_2}{2})] \quad (73)$$

and

$$v = \frac{1}{2\mu\sqrt{\pi a}}[2(1-\nu)K_I\sqrt{r_1r_2}\sin\frac{\theta_1+\theta_2}{2} - \frac{K_Ir^2}{\sqrt{r_1r_2}}\sin\theta\cos(\theta-\frac{\theta_1+\theta_2}{2}) - $$
$$-(1-2\nu)K_{II}\sqrt{r_1r_2}\cos\frac{\theta_1+\theta_2}{2} - \frac{K_{II}r^2}{\sqrt{r_1r_2}}\sin\theta\sin(\theta-\frac{\theta_1+\theta_2}{2})] \qquad (74)$$

The J-integral computations were carried out with 30 internal points and the trapezoidal integration rule. With such a coarse mesh, the results obtained are remarkably accurate. For mode II, the stability of the J-integral results is slightly lower than for mode I. This means that, for the mode II deformation, the variation of the elastic field along the contour paths is not properly approximated with the trapezoidal integration rule. Improved stability could be obtained by considering either more internal points along each path or a higher order integration rule.

Fig. 11 shows the initial and deformed boundary element meshes for the case $a/w = 0.6$.

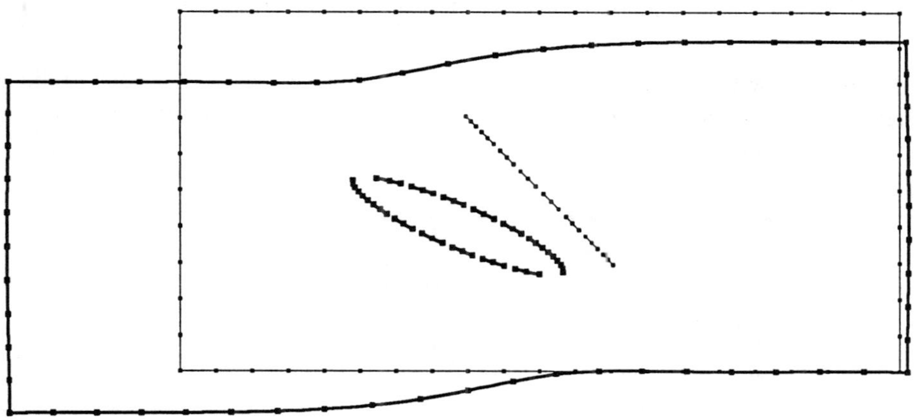

Figure 11: Initial and deformed boundary element meshes for the centre crack problem ($a/w = 0.6$)

8.3 The general crack problem

As a final test consider the analysis of a rectangular plate with an internal kinked crack, represented in Fig. 12. The ratio between the height and the width of the plate is given by $h/w = 2$. The segment of the crack of length a makes an angle of 90° with the direction of the applied traction, while the other segment makes an angle of 45° with the same direction and has a length b. The projection of the total crack in the direction perpendicular to the traction is given by $2c = a + b/\sqrt{2}$. The kink of the crack is at the center of the plate which is loaded with a uniform traction $\bar{t}$, symmetrically applied at the ends of the plate. Three cases were considered, $b/a = 0.2$, 0.4 and 0.6, with $a/w = 0.1$. The stress intensity factors were obtained for both tips A and B with a boundary element mesh of 48 quadratic elements. The crack segment of length

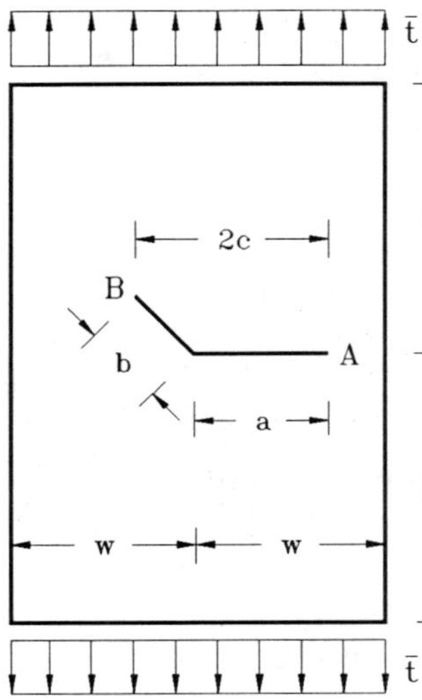

Figure 12: Rectangular plate with an internal kinked crack ($h/w = 2$, $a/w = 0.1$)

a was discretized with 5 discontinuous quadratic elements, on each crack face, while the other crack segment was discretized with 4 discontinuous quadratic elements on each crack face. Accurate results for comparison are published by Murakami.[42]

b/a	$K_I/(t\sqrt{\pi c})$					
	J-Integral Contour Path					Ref.
	2	3	4	5	8	42
0.2	0.998	0.995	0.997	0.996	0.993	0.995
0.4	0.993	0.990	0.992	0.991	0.989	0.990
0.6	0.990	0.987	0.989	0.988	0.987	0.986

Table 4: Mode I stress intensity factors for tip A; kinked crack problem

b/a	$K_{II}/(t\sqrt{\pi c})$					
	J-Integral Contour Path					Ref.
	2	3	4	5	8	42
0.2	0.029	0.029	0.030	0.030	0.030	0.028
0.4	0.034	0.035	0.035	0.035	0.036	0.033
0.6	0.031	0.032	0.032	0.032	0.032	0.030

Table 5: Mode II stress intensity factors for tip A; kinked crack problem

The results obtained for this problem, presented in Tables 4 to 7, where obtained by the J-integral only and are in good agreement with those of the Ref. 42, except where

K_{II} is very small.

A singular particular solution, valid in the global problem domain, is not available for an analysis of this problem with the singularity subtraction technique. Therefore, partition of the problem domain would be necessary in order to regularize locally the near-tip field regions.

b/a	$K_I/(t\sqrt{\pi c})$					
	J-Integral Contour Path					Ref.
	2	3	4	5	8	42
0.2	0.605	0.603	0.603	0.604	0.604	0.598
0.4	0.578	0.576	0.576	0.576	0.576	0.574
0.6	0.572	0.570	0.570	0.570	0.570	0.568

Table 6: Mode I stress intensity factors for tip B; kinked crack problem

b/a	$K_{II}/(t\sqrt{\pi c})$					
	J-Integral Contour Path					Ref.
	2	3	4	5	8	42
0.2	0.556	0.556	0.555	0.555	0.556	0.557
0.4	0.602	0.602	0.601	0.602	0.603	0.607
0.6	0.623	0.623	0.623	0.623	0.624	0.627

Table 7: Mode II stress intensity factors for tip B; kinked crack problem

9 Incremental crack-extension analysis

The incremental crack-extension analysis assumes a piece-wise linear discretization of the unknown crack path. For each increment of the crack extension, the dual boundary element method is applied to carry out a stress analysis of the cracked structure and the J-integral is the technique used for the evaluation of the stress intensity factors. The steps of this basic computational cycle, repeatedly executed for any number of crack-extension increments, are summarized as follows:

- Carry out a dual boundary element method stress analysis of the structure.
- Compute the stress intensity factors with the J-integral technique.
- Compute the direction of the crack-extension increment.
- Extend the crack one increment along the direction computed in the previous step.
- Repeat all the above steps sequentially until a specified number of crack-extension increments is reached.

For the sake of simplicity, the increment of crack extension is discretized with a fixed number of new boundary elements. In order to avoid numerical problems, concerned with the relative size of neighbouring elements, the crack increment length is kept between convenient limiting bounds, defined in terms of the size of the crack-tip element. Apart from this constraint, the length of the crack extension increment may be defined as the result of a compromise between accuracy and computational cost; the smaller the crack increment the more accurate and expensive is the analysis.

The results obtained from an incremental crack-extension analysis are a crack path diagram and diagrams of the stress intensity factor variation along the crack path.

10 Crack-extension criterion

Several criteria have been proposed to describe the mixed-mode crack growth. Among them, one of the most commonly used is based on the maximum principal stress at the crack tip, Erdogan & Sih.[44] The maximum principal stress criterion postulates that the growth of the crack will occur in a direction perpendicular to the maximum principal stress. Thus, the local crack-growth direction θ_t is determined by the condition that the local shear stress is zero, that is

$$K_I \sin\theta_t + K_{II}(3\cos\theta_t - 1) = 0 \tag{75}$$

where θ_t is an angular coordinate centered at the crack tip and measured from the crack axis ahead of the tip. As a continuous criterion, the maximum principal stress does not take account of the discreteness of the crack extension modelling procedure. In other words, the crack path is defined continuously by the trajectory of the maximum principal stresses, evaluated locally by eqn (75). Therefore, the incremental extension of a crack in a general mixed-mode deformation field, computed by eqn (75), is always defined locally in the same direction, whatever length of crack extension, Δa is considered. As a consequence, uniqueness of the crack path cannot be assured with different sizes of the crack-extension increment. Therefore, in an incremental analysis, the tangent direction of the crack-path, predicted by eqn (75), must be corrected to give the direction of the actual crack-extension increment.

The correct direction of the crack-extension increment ensures that, for different analysis of the same problem with different crack-extension increments, a unique final crack path is achieved. The procedure applied to define the direction of the n-th crack-extension increment introduces a correction angle β to the tangent direction $\theta_{t(n)}$ predicted by the maximum principal stress criterion, as shown in Fig. 13. Using geometric relationships, this correction angle is given by $\beta = \theta_{t(n+1)}/2$, in which $\theta_{t(n+1)}$ is the direction of the next crack-extension increment, also computed with the maximum principal stress criterion. This predictor-corrector procedure can be applied while each correction is smaller than the previous one. For the current n-th crack-extension increment, the i-th iteration can be summarized as follows:

- for the first iteration only, evaluate the crack-path tangent direction $\theta^i_{t(n)}$ with the maximum principal stress criterion, eqn (75);

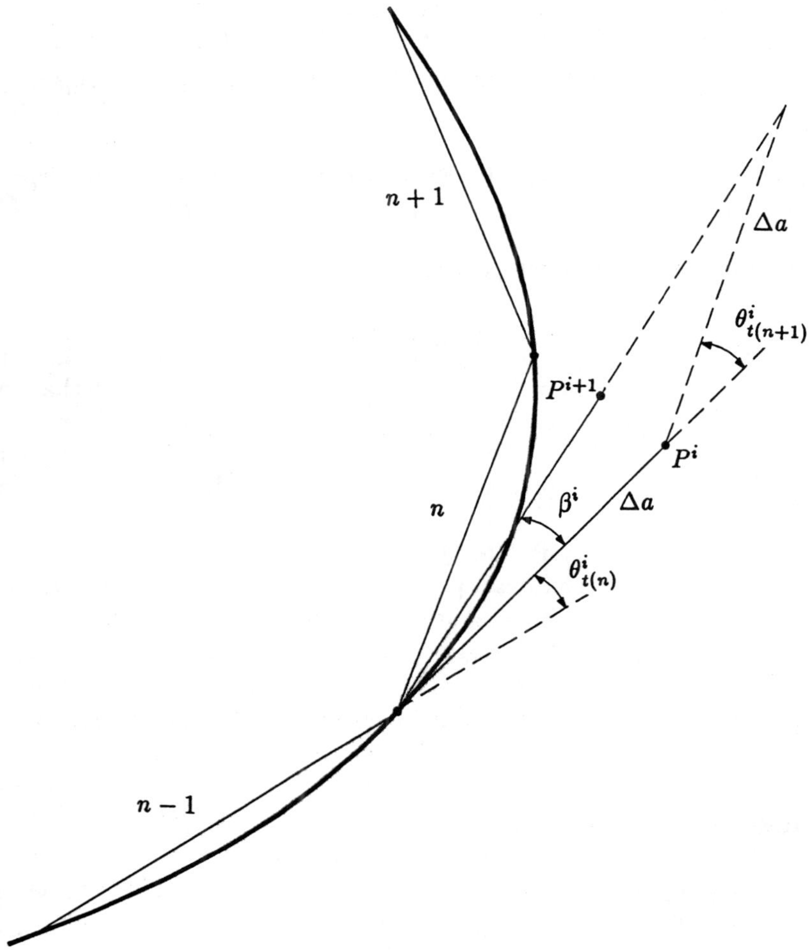

n - current crack-extension increment
Δa - increment size
i - iteration number
θ_t - increment direction computed with the tangent criterion
β - correction angle ($\beta = \theta_{t(n+1)}/2$)
P - crack tip location

Figure 13: Incremental crack-extension direction

- along the direction computed in the previous step, extend the crack one increment Δa to P^i and evaluate the new stress intensity factors;
- with the new stress intensity factors and the maximum principal stress criterion, eqn (75), evaluate the next crack-path direction $\theta^i_{t(n+1)}$;
- define the correction angle $\beta^i = \theta^i_{t(n+1)}/2$, measured from the increment Δa defined in the second step;
- correct the crack-extension increment, defined in the second step, to its new position P^{i+1} in the direction given by $\theta^{i+1}_{t(n)} = \theta^i_{t(n)} + \beta^i$;
- starting from the second step, repeat the above steps sequentially while $|\beta^{i+1}| < |\beta^i|$.

When the size of the crack-extension increment, Δa tends to zero, the angle $\theta_{t(n+1)}$ also tends to zero and so does the correction angle, β. This means that in the limit, the direction of the incremental crack-extension tends to the direction of the tangent of the continuous crack path. The present procedure is simple, but more research may be needed to define an efficient criterion for the prediction of the direction of the crack-extension increment.

In a mixed-mode analysis, an equivalent mode I stress intensity factor can be defined, Broek.[45] For the maximum principal stress criterion , it is given by

$$K_{I\text{eq}} = K_I \cos^3 \frac{\theta_t}{2} - 3K_{II} \cos^2 \frac{\theta_t}{2} \sin \frac{\theta_t}{2} \tag{76}$$

The fracture condition then follows from $K_{I\text{eq}} = K_{Ic}$, in which K_{Ic} is the fracture toughness.

11 Crack extension applications

Two applications of the incremental crack extension analysis will be studied in the present section. The first application characterizes the behaviour of a crack that initially deforms under pure mode I or pure mode II. The second application makes the analysis of a crack that initially deforms under mixed mode.

For the first application, consider a square plate of height h, with an edge crack of length $a = h/3$, subjected to either pure mode I or pure mode II loading, as represented in Fig. 14. Results of the incremental crack extension analysis of these problems were obtained with the maximum principal stress criterion and a crack extension increment equal to three times the length of the initial crack-tip element. Fig.s 15 and 16 present these results in terms of crack paths and stress intensity factor diagrams, for each one of the problems. The stress intensity factor diagrams are normalized against the mode *I* equivalent stress intensity factor computed at the initial crack length.

The second application, where the initial crack is under mixed-mode deformation, is the cruciform plate represented in Fig. 17. The plate has square arms of length h which are loaded with uniform tractions $\bar{t}_1$ and $\bar{t}_2$, applied at the ends of each arm,

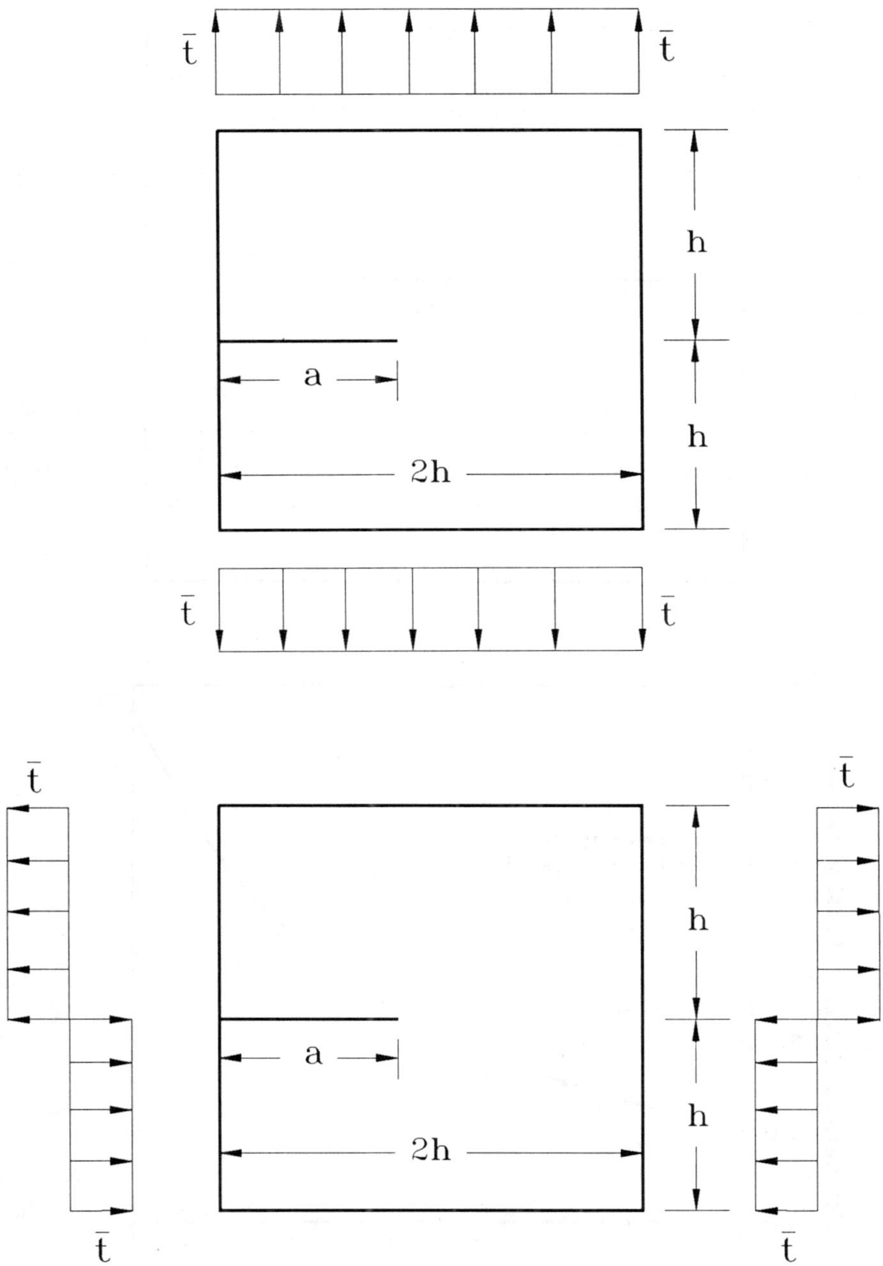

Figure 14: Square plate with an edge crack under pure mode I and pure mode II

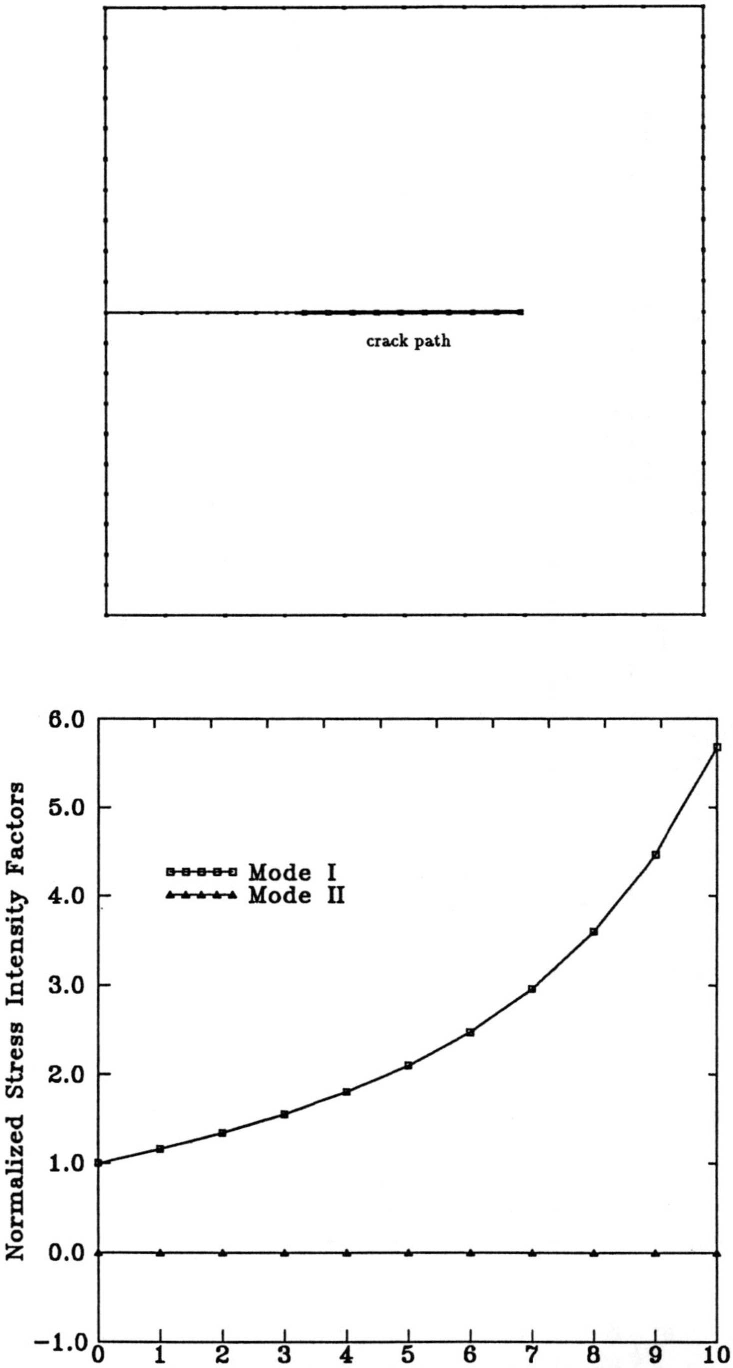

Figure 15: Crack-extension analysis of a crack initially under pure mode I

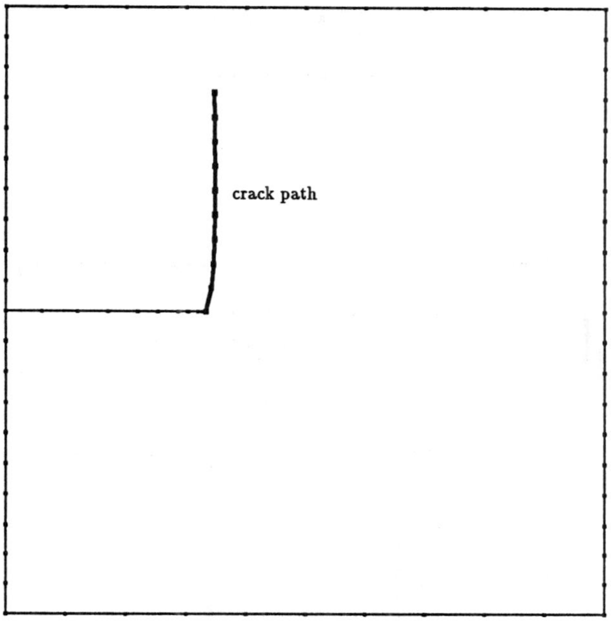

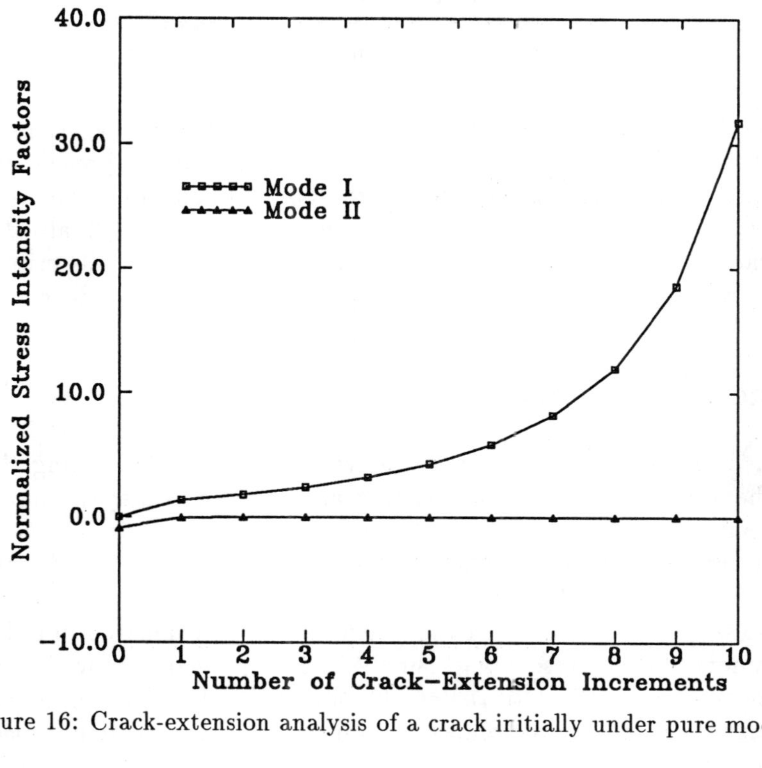

Figure 16: Crack-extension analysis of a crack initially under pure mode II

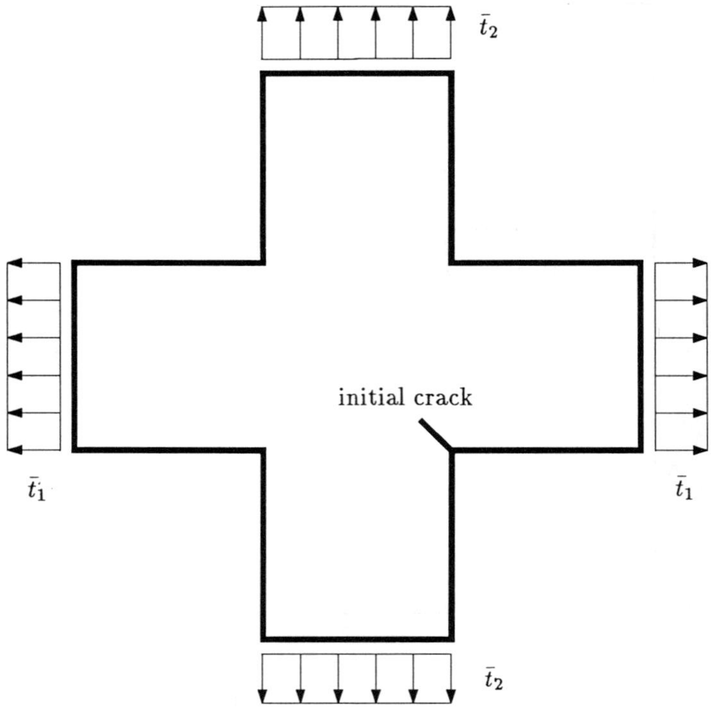

Figure 17: Cruciform cracked plate

respectively. At one of the interior corners, a crack of length $a = h/4$ was introduced in the direction of the diagonally opposite corner of the plate. In the present application, incremental crack extension analyses were carried out for four different cases in which $\bar{t}_1 = 0$ and $\bar{t}_2 = p$, $\bar{t}_1 = p$ and $\bar{t}_2 = 0$, $\bar{t}_1 = p$ and $\bar{t}_2 = p$ and finally $\bar{t}_1 = p$ and $\bar{t}_2 = 2p$. Crack paths, obtained with the maximum principal stress criterion and a crack increment equal to three times the length of the initial crack-tip element, are presented in Fig. 18.

12 Fatigue crack-growth

In general, fatigue crack-growth is driven by variable amplitude loading. In the simplest case, the loading cycles have a constant amplitude and may be described by a constant amplitude load with a constant maximum/minimum stress ratio. The aim of the fatigue crack-growth analysis is to obtain residual-strength and fatigue-life diagrams for each crack. From these results, both the maximum permissible size of each crack and the safe operational life of the structure can be evaluated in a damage tolerance assessment. In the present work, analysis of fatigue crack growth is introduced as a post-processing procedure on the results of each crack-extension increment.

A residual-strength diagram shows the variation in the maximum load that the cracked structure can sustain, that is, the load that causes fracture instability, as the crack length varies. Thus, for a given crack length a, the residual strength, σ_c is given

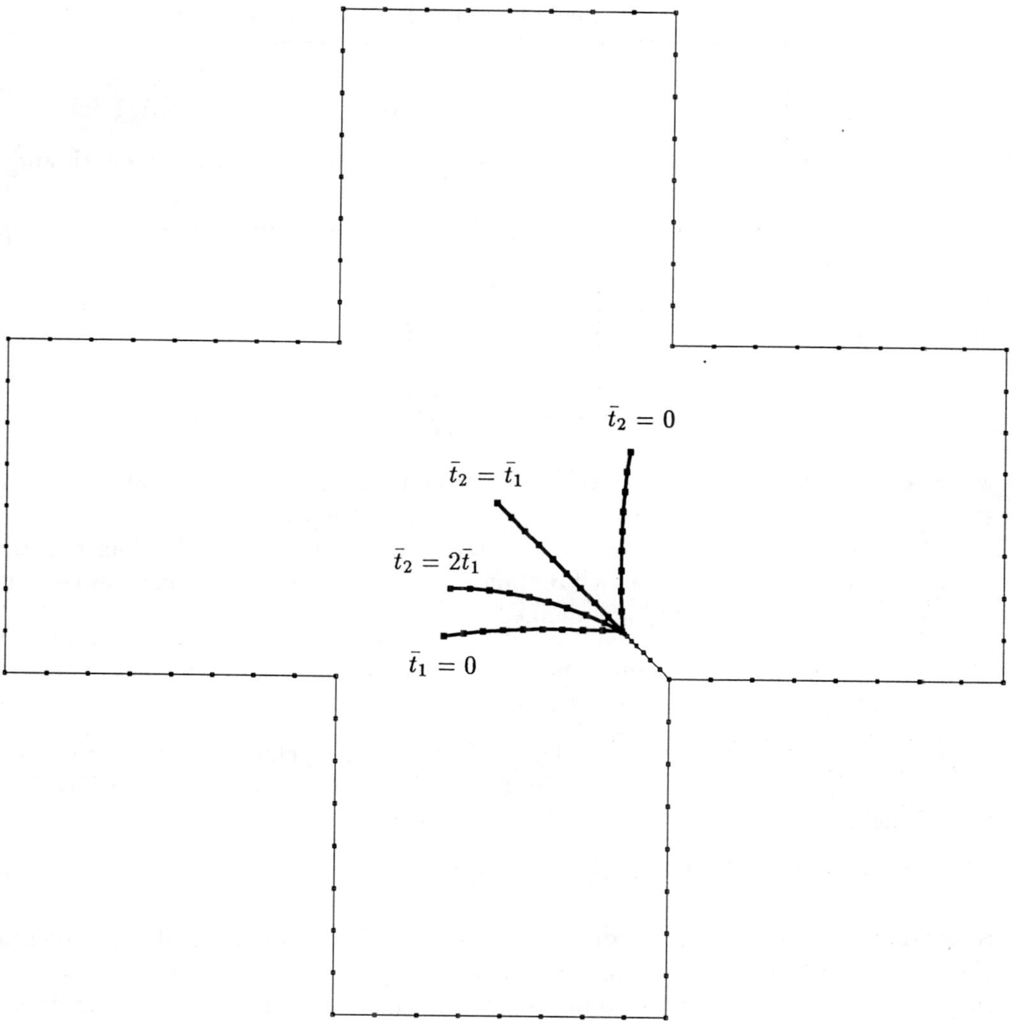

Figure 18: Crack paths for the cruciform cracked plate problem

by the critical load which is defined as

$$\sigma_c = \frac{K_{Ic}}{Y\sqrt{\pi a}} \tag{77}$$

where Y is the geometry factor. For the same crack length, the geometry factor is constant and hence, can be defined in terms of the applied reference stress, σ_r and the corresponding mode I equivalent stress intensity factor, $K_{I\text{eq}}$ as

$$Y = \frac{K_{Ic}}{\sigma_c\sqrt{\pi a}} = \frac{K_{I\text{eq}}}{\sigma_r\sqrt{\pi a}} \tag{78}$$

From this equation, the following relationship between the residual strength and the applied reference stress holds:

$$\sigma_c = \frac{K_{Ic}}{K_{I\text{eq}}}\sigma_r \tag{79}$$

At each step of an incremental analysis, the residual strength is represented conveniently in a normalized form, derived from eqn (79) as

$$\frac{\sigma_c}{\sigma_{c*}} = \frac{K_{I\text{eq}*}}{K_{I\text{eq}}} \tag{80}$$

where σ_{c*} and $K_{I\text{eq}*}$ are respectively the residual strength and the mode I equivalent stress intensity factor computed at the initial crack length.

A fatigue-life diagram shows the variation in the number of loading cycles, required to extend the crack, as a function of the crack length. It can be computed from the generalized Paris model defined as

$$\frac{\mathrm{d}a}{\mathrm{d}N} = C(\Delta K_{\text{eff}})^m \tag{81}$$

where a is the crack length, N is the number of load cycles, C and m are material dependent constants and ΔK_{eff} is the range of the effective stress intensity factor. The model of Tanaka,[46] in which ΔK_{eff} is defined by

$$\Delta K_{\text{eff}}^2 = \Delta K_I^2 + 2\Delta K_{II}^2 \tag{82}$$

is usually applied in mixed-mode analyses, since it has been verified experimentally. Thus, although $\Delta K_{\text{eff}} = \Delta K_{I\text{eq}}$, when the maximum principal stress criterion is used, the model of Tanaka was applied in the present work. In eqn (82), the stress intensity factor range of the individual modes is given by $\Delta K = K_{\max} - K_{\min} = K_{\max}(1-R)$ in which $R = K_{\min}/K_{\max} = \sigma_{\min}/\sigma_{\max}$ is the load ratio of the loading cycle. The number of loading cycles required to extend the crack a given increment is evaluated by integration of eqn (81) with the trapezoidal rule, applied for each increment of the analysis.

Results of fatigue crack-growth analysis were obtained for the cruciform plate represented in Fig. 17, for the case in which $\bar{t}_1 = \bar{t}_2$, with a crack-extension increment equal to three times the length of the initial crack-tip element. The residual strength diagram is presented in Fig. 19. The fatigue-life diagram, obtained for the case in

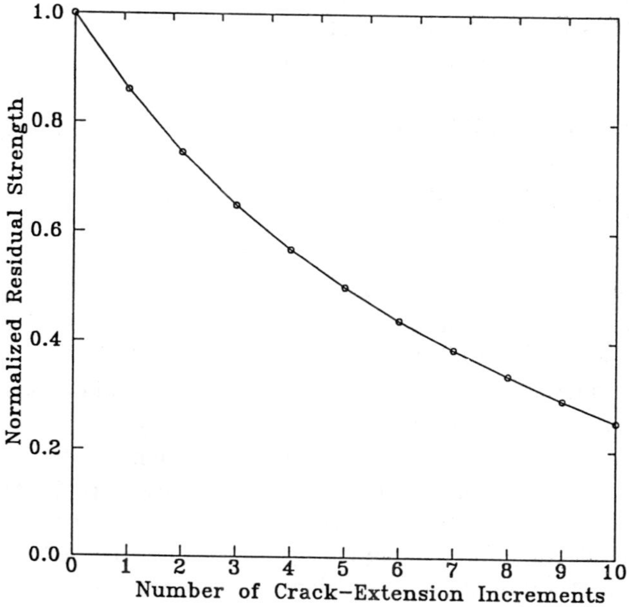

Figure 19: Residual-strength diagram for the cruciform cracked plate problem with $\bar{t}_1 = \bar{t}_2 = p$

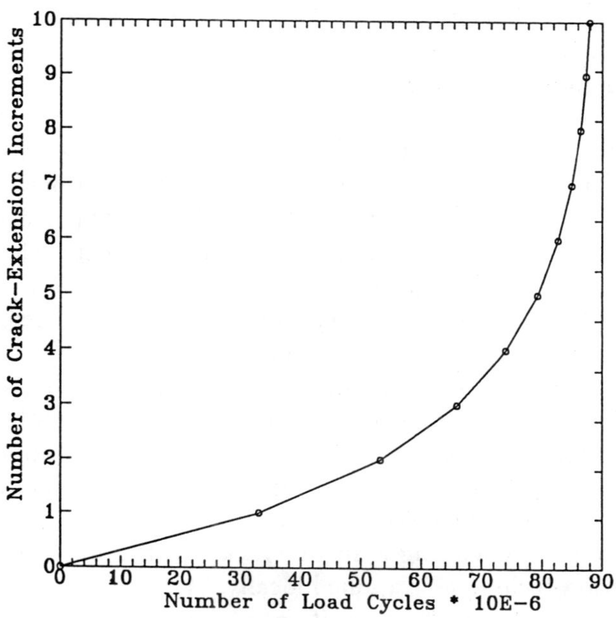

Figure 20: Fatigue-life diagram for the cruciform cracked plate problem with $\bar{t}_1 = \bar{t}_2 = p$

which the stress ratio=0.66666, C =4.624E-12 and m =3.3, is presented in Fig. 20. With these diagrams and a knowledge of the fracture toughness of the material, a complete damage tolerance analysis can be carried out.

13 Discussion and conclusions

In the present work, the dual boundary element method is applied to the incremental linear elastic analysis of crack extension problems. For each increment of the crack extension, a stress analysis is carried out and the stress intensity factors are evaluated with the J-integral technique. This basic computational cycle is repeated for each crack extension in a direction predicted by the maximum principal stress criterion, corrected to account for the discreteness of the analysis.

The dual boundary element method incorporates two independent boundary integral equations; one is the usual displacement boundary integral equation and the other one is the traction boundary integral equation. When the displacement equation is applied for collocation on one of the crack boundaries and the traction equation is applied for collocation on the other, general crack problems can be solved in a single-region formulation. This feature constitutes a practical advantage of the dual boundary element method over the finite element and multi-region boundary element methods, because remeshing, extensively used in these methods, is no longer required when the crack extension increment is modelled with new elements.

The existence of the principal-value integrals of the traction equation requires continuity of the displacement derivatives at the collocation node, on a smooth boundary. Because of this requirement, both crack boundaries are discretized with standard discontinuous quadratic boundary elements. In addition, the discontinuous elements circumvent the problem of collocation at crack tips, crack kinks and crack-edge corners. The effective treatment of the hypersingular integrals that appear in the traction equation is of fundamental importance. For curved boundary elements, the natural definition of ordinary finite-part integrals is proposed in the present work. For flat boundary elements, the direct analytic integration is the most effective method to deal with such integrations. At a crack node, singular integrations occur twice, once on the self-point element and again on the opposite one. This feature prevents the use of the standard rigid body condition to evaluate indirectly the diagonal terms at crack nodes. Several cracked geometries were analyzed with the dual boundary element method. It was demonstrated that the dual boundary element method can be used for solving general crack problems efficiently.

The singularity subtraction technique uses a singular particular solution to introduce the stress intensity factors as primary unknowns in the problem. In the basic formulation, these particular solutions are applied globally in the problem domain. The modelling strategy presented to solve a general crack problem considers an automatic partition of the problem domain into near-tip and far-tip field regions. In each near-tip field region, enclosing a single crack tip, the basic formulation of the singularity subtraction technique is applied locally.

A procedure based on the decomposition of the elastic field into its respective symmetric and antisymmetric components was used to decouple the stress intensity factors of mixed-mode problems, in the J-integral technique. This procedure is im-

plemented efficiently in the boundary element method because the integration path is defined with internal points, where the elastic field is accurately determined. This facility is not shared by the domain methods, like the finite element method, where the field variation is pre-assumed.

The assessment of an efficient technique for the evaluation of the stress intensity factors in general mixed-mode crack problems, was based on the performance of the analysis of standard benchmark problems. The accuracy of both the singularity subtraction technique and the J-integral was shown with the edge and centre crack problems. The analysis of general crack problems with the singularity subtraction technique has the disadvantage of the presence of artificial boundaries. As a final test the J-integral was applied in the analysis of a kinked-crack problem. The accuracy as well as the stability of the results, for any contour path, was excellent. In general, the J-integral was found to be an efficient tool for the stress intensity factor evaluation, in the analysis of crack problems with the dual boundary element method.

The new boundary elements, introduced for the discretization of each crack extension increment, generate new algebraic equations and introduce new unknowns in the equations already existing. An *LU* decomposition method is used to solve the system of algebraic equations. The *LU* decomposition is carried out incrementally over the new rows and new columns of the system matrix. The remaining part of the system matrix is carried over, already decomposed, from the previous iteration of the analysis.

The reliability of the whole incremental analysis process lies in the accuracy of the stress intensity factors evaluation, as well as on the accuracy of the prediction of the crack-extension increment direction. In this work, the stress intensity factors are evaluated by the J-integral technique, applied with a circular contour around each crack tip. Since the exact variation of the interior elastic field is built into the fundamental solution of the boundary element method, this technique is very effective. In contrast to the finite element method, where the field is approximated by a polynomial variation in each element.

A simple prediction-correction procedure was adopted for the definition of the direction of each crack-extension increment. It predicts the tangent direction of the crack path using the maximum principal stress criterion and then introduces a correction to this direction with information derived from one step ahead of the current crack-extension increment.

The results of incremental crack-extension analyses of several structures are presented. A fatigue crack-growth analysis is also presented for the simplest case of constant amplitude loading cycles.

References

1. Gallagher, R.H. A review of finite element techniques in fracture mechanics, *Proc. of the First Conference on Numerical Methods in Fracture Mechanics*, A.R. Luxmoore & D.R.J. Owen (ed), University college of Swansea, 1978, pp. 1-25.

2. Rice, J.R. & Tracey, D.M. Computational fracture mechanics, in *Numerical and Computer Methods in Structural Mechanics*, Academic Press, New York, 1973.

3. Shephard, M.S., Yehia, N.A.B., Burd, G.S. & Weidner, T.J. Automatic crack propagation tracking, *Comp. and Struc.*, 1985, **20**, 211–223.

4. Valliappan, S. & Murti, V. Automatic remeshing technique in quasi-static and dynamic crack propagation, *Proc. NUMETA'85 Conf.*, Swansea, 1985, pp. 107-116.

5. Swenson, D.V. & Ingraffea, A.R. Modelling mixed-mode dynamic crack propagation using finite elements: theory and applications, *Comp. Mech.*, 1988, **3**, 381-397.

6. Remzi, E.M. & Blackburn, W.S. Automatic crack propagation studies in T-junctions and cross bars, *Eng. Comp.*, 1990, **7**, 116-124.

7. Wawrzynek, P.A., Martha, L.F. & Ingraffea, A.R. A computational environment for the simulation of fracture processes in three dimensions, in *Anal. Numer. Exper. Aspects of Three–Dimensional Fracture Process*, A.J. Rosakis *et al.* (ed), ASME, AMD 91, 1988, pp. 321-327.

8. Aliabadi, M.H. & Rooke, D.P. *Numerical Fracture Mechanics*, Computational Mechanics Publications, Southampton and Kluwer Academic Publishers, Dordrecht, 1991.

9. Ingraffea, A.R., Blandford, G.E. & Ligget, J.A. Automatic modelling of mixed-mode fatigue and quasi-static crack propagation using the boundary element method, *Proc. of Fracture Mechanics: Fourteenth Symposium*, ASTM STP 791, J.C. Lewis & G. Sines (ed), ASTM, 1983, I 407–I 426.

10. Brebbia, C.A. & Dominguez, J. *Boundary Elements - an Introductory Course*, Computational Mechanics Publications, Southampton, UK, 1989.

11. Snyder, M.D. & Cruse, T.A. Boundary integral equation analysis of cracked anisotropic plates, *Int. Journal of Fracture*, 1975, **11**, 315-328.

12. Crouch, S.L. Solution of plane elasticity problems by the displacement discontinuity method, *Int. Journ. Num. Meth. Engng.*, 1976, **10**, 301-342.

13. Blandford, G.E., Ingraffea, A.R. & Liggett, J.A. Two-dimensional stress intensity factor computations using the boundary element method, *Int. Journ. Num. Meth. Engng.*, 1981, **17**, 387-404.

14. Portela, A., Aliabadi, M.H. & Rooke, D.P. The dual boundary element method: effective implementation for crack problems, *Int. Journ. Num. Meth. Engng.*, 1991, **33**, 1269-1287.

15. Sladek, V., Sladek, J. & Balas, J. Boundary integral formulation of crack problems, *ZAMM*, 1986, **66**, 83-94.

16. Cruse, T.A. *Boundary Element Analysis in Computational Fracture Mechanics*, Kluwer Academic Publishers, 1989.

17. Bueckner, H.F. Field singularities and related integral representations, *Mechanics of Fracture*, G.C. Sih (ed), Volume 1, Nordhoff, 1973.

18. Watson, J.O. Hermitian cubic boundary elements for the analysis of cracks of arbitrary geometry, in *Advanced Boundary Element Methods*, T.A. Cruse (ed), Springer-Verlag, 1988.

19. Hong, H. & Chen, J. Derivations of integral equations of elasticity, *Journ. Engng. Mech.*, 1988, **114**(6), 1028-1044.

20. Gray, L.J. *Boundary Element Method for Regions with Thin Internal Cavities*, IBM Bergen Scientific Centre, Allegaten 36, 5007 Bergen, Norway, 1987.

21. Gray, L.J., Martha, L.F. & Ingraffea, A.R. Hypersingular integrals in boundary element fracture analysis, *Int. Journ. Num. Meth. Engng.*, 1990, **29**, 1135-1158.

22. Rudolphi, T.J., Krishnasamy, G., Schmerr, L.W. & Rizzo, F.J. On the use of strongly singular integral equations for crack problems, *Proc. Boundary Elements X*, Computational Mechanics, Southampton, 1988.

23. Portela, A., Aliabadi, M.H. & Rooke, D.P. Dual boundary element incremental analysis of crack propagation, *Int. Journ. Computers and Structures*, to appear.

24. Portela, A. *Dual Boundary Element Analysis of Fatigue Crack Growth*, Computational Mechanics Publications, Southampton, 1992.

25. Aliabadi, M.H., Rooke, D.P. & Cartwright, D.J. An improved boundary element formulation for calculating stress intensity factors: application to aerospace structures, *J. Strain Anal.*, 1987, **22**(4), 1-5.

26. Portela, A., Aliabadi, M.H. & Rooke, D.P. Boundary element analysis of V-notched plates, *Proc. of the Fourth International Conference on Boundary Element Technology*, Windsor, Canada, C.A. Brebbia (ed), Computational Mechanics Publications, Southampton, UK, 1989.

27. Aliabadi, M.H. An enhanced boundary element method for determining fracture parameters, *Proc. 4th Int. Conf on Num. Methods in Fracture Mechanics*, San Antonio, Texas, Pineridge Press, 1987.

28. Portela, A., Aliabadi, M.H. & Rooke, D.P. Efficient boundary element analysis of sharp notched plates, *Int. Journ. Num. Meth. Engng.*, 1991, **32**, 445-470.

29. Portela, A., Aliabadi, M.H. & Rooke, D.P. Dual boundary element analysis of cracked plates: singularity subtraction technique, *Int. Journ. Fracture*, to appear.

30. Rice, J.R. A path independent integral and the approximate analysis of strain concentration by notches and cracks, Trans. ASME, *J. Appl. Mech.*, 1968, **35**, 379-386.

31. Hadamard, J. *Lectures on Cauchy's Problem in Linear Partial Differential Equations*, New Haven, Yale University Press, 1923.

32. Walters, H.G., Ortiz, J.C., Gipson, G.S. & Brewer, J.A. Overhouser boundary elements in potential theory and linear elastostatics, in *Advanced Boundary Element Methods*, T.A. Cruse (ed), Springer-Verlag, 1988.

33. Kutt H.R. On the numerical evaluation of finite-part integrals involving an algebraic singularity, CSIR Special Report WISK 179, National Research Institute for Mathematical Sciences, P.O. Box 395, Pretoria 0001, South Africa, 1975.

34. Ossicini, A. Alcune formule di quadratura per il calcolo della parte finita e del valore principale di integrali divergenti, *Rend. Matem.*, 1969, Ser. VI, 2, 385-401.

35. Kutt, H.R. Quadrature formulae for finite-part integrals, Special Report WISK 178, National Research Institute for Mathematical Sciences, Pretoria, South Africa, 1975. Available also from the National Technical Information Service, USA.

36. Ioakimidis, N.I. On the numerical evaluation of derivatives of Cauchy principal-value integrals, *Computing*, 1981, **27**, 81-88.

37. Ladopoulos, E.G. On the numerical evaluation of the general type of finite-part singular integrals and integral equations used in fracture mechanics, *Engng. Frac. Mech.*, 1988, **31**(2), 315-337.

38. Williams, M.L. Stress singularities resulting from various boundary conditions in angular corners of plates in extension, *J. Appl. Mech.*, 1952, **19**, 526-528.

39. Westergaard, H.M. Bearing pressures and cracks, *J. Appl. Mech.*, 1939, **6**, A49-53.

40. Civelek, M.B. & Erdogan, F. Crack problems for a rectangular sheet and an infinite strip, *Int. Journ. Frac.*, 1982, **19**, 139-159.

41. Krishnasamy, G., Rudolphi, T.J., Schmerr, L.W. & Rizzo F.J. Discretization considerations with hypersingular integral formulas for crack problems, in *Discretization Methods in Structural Mechanics*, G. Kuhn & H. Mang (ed), Springer-Verlag, 1990.

42. Murakami, Y. *Stress Intensity Factors Handbook*, Pergamon Press, 1987.

43. Davis, P.J. & Rabinowitz, P. *Methods of Numerical Integration*, Academic Press, 1984.

44. Erdogan, F. & Sih, G.C. On the crack extension in plates under plane loading and transverse shear, *J. Basic Eng.*, 1963, **85**, 519-527.

45. Broek, D. *Elementary Engineering Fracture Mechanics*, Martinus Nijhoff Publishers, Dordrecht, 1986.

46. Tanaka, K. Fatigue crack propagation from a crack inclined to the cyclic tensile axis, *Eng. Frac. Mech.*, 1974, **6**, 493-507.

Chapter 2

BEM computations of elastodynamic fields in bodies containing internal, near-surface and surface-breaking cracks

Z.L. Li & J.D. Achenbach

Center for Quality Engineering and Failure Prevention, Northwestern University, Evanston, Illinois 60208, USA

Summary

The formulation of boundary integral equations and their numerical solution by the boundary element method are discussed for elastodynamic fields in bodies containing interior, surface-breaking and near-surface cracks. Special attention is devoted to the 'crack-opening displacement' (COD) approach and the 'multi-region' approach. Techniques to deal with the higher-order singularities that occur in the COD approach are discussed in some detail. Applications of the multi-region approach to bodies containing interphase failures and interface cracks are also discussed. Methods to obtain solutions for surface-breaking and near-surface cracks are given detailed attention.

Introduction

The boundary element method has become a well established technique for the numerical solution of the boundary integral equations that govern the displacements of the crack faces in a cracked body. Once the relative separations of the corresponding points on the crack faces (i.e. the crack-opening displacements) have been calculated by the boundary element method, other quantities of interest can be obtained by the use of appropriate integral representations. For the static problem quite a large amount of published work is available. For a recent review paper we refer to Cruse.[6] The body of published work on elastodynamic problems for cracked bodies is much smaller. Although there have been review articles on the application of the boundary element method to elastodynamics, see e.g. the papers by Kobayashi[12] and Beskos,[3] the specific area of elastodynamic fields in cracked bodies has, to the authors' knowledge, not been reviewed. It is the purpose of this paper to provide an overview of recent contributions in that area.

The general category of problems that are discussed in this paper is of interest when a body containing a crack is subjected to rapid excitation. When the excitation is generated by a service load, the quantities of interest are ones related to the mechanics of fracture, such as elastodynamic stress intensity factors. A second application, which requires essentially the same mathematical formalisms, is in the field of quantitative non-destructive evaluation. Here the general purpose is to detect and characterize a crack, by introducing a probing field of ultrasound into the cracked body. Now the quantities of interest are the scattered wavefields that are generated by the interaction of the crack with the externally excited fields of ultrasound.

The formulation of a transient elastodynamic problem for a cracked body can be reduced to a system of singular integral equations whose numerical solution can be obtained by a direct time-stepping approach in conjunction with a boundary element discretization over the crack surface. For a planar crack this approach has been pursued by Hirose & Achenbach.[9] An alternate approach, which presently seems to be preferred, first disposes of the time dependence by the use of a Fourier integral approach. The resulting problem in the frequency domain only requires a boundary element discretization over the crack surface. A Fast-Fourier Transform can subsequently be employed to construct the time-domain solution. Naturally it may be necessary to calculate frequency-domain results for a fairly large range of frequencies, which may require a considerable investment in computer time. Nevertheless the frequency domain approach has the advantage of being straight forward and for elastodynamic crack problems it is, so far, the most frequently used method. In another paper by Kitahara, Hirose & Achenbach[10] the direct-time approach will be discussed. Here we will restrict our attention to the frequency-domain approach.

With regard to the mathematical formulation of the frequency-domain boundary integral equations, elastodynamic problems suffer from the same problem of degeneracy as static problems, because the two crack faces occupy the same region in a mathematical surface. The three approaches that have been established to circumvent the degeneracy are the multi-region approach, the special Green's function approach and the crack-opening displacement approach. The crack-opening displacement approach has received most attention in recent years. However, as is well-known, the boundary integral equations for this formulation are hypersingular. These equations require special techniques to deal with these singularities, and these techniques are discussed in some detail in this paper. There are, however, also configurations where the multi-region approach is the optimal one, the primary example being the configuration of a crack in an interface between two solids of different mechanical properties. We discuss an example of the multi-region approach in some detail.

A special class of applications that is discussed in this paper, concerns a crack which breaks a free surface or is located in the vicinity of a free surface. Since elastodynamic Green's functions for bounded bodies are exceedingly complicated, it is generally advantageous to use full space Green's functions. The use of such Green's functions, however, imply that the system of boundary integral equations includes equations over the external surface(s) of the body. Examples of surface-breaking and near surface cracks in an elastic half-space are also discussed.

1 Elastodynamic theory

Consider a region, V, of a homogeneous, isotropic, linearly elastic solid, with Lamé constants λ, μ and mass density ρ. Let S be the boundary of V and n_j the outward normal vector of S. In this paper the region V may be in three-dimensional or two-dimensional space. The region V is in time-harmonic motion, with time dependence $\exp(-i\omega t)$, where ω is the angular frequency. Since the time-harmonic factor $\exp(-i\omega t)$ is common to all field variables it is omitted in the sequel.

The time-harmonic governing equations of elastodynamic motion are given by (see, for example, Achenbach[1])

$$\sigma_{ij,j} + \rho\omega^2 u_i = 0 \tag{1}$$

where σ_{ij} and u_i are the components of the stress tensor and the particle displacement vector, respectively. In eqn (1) body forces are not considered and the summation convention is implied. The relation between stress and displacement gradients is expressed by Hooke's law:

$$\sigma_{ij} = C_{ijk\ell} u_{k,\ell} \tag{2}$$

where $C_{ijk\ell}$ denotes the stiffness tensor of the isotropic material which can be written as

$$C_{ijk\ell} = \lambda\delta_{ij}\delta_{k\ell} + \mu(\delta_{ik}\delta_{j\ell} + \delta_{i\ell}\delta_{jk}) \tag{3}$$

where δ_{ij} is the Kronecker delta.

2 Integral representations

An application of the divergence theorem to the Betti-Rayleigh reciprocal relations for the elasto-dynamic field and the full-space elastodynamic Green's function gives the following well-known representation integral:

$$\epsilon(\mathbf{x}) u_k \mathbf{x}) = \int_S [u_{ik}^{\mathrm{G}}(\mathbf{y}-\mathbf{x}) t_i(\mathbf{y}) - \sigma_{ijk}^{\mathrm{G}}(\mathbf{y}-\mathbf{x}) n_j(\mathbf{y}) u_i(\mathbf{y})] \mathrm{d}S(\mathbf{y}) \tag{4}$$

where

$$\epsilon(x) = \begin{cases} 1\,, & \mathbf{x}\epsilon V\,, \\ (2.2)0\,, & \mathbf{x}\notin V \end{cases} \tag{5}$$

and

$$t_i(\mathbf{y}) = \sigma_{ij}(\mathbf{y}) n_j(\mathbf{y}) \tag{6}$$

In eqn (4), $u_{ik}(\mathbf{y}-\mathbf{x})$ is the full-space elastodynamic displacement Green's function, i.e. the displacement at $\mathbf{y}$ in the i-direction due to a unit time harmonic point load applied at $\mathbf{y}$ in the k-direction. The corresponding stress Green's function, which is denoted $\sigma_{ijk}^{\mathrm{G}}(\mathbf{y}-\mathbf{x})$ is related to $u_{ik}^{\mathrm{G}}(\mathbf{y}-\mathbf{x})$ by eqn (2). Equation (4) states that the displacement $u_k(\mathbf{x})$ at any point $\mathbf{x}$ in V can be expressed in terms of the displacement $u_i(\mathbf{y})$ and traction $t_i(\mathbf{y})$ on the boundary S. It is also noted here that boundary displacements and tractions satisfy a constraint equation when $\mathbf{x}$ is outside the volume

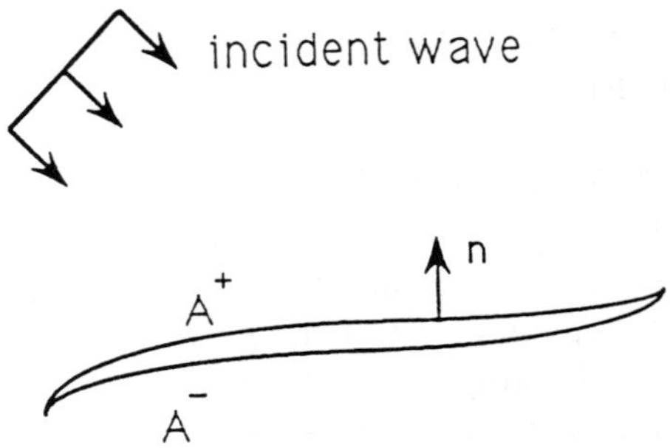

Figure 1: A plane wave incident on a crack of arbitrary shape

V. The explicit functional forms of $u^G_{ik}(\mathbf{y}-\mathbf{x})$ and $\sigma^G_{ijk}(\mathbf{y}-\mathbf{x})$ are given in Appendix A.

If V is composed of multiply connected regions, and some of these extend to infinity, eqn (4) is still valid provided that $u_k(x)$ satisfies the well-known Sommerfeld radiation condition for $|x| \to \infty$. Consequently, for an infinite solid containing a void we express the total fields, u_i, as the sum of the field in the absence of the void and the field generated by the presence of the void. For the important example of scattering of an incident wave by a void, these two fields correspond to the incident field, u_i^{in}, and the scattered field, u_i^{sc}, respectively, and we write

$$u_i = u_i^{\text{in}} + u_i^{\text{sc}} \tag{7}$$

Since the scattered field satisfies the Sommerfeld radiation condition for $|\mathbf{x}| \to \infty$, the integral representation for the scattered field follows from (4) as

$$\epsilon(\mathbf{x})u_k^{\text{sc}}(\mathbf{x}) = \int_A [-u^G_{ik}(\mathbf{y}-\mathbf{x})t_i^{\text{sc}}(\mathbf{y}) + \sigma^G_{ijk}(\mathbf{y}-\mathbf{x})n_j(\mathbf{y})u_i^{\text{sc}}(\mathbf{y})]\mathrm{d}A(\mathbf{y}) \tag{8}$$

where A is the surface of the void and the direction of the normal $n_j(\mathbf{y})$ is pointed to the interior of the solid. If the surface of the void is traction-free, we obtain

$$u_k^{\text{sc}}(\mathbf{x}) = \int_A \sigma^G_{ijk}(\mathbf{y}-\mathbf{x})n_j(\mathbf{y})u_i^{\text{sc}}(\mathbf{y})\mathrm{d}A, \quad \mathbf{x}\epsilon V \tag{9a}$$

$$0 = \int_A \sigma^G_{ijk}(\mathbf{y}-\mathbf{x})n_j(\mathbf{y})u_i^{\text{sc}}(\mathbf{y})\mathrm{d}A, \quad \mathbf{x}\notin V \tag{9b}$$

3 Integral equations for crack problems

When external loads are applied to a solid, a crack becomes a surface of displacement discontinuity. The faces of a mathematical crack are infinitesimally close prior to loading, and they do not interact when loads are applied. This is an acceptable approximation for real cracks whose faces will not touch when the solid is disturbed. Geometrically, a crack is a limiting case of a void. For a crack of arbitrary shape, we write $A = A^+ + A^-$, where, for the scattering problem, A^+ is the insonified side of the crack and A^- is the shadow side. Also n_j is the unit normal vector of A with direction shown in Fig. 1.

For a traction-free void the limit of $\mathbf{x}\epsilon V \to \mathbf{x}\epsilon A$ in eqn (9a), yields the integral equation,

$$u_k^{sc}(\mathbf{x}^A) = \lim_{\mathbf{x}\to\mathbf{x}^A} \sigma_{ijk}^{G}(\mathbf{y}-\mathbf{x})n_j(\mathbf{y})u_i^{sc}(\mathbf{y})\mathrm{d}A$$

$$= \int_A \sigma_{ijk}^{G}(\mathbf{y}-\mathbf{x}^A)n_j(\mathbf{y})u_i^{sc}(\mathbf{y})\mathrm{d}A \qquad (10)$$

where $\mathbf{x}^A$ denotes a point on the surface of the void.

For a crack, the integral representation for the scattered field eqn (9a) can be rewritten as

$$u_k^{sc}(\mathbf{x}) = \int_{A+} \sigma_{ijk}^{G}(\mathbf{y}-\mathbf{x})n_j(\mathbf{y})\Delta u_i^{sc}(\mathbf{y})\mathrm{d}A, \quad \mathbf{x}\epsilon V,\ \mathbf{x}\notin A^+ \qquad (11)$$

where $\Delta u_i^{sc}(\mathbf{y})$ defines the displacement discontinuity across the faces of the crack,

$$\Delta u_i^{sc}(\mathbf{y}) = u_i^{sc}(\mathbf{y}^+) - u_i^{sc}(\mathbf{y}^-) \qquad (12)$$

here $\mathbf{y}^+\epsilon A^+, \mathbf{y}^-\epsilon A^-$. A system of integral equations cannot be obtained directly from eqn (11) because the limiting process of $\mathbf{x}\to\mathbf{x}^A$ will produce a degenerate system of boundary integral equations.

Substituting eqn (11) into Hooke's law (2) and using $t_p^{sc}(\mathbf{x}) = -\sigma_{pq}^{sc}n_q(\mathbf{x})$, we obtain

$$t_p^{sc}(\mathbf{x}) = -C_{pqk\ell}n_q(\mathbf{x}) \int_{A+} \sigma_{ijk,\ell}^{G}(\mathbf{y}-\mathbf{x})n_j(\mathbf{y})\Delta u_i^{sc}(\mathbf{y})\mathrm{d}A(\mathbf{y}), \mathbf{x}\epsilon V,\ \mathbf{x}\notin A^+ \qquad (13)$$

A system of boundary integral equations without degeneracy can be derived from eqn (13) by taking the limit $\mathbf{x}\to A^+$, and by applying the traction-free boundary conditions. We obtain

$$\lim_{\mathbf{x}\to\mathbf{x}^A} t_p^{sc}(\mathbf{x}) = t_p^{sc}(\mathbf{x}^A) = -t_p^{in}(\mathbf{x}^A)$$

$$= \lim_{\mathbf{x}\to\mathbf{x}^A}\left\{-C_{pqk\ell}n_q(\mathbf{x})\int_{A+}\sigma_{ijk,\ell}^{G}(\mathbf{y}-\mathbf{x})n_j(\mathbf{y})\Delta u_i^{sc}(\mathbf{y})\mathrm{d}A(\mathbf{y})\right\}$$

$$= -C_{pqk\ell}n_q(\mathbf{x}^A)\lim_{\mathbf{x}\to\mathbf{x}^A}\int_{A+}\sigma_{ijk,\ell}^{G}(\mathbf{y}-\mathbf{x})n_j(\mathbf{y})\Delta u_i^{sc}(\mathbf{y})\mathrm{d}A(\mathbf{y})$$

$$= -C_{pqk\ell}n_q(\mathbf{x}^A)\int_{A+}\sigma_{ijk,\ell}^{G}(\mathbf{y}-\mathbf{x}^A)n_j(\mathbf{y})\Delta u_i^{sc}(\mathbf{y})\mathrm{d}A(\mathbf{y}) \qquad (14)$$

In the last step of eqn (14), the limit $\mathbf{x}\to\mathbf{x}^A$ has been brought inside the integration. In a strict sense this interchange is not valid. In fact the system of boundary integral equations obtained in this manner is hypersingular when the observation point $\mathbf{y}$ and the source point $\mathbf{x}$ coincide, since the terms $\sigma_{ijk,\ell}^{G}(\mathbf{y}-\mathbf{x}^A)$ behaves as

$$\sigma_{ijk,\ell}(\mathbf{y}-\mathbf{x}^A) \sim \begin{cases} \frac{1}{r^2}, & \text{two-dimensional,} \\ \frac{1}{r^3}, & \text{three-dimensional,} \end{cases} \quad \text{as } r\to 0 \qquad (15)$$

here $r = |\mathbf{y} - \mathbf{x}^A|$, thus the expression in the right-hand side of (14) is not integrable.

Several methods have been proposed to overcome this difficulty, but all of them have the same basic idea: before taking the limit, the integral with the hypersingular integrand is transformed to several integrals with singular integrands of lower order which are integrable after the limiting process $\mathbf{x} \to \mathbf{x}^A$. Two strategies have been employed. For the first strategy, before taking the limit $\mathbf{x} \to \mathbf{x}^A$ integration-by-parts is used to reduce the higher-order singularities and then the limit $\mathbf{x} \to \mathbf{x}^A$ is taken. In this manner a system of boundary integral equations is obtained but for both the crack-opening displacements $\Delta u_i^{\mathrm{sc}}(\mathbf{y})$ and the dislocation densities (derivatives of the crack-opening displacement) $\Delta u_{i,j}^{\mathrm{sc}}(\mathbf{y})$ as unknowns. The process is referred to as a regularization of the (higher-order singular) integrals. This method was proposed by Budiansky & Rice[4] for solving a 3-D flat crack scattering problem. The integration-by-parts regularization procedure has also been used by Sládek & Sládek,[21] Nishimura & Kobayashi[17] and Budreck & Achenbach.[5] For a 2-D crack configuration, analogous formulations have been proposed by Tan,[22] Schmerr[19] and Zhang & Achenbach.[23,24] The integration-by-parts regularization procedure is easily implemented for a flat (3-D problem) or straight (2-D problem) crack, but it becomes quite cumbersome for curved cracks. Furthermore, different forms of the regularized BIEs are obtained through the nonunique integration-by-parts procedure, though they are equivalent (see Cruse[6]).

Zhang & Achenbach[25] proposed a new approach to deal with this problem. By employing an elastodynamic conservation integral, instead of using the Betti-Rayleigh reciprocity relation, they derived the boundary integral equations for crack configurations very naturally without any integration-by-parts procedure. The result is

$$t_p^{\mathrm{in}}(\mathbf{x}) = C_{pqk\ell} n_q(\mathbf{x}) \int_{A^+} \left\{ \left[\Delta u_{m,n}^{\mathrm{sc}}(\mathbf{x}) \sigma_{mn\ell}^{\mathrm{G}}(\mathbf{y}-\mathbf{x}) - \rho\omega^2 \Delta u_i^{\mathrm{sc}}(\mathbf{y}) u_{i\ell}^{\mathrm{G}}(\mathbf{y}-\mathbf{x}) \right] \delta_{jk} - \Delta u_{i,k}^{\mathrm{sc}}(\mathbf{y}) \sigma_{ij\ell}^{\mathrm{G}}(\mathbf{y}-\mathbf{x}) \right\} n_j(\mathbf{y}) \mathrm{d}A(\mathbf{y}), \quad \mathbf{x} \epsilon A^+ \tag{16}$$

Equation (16) is valid for a crack of arbitrary shape and it does not involve a hypersingularity. All integral equations obtained by various integration-by-parts procedures are special cases of eqn (16). The disadvantages of eqn (16) is that it requires not only the computation of the crack-opening displacements, but also of the generally irrelevant displacement derivatives.

The main idea of the alternate strategy can be explained as follows. First, A^+ is divided into two parts: $A^+ = A_c^+ + A_s^+$, where A_s^+ is a small area on A^+ with $\mathbf{x}^A$ inside it, for example with $\mathbf{x}^A$ located at its geometrical centroid. In actual numerical calculations, A_s^+ can be chosen as one element. We can write (see (14)).

$$t_p^{\mathrm{sc}}(\mathbf{x}^A) = -C_{pqk\ell} n_q(\mathbf{x}^A) \int_{A_c^+} \sigma_{ijk,\ell}^{\mathrm{G}}(\mathbf{y}-\mathbf{x}^A) n_j(\mathbf{y}) \Delta u_i^{\mathrm{sc}}(\mathbf{y}) \mathrm{d}A(\mathbf{y}) - C_{pqk\ell} n_q(\mathbf{x}^A) \lim_{\mathbf{x} \to \mathbf{x}^A} I \tag{17}$$

where

$$I = \int_{A_s^+} \sigma_{ijk,\ell}^{\mathrm{G}}(\mathbf{y}-\mathbf{x}) n_j(\mathbf{y}) \Delta u_i^{\mathrm{sc}}(\mathbf{y}) \mathrm{d}A(\mathbf{y}), \quad \mathbf{x} \notin A_s^+ \tag{18}$$

It is well-known that the elastodynamic Green's function can be expressed as

$$\sigma_{ijk}^{\mathrm{G}} = \sigma_{ijk}^{GD} + \sigma_{ijk}^{GS} \tag{19}$$

where σ_{ijk}^{GS} is the corresponding elastostatic Green's function which is hypersingular but the dynamic part of σ_{ijk}^{GD} is regular. Now assuming that $\Delta u_i^{sc}(\mathbf{y})$ is differentiable and its derivatives, $\Delta u_{i,m}^{sc}(\mathbf{y})$, satisfy a Hölder condition at $\mathbf{y} = \mathbf{x}^A, \Delta u_i^{sc}(\mathbf{y})$ can be represented by the following expansions

$$\Delta u_i^{sc}(\mathbf{y}) = \Delta u_i^{sc}(\mathbf{x}^A) + \Delta u_{i,m}^{sc}(\mathbf{x}^A)(y_m - x_m^A) + O(r^{1+\alpha}) \tag{20}$$

where $r = |\mathbf{y} - \mathbf{x}^A|$ and $\alpha > 0$. Substitution of (19) and (20) into (18) yields

$$I = \int_{A_s^+} \sigma_{ijk,\ell}^{GD}(\mathbf{y}-\mathbf{x}) n_j(\mathbf{y}) \Delta u_i^{sc}(\mathbf{y}) \mathrm{d}A(\mathbf{y})$$

$$+\Delta u_{i,m}^{sc}(\mathbf{x}^A) \int_{A_s^+} \sigma_{ijk,\ell}^{GS}(\mathbf{y}-\mathbf{x}) n_j(\mathbf{y})(y_m - x_m) \mathrm{d}A(\mathbf{y})$$

$$+ \Delta u_i^{sc}(\mathbf{x}^A) \int_{A_s^+} \sigma_{ijk,\ell^{GS}}(\mathbf{y}-\mathbf{x}) n_j(\mathbf{y}) \mathrm{d}A(\mathbf{y}), \quad \mathbf{y} \notin A_s^+ \tag{21}$$

In the limit $\mathbf{x} \to \mathbf{x}^A$, the first integral of (21) is regular, the second integral is found to be integrable in the sense of a Cauchy Principle Value (CPV) integral. By using the equation of static equilibrium, $\sigma_{ijk,j}^{GS} = 0$, the third integral of (21) can be rewritten as

$$\int_{A_s^+} \sigma_{ijk,\ell}^{GS}(\mathbf{y}-\mathbf{x}) n_j(\mathbf{y}) \mathrm{d}A(\mathbf{y})$$

$$= \int_{A_s^+} [\sigma_{ijk,\ell}^{GS}(\mathbf{y}-\mathbf{x}) n_j(\mathbf{y}) - \sigma_{ijk,j}^{GS}(\mathbf{y}-\mathbf{x}) n_\ell(\mathbf{y})] \mathrm{d}A(\mathbf{y})$$

$$= \int_{\partial A_s^+} e_{mj\ell} \sigma_{ijk}^{GS}(\mathbf{y}-\mathbf{x}) \nu_m(\mathbf{y}) \mathrm{d}S(\mathbf{y}) \tag{22}$$

where $e_{mj\ell}$ is the alternating symbol, ∂A_S^+ denotes the edge of A_S^+ and ν_m is the unit normal vector of ∂A_S^+. In the last step of deriving (22), Stokes' integral theorem has been used. Since $\mathbf{x}^A$ is not on the edge of A_S^+, the line integral on the right-hand side of (21) will remain regular in the limit $\mathbf{x} \to \mathbf{x}^A$. By using Taylor series expansion of the Green's function, the first two integrals of (21) and the line integral (22) can be evaluated analytically for simple shape of A_S^+.

This method was used by Budreck & Achenbach[5] to solve the problem of intermediate and short wave length scattering from a penny-shaped and elliptic crack. A similar method was used by Lin & Keer.[15]

The key point of this method, the derivation of (22), is to convert the surface integral on A_S^+ which is hypersingular when $\mathbf{x} \to \mathbf{x}^A$ where $\mathbf{x} \epsilon V$ and $\mathbf{x}^A \notin \partial A_S^+$ to a line integral around A_S^+ which is regular as $\mathbf{x} \to \mathbf{x}^A$ by the use of Stokes' theorem. Krishnaswamy *et al.*[13] recently realized that this procedure can also be performed on the total crack surface A^+ regardless of how the surface is discretized and whether the surface is open or closed. They presented an integral equation which includes a sum of surface integrals on A^+, and line integrals around ∂A^+, which are all regular or at most weakly singular.

Thus, in the approach of Budreck & Achenbach,[5] Stokes' theorem is used not only to reduce the order of singularity by transforming a derivative on the fundamental solution to the density function, i.e. the usual regularization as in the integration-by-parts method, but also to convert surface integrals to line integrals. Instead of using Stokes' theorem, Guiggiani *et al.*,[8] have recently obtained similar results by careful analysis of the limiting process and by expanding the hypersingular integrand in Laurent series in the local plane of intrinsic coordinates. Following Guiggiani *et al.*,[8] we differentiate (9b) with respect to x_i and then multiply the results by $C_{pqk\ell}n_q(\mathbf{x})$, to obtain,

$$C_{pqk\ell}n_q(\mathbf{x}) \int_{A+} \sigma^{\mathrm{G}}_{ijk,\ell}(\mathbf{y}-\mathbf{x})n_j(\mathbf{y})\Delta u^{\mathrm{sc}}_i(\mathbf{y})\mathrm{d}A(\mathbf{y}) = 0, \mathbf{x}\epsilon V \tag{23}$$

The traction BIE can be expressed as following

$$C_{pqk\ell}n_q(\mathbf{x}^A) \lim_{\epsilon\to 0} \int_{A+-A^+_\epsilon+E_\epsilon} \sigma^{\mathrm{G}}_{ijk,\ell}(\mathbf{y}-\mathbf{x}^A)n_j(\mathbf{y})\Delta u^{\mathrm{sc}}_i(\mathbf{y})\mathrm{d}A(\mathbf{y}) = 0 \tag{24}$$

The meaning of A^+_ϵ and E_ϵ is shown in Fig. 2, here for simplicity, it is assumed that $\mathbf{x}^A$ is on a smooth boundary point. Consider an element A^+_s which includes $\mathbf{x}^A$ inside it. The contribution of A^+_s to the left-hand side of (24) can be rewritten as

$$\begin{aligned} &C_{pqk\ell}n_q(\mathbf{x}^A) \lim_{\epsilon\to 0} \int_{A^+_s-A^+_\epsilon+E_\epsilon} \sigma^{\mathrm{G}}_{ijk,\ell}(\mathbf{y}-\mathbf{x}^A)n_j(\mathbf{y})\Delta u^{\mathrm{sc}}_i(\mathbf{y})\mathrm{d}A(\mathbf{y}) \\ &\qquad = C_{pqk\ell}n_q(\mathbf{x}^A) \lim_{\epsilon\to 0}(I_1+I_2+I_3+I_4) \end{aligned} \tag{25}$$

where

$$\begin{aligned} I_1 &= \int_{A^+_s-A^+_\epsilon} \sigma^{\mathrm{G}}_{ijk,\ell}(\mathbf{y}-\mathbf{x}^A)n_j(\mathbf{y})\Delta u^{\mathrm{sc}}_i(\mathbf{y})\mathrm{d}A(\mathbf{y}) \\ I_2 &= \Delta u^{\mathrm{sc}}_i(\mathbf{x}^A) \int_{E_\epsilon} \sigma^{\mathrm{G}}_{ijk,\ell}(\mathbf{y}-\mathbf{x}^A)n_j(\mathbf{y})\mathrm{d}A(\mathbf{y}) \\ I_3 &= \int_{E_\epsilon} \sigma^{\mathrm{G}}_{ijk,\ell}(\mathbf{y}-\mathbf{x}^A)n_j(\mathbf{y})[\Delta u^{\mathrm{sc}}_i(\mathbf{y}) - \Delta u^{\mathrm{sc}}_i(\mathbf{x}^A) - \Delta u^{\mathrm{sc}}_{i,m}(\mathbf{x}^A)(y_m - x^A_m)]\mathrm{d}A(\mathbf{y}) \\ I_4 &= \Delta u^{\mathrm{sc}}_{i,m}(\mathbf{x}^A) \int_{E_\epsilon} \sigma^{\mathrm{G}}_{ijk,\ell}(\mathbf{y}-\mathbf{x}^A)n_j(\mathbf{y})(y_m - x^A_m)\mathrm{d}A(\mathbf{y}) \end{aligned}$$

By considering eqn (20), we find that in the limit $\epsilon \to 0$, $I_3 \to 0$, the integral I_4 can be evaluated analytically to yield

$$I_4 = 0.5\Delta u^{\mathrm{sc}}_{i,m}(\mathbf{x}^A)\delta_{\ell k}\delta_{m\ell} = 0.5\Delta u^{\mathrm{sc}}_{k,\ell}(\mathbf{x}^A) \tag{26}$$

The integrals I_1 and I_2 are both unbounded as $\epsilon \to 0$ because of the hypersingularity of the integrands. Guiggiani *et al.*[8] have however pointed out, that the two unbounded terms will cancel each other, and the sum $I_1 + I_2$ is bounded. Actually, expanding the integrands of I_1 and I_2 in Laurent series in the local plane of intrinsic coordinates, with special care being taken to preserve the features of the limiting process, the hypersingular integrals $I_1 + I_2$ are transformed to a sum of double and simple regular integrals:

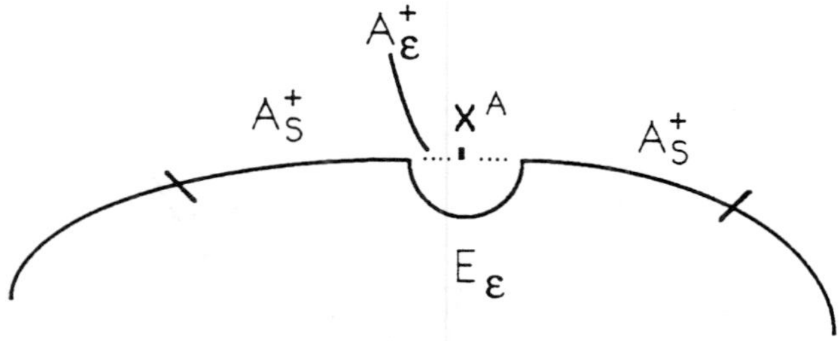

Figure 2: Integration of singular integrals near a point on a smooth boundary

$$I_1 + I_2 = \int_o^{2\pi}\int_o^{S(\theta)} \left\{ F_{k\ell}(\rho,\theta) - \left[\frac{F_{k\ell}^{(-2)}(\theta)}{\rho^2} + \frac{F_{k\ell}^{(-1)}}{\rho} \right] \right\} \mathrm{d}\rho \mathrm{d}\theta$$

$$+ \int_o^{2\pi} | F_{k\ell}^{(-1)}(\theta) \ell n | \frac{s(\theta)}{\beta(\theta)} - F_{k\ell}^{(-2)}(\theta) | \frac{\gamma(\theta)}{\beta^2(\theta)} + \frac{1}{s(\theta)} | \Big\} \mathrm{d}\theta \tag{27}$$

where

$$F_{k\ell}(\rho,\theta) = \sigma_{ijk,\ell}^{\mathrm{G}}[\mathbf{y}(\rho,\theta) - \mathbf{x}^A] n_j(\mathbf{y}(\rho,\theta))] N_i(\rho,\theta)\rho$$

$$= \frac{F_{k\ell}^{(-2)}(\theta)}{\rho^2} + \frac{F_{k\ell}^{(-1)}(\theta)}{\rho} + 0(1) \tag{28}$$

Here, $N_i(\rho,\theta)$ are shape functions, and ρ and θ are the local polar intrinsic coordinates centered at $\mathbf{x}^A$. Also $\rho = s(\theta)$ is the equation of the external contour of A_s and

$$\rho = \epsilon\beta(\theta) + \epsilon^2\gamma(\theta) + 0(\epsilon^3) \tag{29}$$

is the equation of the edge of A_ϵ^+.

Actually the one-dimensional integrals in the right-hand side of (27) which take account of the external shape of the element, are the counterparts of the line integrals in (22) which were obtained through the use of Stokes' theorem. The advantage of this method is that since all integrals in eqn (27) are expressed in the local plane of intrinsic coordinates, the actual computation can be systematically applied in any BEM analysis, either with an open or a closed surface, and with curved boundary elements of any kind and order which meet necessary smoothness requirements, and the computation requires only a straightforward application of standard quadrature rules.

4 Shape functions for crack opening displacements

Special care must be taken in the numerical implementation to account for the square root behavior of the crack-face displacements near the edges of cracks. Consider a typical element which shares a boundary with the crack edge. In a local coordinate system $(\xi_1\xi_2\xi_3)$ the element is shown in Fig. 3. The element lies in the $\xi_1\xi_3$-plane, where the ξ_1 axis is normal to the crack edge. The shape functions must have the following form:

$$\Delta u_i^{\mathrm{sc}} \rightarrow C_i\rho^{\frac{1}{2}}, \quad \text{as} \quad \rho = \overline{0T} \rightarrow 0 \tag{30}$$

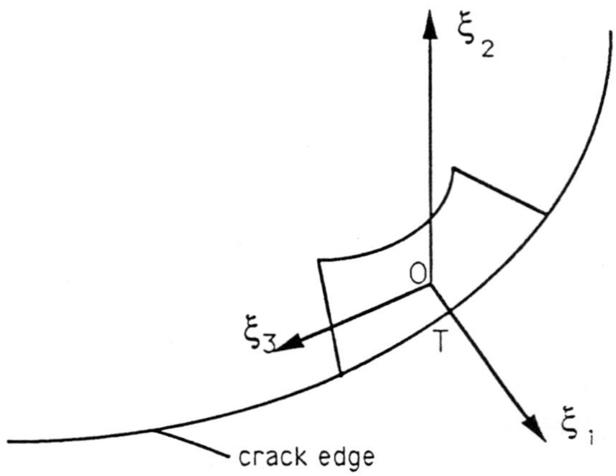

Figure 3: Local coordinates for an element on the crack edge

where C_i is a constant proportional to the local stress intensity factors.

For a flat elliptical crack lying in the plane $y_3 = 0$, with semi-major and minor axes a_1 and a_2, the following crack opening displacement was used by Budreck & Achenbach[5]

$$\Delta u_m^{sc}(\mathbf{y}) = f(\mathbf{y})\Delta u_m(\mathbf{y}_g) \tag{31}$$

Here **yg** denotes the geometrical centroid of the q-th element. The shape function $f(\mathbf{y}) \equiv 1$ was used for all elements except those sharing a border with the crack edge. For edge elements, the shape function was taken as

$$f(\mathbf{y}) = \{[1-(y_1/a_1)^2-(y_2/a_2)^2]/[1-(y_{1_q}/a_1)^2-(y_{2_q}/a_2)^2]\}^{\frac{1}{2}} \tag{32}$$

For a flat crack of arbitrary shape, the shape function

$$f(\mathbf{y}) = [2a\mathrm{d}(\mathbf{y}) - \mathrm{d}^2(\mathbf{y})]^{\frac{1}{2}} \tag{33}$$

was used by Lin & Keer.[15] Here $2a$ is a parameter related to the crack size, which is taken as the maximum distance along the edge and $\mathrm{d}(\mathbf{y})$ is the minimum distance from $\mathbf{y}$ to the edge. It should be noted that (31)-(33) satisfy the property stated by eqn (30).

For BIEs involving both the crack-opening displacement $\Delta u_i^{sc}(y)$ and its derivatives, the dislocation density $\Delta u_{i,j}^{sc}(y)$, it is particularly useful to employ an *a-priori* assumed functional form of $\Delta u_j^{sc}(y)$, so that its differentiation can be carried out explicitly. This methodology was essentially used by Nishimura & Kobayashi,[17] who employed cubic spline functions over a circular crack.

Of special importance in fracture mechanics is the determination of physically relevant parameters which characterize the crack-tip features. In linear fracture mechanics the most relevant quantities are the stress intensity factors. In the local coordinate system $(\xi_1\xi_2\xi_3)$ shown in Fig. 3 the crack opening displacements Δu_i^{sc} may be expressed as (see Erdogan[7])

$$K_I = \frac{\mu\sqrt{2\pi}}{4(1-\nu)}\lim_{\rho\to 0}\frac{\Delta u_2^{sc}}{\sqrt{\rho}} = \frac{\mu\sqrt{2\pi}}{4(1-\nu)}C_2 \tag{34}$$

$$K_{II} = \frac{\mu\sqrt{2\pi}}{4(1-\nu)} \lim_{\rho\to 0} \frac{\Delta u_1^{sc}}{\sqrt{\rho}} = \frac{\mu\sqrt{2\pi}}{4(1-\nu)} C_1 \tag{35}$$

$$K_{III} = \mu\sqrt{2\pi} \lim_{\rho\to 0} \frac{\Delta u_3^{sc}}{\sqrt{\rho}} = \mu\sqrt{2\pi} C_3 \tag{36}$$

where eqn (30) has been used, and ν is the Poisson's ratio.

By using eqns (34)-(36), the stress intensity factors can be directly determined, provided that the components Δu_i^{sc} have been evaluated as part of the BEM procedure. It should be noted here that it is important to use 'crack-edge elements' such as defined by eqn (30) to calculate the stress intensity factors since they represent the proper local behavior of Δu_i^{sc} at the crack edge.

Actual results will not be presented here since they are available in the original papers listed among the References.

5 Surface-breaking and near-surface cracks

The elastodynamic fields generated by cracks are significantly affected by near-by free surfaces or interfaces. In this section we consider surface-breaking and near-surface cracks. The presence of another surface in addition to the surface compressed by the crack faces presents a considerable additional complication to the application of the boundary element method.

The configuration to be discussed in some detail concerns a homogeneous, isotropic, linearly elastic half-plane containing an inclined surface-breaking or near surface crack of length a. The geometry is shown in Fig. 4. The geometry of the problem is two-dimensional, and the deformation is in plane strain. The inclination of the crack with respect to the normal to the stress-free boundary of the half-plane is given by the angle α. For convenience we introduce a local coordinate system x_1, x_2 and a global coordinate system x_1', x_2'. The coordinate systems are related by

$$\begin{Bmatrix} x_1' \\ x_2' \end{Bmatrix} = \begin{Bmatrix} \sin\alpha & -\cos\alpha \\ \cos\alpha & \sin\alpha \end{Bmatrix} \begin{Bmatrix} x_1 \\ x_2 \end{Bmatrix}$$

The boundary conditions on the faces of the crack, A, and on the stress-free boundaries of the half-plane, Γ, are:

$$t_i = \sigma_{ij} n_j = 0 \quad , \quad \mathbf{x}\epsilon\Gamma, A^+ \text{ or } A^- \tag{37}$$

in which Γ is the free boundary of the half-plane, A^+ is the insonified side of the crack and A^- is the other side, and n_j denotes the components of the normal vectors of Γ, A^+ and A^- in the local coordinate system x_1, x_2.

The displacement components of an incident Rayleigh wave which propagates in the positive x_1'-direction along the surface of the half-plane may be written as

$$u_i^{in}(\mathbf{x}') = U\left\{ \binom{k_R}{i\gamma_L} \frac{1}{k_L} \exp(-\gamma_L x_2') - \frac{2ik_R\gamma_L}{(k_T^2 - 2k_R^2)k_L} \binom{i\gamma_T}{-k_R} \exp(-\gamma_T x_2') \right\} \exp(ik_R x_1') \tag{38}$$

where U is an amplitude factor, and

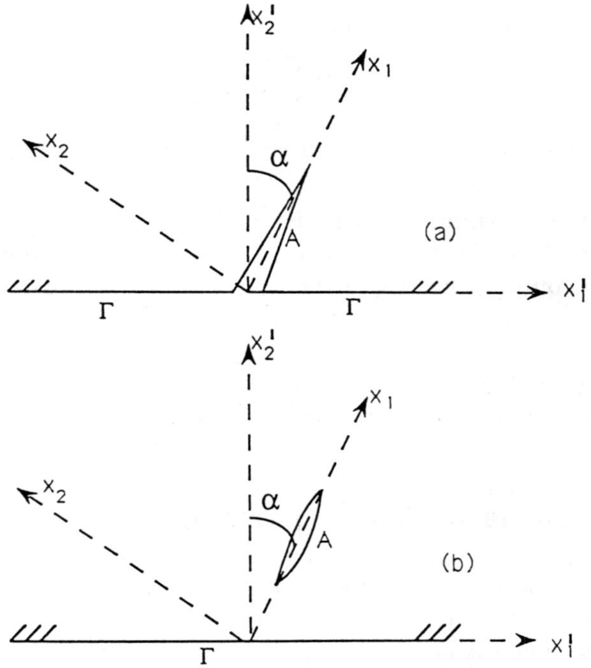

Figure 4: (a) Surface-breaking crack; (b) sub-surface crack

$$k_{\mathrm{R}} = \omega/c_{\mathrm{R}}, k_{\mathrm{L}} = \omega/c_{\mathrm{L}}, k_{\mathrm{T}} = \omega/c_{\mathrm{T}} \tag{39a,b,c}$$

$$\gamma_{\mathrm{L}} = (k_{\mathrm{R}}^2 - k_{\mathrm{L}}^2)^{\frac{1}{2}} \quad , \quad \gamma_{\mathrm{T}} = (k_{\mathrm{R}}^2 - k_{\mathrm{T}}^2)^{\frac{1}{2}} \tag{40a,b}$$

In (39a,b,c), $c_{\mathrm{L}}, c_{\mathrm{T}}$ and c_{R} are the velocities of longitudinal, transverse and Rayleigh waves, respectively. The corresponding expressions for the stress components $\sigma_{ij}^{\mathrm{in}}$ can be obtained by substitution of (38) into Hooke's law, eqn (2).

For the present half-plane problem, a term for the free boundary Γ is added to the integral representation for the scattered field eqn (11). In the local coordinate system x_1, x_2 we have

$$u_k^{\mathrm{sc}}(\mathbf{x}) = \int_{\Gamma} \sigma_{ijk}^{\mathrm{G}}(\mathbf{y}-\mathbf{x}) u_i^{\mathrm{sc}}(\mathbf{y}) n_j(\mathbf{y}) \mathrm{d}s$$

$$+ \int_{A^+} \sigma_{i2k}^{\mathrm{G}}(\mathbf{y}-\mathbf{x}) \Delta u_i^{\mathrm{sc}}(\mathbf{y}) \mathrm{d}y_1 \, , \quad x_2' > 0, \mathbf{x} \notin \Gamma, A^+ \tag{41}$$

Substituting eqn (41) into eqn (2) and using the relation between surface tractions and the stress components eqn (6) we obtain the following representation formula for the traction components t_i^{sc} at $\mathbf{x}$:

$$t_i^{\mathrm{sc}}(\mathbf{x}) = -n_j(\mathbf{x}) \left\{ \int_{\Gamma} K_{ijk\ell}^{\mathrm{G}}(\mathbf{y}-\mathbf{x}) u_k^{\mathrm{sc}}(\mathbf{y}) n_\ell(\mathbf{y}) \mathrm{d}s \right.$$

$$\left. + \int_{A^+} K_{ijk2}^{\mathrm{G}}(\mathbf{y}-\mathbf{x}) \Delta u_k^{\mathrm{sc}}(\mathbf{y}) \mathrm{d}y_1 \right\} , \quad x_2' > 0, \quad \mathbf{x} \notin \Gamma, A^+ \tag{42}$$

The function $K^{G}_{ijk\ell}$ is defined by

$$K^{G}_{ijk\ell} = \lambda\delta_{ij}\sigma^{G}_{k\ell m,m} + \mu(\sigma^{G}_{k\ell i,j} + \sigma^{G}_{k\ell j,i}) \tag{43}$$

Next, the integration-by-parts regularization procedure is used (see Section 3) to obtain the BIE from (42). First Γ is divided into N_1, and A^+ into N_2 elements, respectively. Then eqn (42) can be written in the following form:

$$t^{sc}_i(\mathbf{x}) = -n_j(x)\left\{\sum_{j=1}^{N_1}\int_{sj}^{s_{j+1}} K^{G}_{ijk\ell}(\mathbf{y}-\mathbf{x})u^{sc}_k(\mathbf{y})n_\ell(\mathbf{y})ds\right.$$

$$\left.+\sum_{j=N_1+1}^{N}\int_{s_j}^{s_{j+1}} K^{G}_{ijk2}(\mathbf{y}-\mathbf{x})\Delta u^{sc}_k(\mathbf{y})dy_1\right\},\ x_2^{'} > 0,\ \mathbf{x}\notin\Gamma, A^+ \tag{44}$$

It is noted here that the original integration paths along Γ in (42) extend to infinity, which is not suitable for a numerical procedure. Hence, Γ is truncated in such a way that a desired accuracy for the crack opening displacement is achieved. This method has been applied in other cases for half-plane related problems, and it yields quite good results.

Through integration-by-parts, (for details, see Zhang & Achenbach[25]) the following relations can be obtained after some manipulations

$$\int_{s_j}^{s_{j+1}} K^{G}_{ijk\ell}(\mathbf{y}-\mathbf{x})u^{sc}_k(\mathbf{y})n_\ell(\mathbf{y})ds$$

$$= H^{1}_{ijk}(\mathbf{y}-\mathbf{x})u^{sc}_\ell(\mathbf{y})|^{s_{j+1}}_{s_j} - \int_{s_j}^{s_{j+1}} H^{1}_{ijk}(\mathbf{y}-\mathbf{x})\epsilon_{mn}u^{sc}_{\ell,m}(\mathbf{y})n_n(\mathbf{y})ds$$

$$-\rho\omega^2\int_{s_j}^{s_{j+1}} H^{2}_{ijk}(\mathbf{y}-\mathbf{x})u^{sc}_k(\mathbf{y})ds \tag{45}$$

in which H^1_{ijk} and H^2_{ijk} are defined by

$$H^{1}_{ijk} = \lambda\delta_{ij}\epsilon_{m\ell}\sigma^{G}_{k\ell m} + \mu(\epsilon_{j\ell}\sigma^{G}_{k\ell i} + \epsilon_{i\ell}\sigma^{G}_{k\ell j}) \tag{46}$$

$$H^{2}_{ijk} = \lambda\delta_{ij}u^{G}_{km}n_m + \mu(u^{G}_{ki}n_j + u^{G}_{kj}n_i) \tag{47}$$

Here ϵ_{ij} is the two-dimensional permutation tensor. Similar results can be obtained for the second integral of eqn (44); they will not be given here in the interest of brevity.

It should be noted that the integrands in (45) possess now a logarithmic singularity $\ell n(r)$ (where $r = |\mathbf{y}-\mathbf{x}|$) as well as a singularity of the type $1/r$ at $\mathbf{x} = \mathbf{y}$. These singularities can be integrated analytically or numerically without difficulties as in a usual BIE formulation (see Banerjee & Butterfield[2]). Using the relation (45), choosing suitable shape functions for u^{sc}_k and Δu^{sc}_k, taking the limit $\mathbf{x}\rightarrow\Gamma + A^-$ and considering the boundary conditions

$$t^{in}_i(\mathbf{x}^\alpha) + t^{sc}_i(\mathbf{x}^\alpha) = 0\ ,\ \mathbf{x}^\alpha\epsilon\Gamma + A^+, \alpha = 1, 2, \cdots N \tag{48}$$

at the N collocation points on Γ, a system of linear, algebraic equations is obtained from eqn (44) for the unknown quantities u_k^{sc} and Δu_k^{sc} at N discrete points. In eqn (48) t_i^{in} denotes the traction components on $\Gamma + A^+$ due to the incident wave.

We mention at this point that in the representation integral (41) the half-plane Green's function which satisfies the stress-free conditions on the boundary of the half-plane can be employed so that a discretization along Γ is not necessary. Unfortunately, the use of half-plane Green's functions increases the computing effort due to the high complexity of these functions (see Kobayashi[11]). Therefore, it is advantageous to use the full-plane Green's function in conjunction with a discretization of the boundary Γ, to obtain a more economical scheme. One disadvantage of this approach is that we have to estimate the required lengths Γ, which depend in general on the crack inclination and on the type, frequency and incidence angle of the incident wave.

To obtain the reflection and transmission coefficients we need the far-field asymptotic approximations for the scattered surface-wave displacements. Equation (41) is not suitable for such an asymptotic analysis since it contains integrals along the infinite length segments Γ. Hence, the half-plane Green's function will be used in the representation integral. In the x_1', x_2' coordinate system this yields the following result

$$u_k^{sc}(\mathbf{x}') = \int_{A+} \overline{\sigma}_{ijk}^{-G}(\mathbf{y}' - \mathbf{x}')\Delta u_i^{sc}(\mathbf{y}')n_j'(\mathbf{y}')\mathrm{d}s \tag{49}$$

Here $\overline{\sigma}_{ijk}^{-G}$ denotes the half-plane Green's function (see Kobayashi[11] and Neerhoff[16]), and n_j' represents the components of the unit normal vector of A^+ in the x_1', x_2' system. The advantage of using $\overline{\sigma}_{ijk}^{-G}$ is that in eqn (49) only an integration along A^+ enters because the Green's function $\overline{\sigma}_{ijk}^{-G}$ satisfies traction-free boundary conditions on Γ.

By substituting the asymptotic expansions of the half-plane Green's function $\overline{\sigma}_{ijk}^{-G}$ into eqn (49) we can write $u_k^{sc}(\mathbf{x}')$ as (see Neerhoff,[16] Zhang & Achenbach[23])

$$u_k^{sc}(\mathbf{x}') = u_k^{b}(\mathbf{x}') + u_k^{s}(\mathbf{x}')\ , \quad \text{if } |\mathbf{x}'| \to \infty \tag{50}$$

in which u_k^b and u_k^s denote the body-wave part and the surface-wave part of the scattered displacements. The body-wave part, u_k^b, decays as $1/|x_1'|^{\frac{1}{2}}$ for fixed x_2' and $|x_1'| \to \infty$. The second term u_k^s represents the outgoing Rayleigh surface waves, which for fixed x_2' and $|x_1'| \to \infty$ propagate in the $\pm x_1'$-directions and do not decay. Hence, the dominant term for the wave motion at a large distance from the crack and near the boundary of the half-plane is the surface-wave term u_k^s, and we can rewrite (50) as

$$u_k^{sc}(\mathbf{x}) \simeq u_k^{s}(\mathbf{x})\ , \quad \text{as } x_1' \to \pm\infty \tag{51}$$

where $u_k^s(\mathbf{x}')$ is given by

$$u_k^{s}(\mathbf{x}') = U_L^{\pm}\left\{\begin{pmatrix} \pm k_R \\ i\gamma_L \end{pmatrix}\frac{1}{k_L}\exp(-\gamma_L x_2') - \right.$$

$$\left. -\frac{U_T^{\pm}}{U_L^{\pm}}\begin{pmatrix} i\gamma_T \\ \pm k \end{pmatrix}\frac{1}{k_T}\exp(-i\gamma_T x_2')\right\}\exp(\pm i k_R x_1')$$

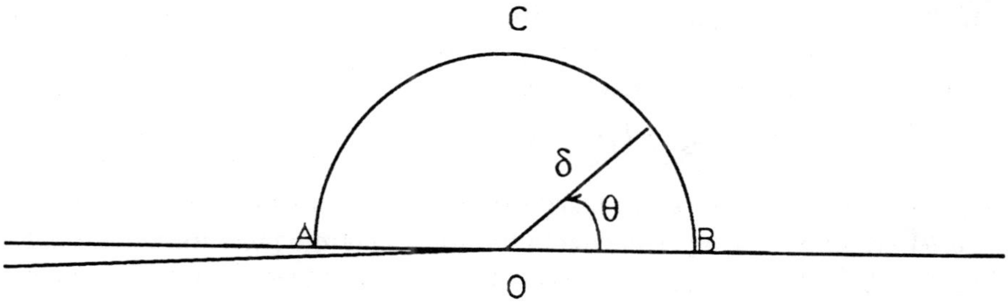

Figure 5: Crack-tip element

$$x_1' \to \pm\infty \tag{52}$$

and

$$U_\xi^{\pm} = \int_{A+} S_{ij}^{\xi R\pm}(\mathbf{y}')\Delta u_i^{sc}(\mathbf{y}')n_j' ds, \quad \xi = L, T \tag{53}$$

in eqns (52) and (53) the '$\pm$' signs refer to Rayleigh waves propagating in the $\pm x_1'$-directions. Explicit expressions for $S_{ij}^{\xi R}$ can be found in Zhang & Achenbach.[24]

Numerous results for reflection and transmission coefficients as well as elastodynamic stress intensity factors can be found in the quoted references.

6 Multi-region method and interface crack problems

The multi-region method avoids the occurrence of hypersingularity of the integral equations for crack problems, because each crack face is modeled as part of the boundary of a separate region. The disadvantage of the multi-region method is that additional boundaries are introduced. For an interface crack problem, i.e. a crack between dissimilar media, this is, however, the only practical method available. In this section we will consider two-dimensional geometries only.

First we consider a crack in a homogeneous solid. In the numerical implementation of the multi-region method, the displacement along the crack line very close to the crack tip is chosen to be proportional to the square root of the distance from the crack tip. On the other hand, along the crack line ahead of the crack tip the stresses are singular, being inversely proportional to the square root of the distance from the crack tip. Thus, along the crack line two different crack tip elements should be used behind and ahead of the crack tip.

Here we propose an alternative way of modeling the behavior of a two-dimensional crack tip region. Let AB be a single crack tip element, as shown in Fig. 5. Here AB is a small length element along the crack line with the crack tip 0 as its midpoint. From eqn (4), we have

$$\int_{AOBCA} [u_{ik}^G(\mathbf{y}-\mathbf{x})t_i^{sc}(\mathbf{y}) - \sigma_{ijk}^G(\mathbf{y}-\mathbf{x})n_j(\mathbf{y})u_i^{sc}(\mathbf{y})]ds(\mathbf{y}) = \epsilon(\mathbf{x})u_k^{sc}(\mathbf{x})$$

or

$$\int_{AOB} - \int_{ACB} [u_{ik}^G(\mathbf{y}-\mathbf{x})t_i^{sc}(\mathbf{y}) - \sigma_{ijk}^G(\mathbf{y}-\mathbf{x})n_j(\mathbf{y})u_i^{sc}(\mathbf{y})]ds(\mathbf{y}) = \epsilon(\mathbf{x})u_k^{sc}(\mathbf{x}) \tag{54}$$

If $\delta = \overline{AB}/2$ is small enough, we can write along ACB

$$u_i^{\rm sc}(\mathbf{y}) \approx u_i^{\rm sc}(\mathbf{y}_o) + K^I T_i^I(\delta,\theta) + K^{II} T_i^I(\delta,\theta) \tag{55}$$

$$t_i^{\rm sc}(\mathbf{y}) \approx [K^I F_{ij}^I(\delta,\theta) + K^{II} F_{ij}^{II}(\delta,\theta)] n_j(\mathbf{y}) \tag{56}$$

where K_I and K_{II} are the stress intensity factors of the crack tip and $T_i^I(\delta,\theta), T_i^{II}(\delta,\theta)$ and $F_{ij}^{II}(\delta,\theta)$ are known analytical functions from dynamic fracture analysis. Explicit expressions for T_i^I, T_i^{II}, F_{ij}^I and F_{ij}^{II} can be found, for example, in Sih & Asaro.[20] It is noted that the traction $t_i^{\rm sc}(\mathbf{y})$ along ACB is not singular for $\delta \neq 0$.

It is well-known that for conditions of perfect contact the stresses near the tip of an interface crack between dissimilar media possess violent oscillations which depend on distance as $r^{-1/2+i\alpha}$, where r is the radial distance from the crack tip and α is a bi-material constant. Related to this behavior the upper and lower faces of the crack should wrinkle and overlap near the crack tip. This is physically impossible. However, as pointed out by Rice,[18] 'while the predicted interpenetration means the solutions must be wrong in detail on the scale of the contact zone, they do nevertheless provide a proper characterizing parameter for the near tip state in typical circumstances when that zone size is much smaller than crack length'. On the other hand the violent oscillatory behavior of the field along the crack line near the crack tip prevents a reliable numerical calculation along the crack line. However, if δ is chosen small enough compared with the crack length and the wave length in the problem considered, but big enough compared with the interpenetration zone behind the crack tip, then eqn (54) can be used for our numerical calculation in solving the BIE. The restrictions on δ are met in many practical cases, particularly when there is some non-negligible tensile component of the loading relative to the crack, for details see Rice. [18]

7 Interaction of a Rayleigh wave with a disbond in a material interphase

As an application of the multi-region method we consider the configuration of an interphase crack in a half-plane. The half plane is composed of two homogeneous, isotropic, linearly elastic quarter planes, A and B. The quarter planes are joined by an interphase which contains a failure region. If the quarter planes are joined perfectly, the failure region becomes an interface crack. The materials of A and B may be the same or they may be different. The half plane occupies the domain $x_2 \geq 0$, and the interphase is located along $x_1 = 0$. Both a surface-breaking and a sub-surface failure region will be considered. The two-dimensional geometry is shown in Fig. 6.

The incident wave is a Rayleigh surface wave which propagates from $x_1 = -\infty$ to $x_1 = 0$ in the region A. The displacement components of the incident wave may be written as

$$u_1^{\rm in}(\mathbf{x}) = U\frac{k_{\rm R}^A}{k_{\rm L}^A}\left\{ {\rm e}^{-\gamma_{\rm L}^A x_2} + \frac{2\gamma_{\rm L}^A\gamma_{\rm T}^A}{(k_{\rm T}^A)^2 - 2(k_{\rm R}^A)^2}{\rm e}^{-\gamma_{\rm T}^A x_2}\right\} {\rm e}^{ik_{\rm R}^A x_1} \tag{57a}$$

$$u_2^{\rm in}(\mathbf{x}) = iU\frac{k_{\rm L}^A}{k_{\rm L}^A}\left\{ {\rm e}^{-\gamma_{\rm L}^A x_2} + \frac{2(k_{\rm R}^A)^2}{(k_{\rm T}^A)^2 - 2(k_{\rm R}^A)^2}{\rm e}^{-\gamma_{\rm T}^A x_2}\right\} {\rm e}^{ik_{\rm R}^A x_1} \tag{57b}$$

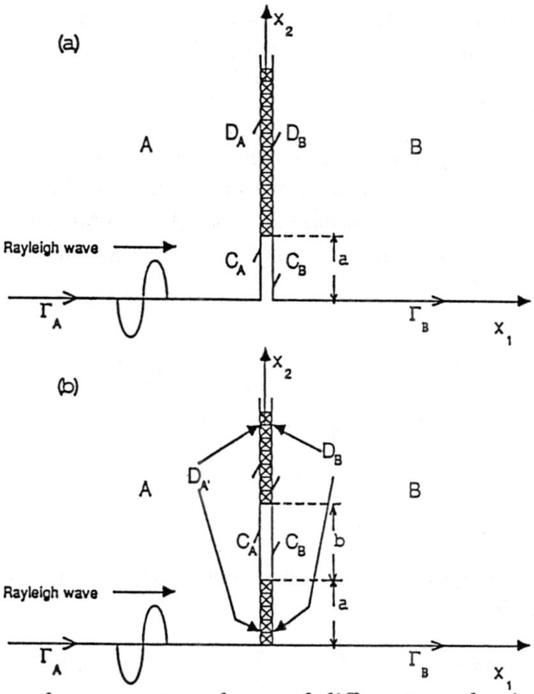

Figure 6: Interphase of two quarter planes of different mechanical properties; (a) a surface-breaking flaw; (b) a sub-surface flaw

where $k_{\rm L}^A$, $k_{\rm T}^A$ and $k_{\rm R}^A$ are the wave numbers of longitudinal, transverse and Rayleigh waves, respectively,

$$k_{\rm L}^A = \omega / c_{\rm L}^A \ , \qquad c_{\rm L}^A = [(\lambda^A + 2\mu^A)/\rho^A]^{\frac{1}{2}} \tag{58a,b}$$

$$k_{\rm T}^A = \omega / c_{\rm T}^A \ , \qquad c_{\rm T}^A = (\mu^A/\rho^A)^{\frac{1}{2}} \tag{59a,b}$$

$$k_{\rm R}^A = \omega / c_{\rm R}^A \tag{60}$$

Here $c_{\rm L}^A$, $c_{\rm T}^A$, $c_{\rm R}^A$ are the longitudinal, transverse and Rayleigh wave velocities, λ^A and μ^A are the Lamé elastic constants, and ρ^A is the mass density of A. Also

$$\gamma_{\rm L}^A = [(k_{\rm R}^A)^2 - (k_{\rm L}^A)^2]^{1/2} \ , \ \gamma_{\rm T}^A = [(k_{\rm R}^A)^2 - (k_{\rm T}^A)^2]^{1/2} \tag{61a,b}$$

In region A, the components of the total displacement field, u_j^A are the sums of the corresponding displacement components, of the incident wave, $u_j^{\rm in}$, and the back-scattered wave, $u_j^{\rm sc}$:

$$u_j^A = u_j^{\rm in} + u_j^{\rm sc} \ , \ \ j = 1, 2 \tag{62}$$

The components of the transmitted displacement field in region B are denoted by u_j^B. According to Hooke's law, eqn (2), the corresponding stress fields are

$$\sigma_{ij}^A = \lambda^A \delta_{ij} u_{k,k}^A + \mu^A (u_{i,j}^A + u_{j,i}^A) \tag{63}$$

$$\sigma_{ij}^B = \lambda^B \delta_{ij} u_{k,k}^B + \mu^B (u_{i,j}^B + u_{j,i}^B) \tag{64}$$

where λ^B and μ^B are the Lamé elastic constants of the material in region B. On a surface with unit outward normal vectors n_j^A or n_j^B, the tractions are

$$t_i^A = \sigma_{ij}^A n_j^A \ , \quad t_i^B = \sigma_{ij}^B n_j^B \tag{65a,b}$$

A spring layer with appropriate spring constants may often be used to model the interphase, and the problem can then be simplified significantly. Let S_{L} and S_{T} denote the extensional and shear constants of the spring layer. The conditions of continuity along the adhered parts of the interphase, D_A and D_B can then be expressed as

$$t_j^A = -t_j^B \tag{66}$$

$$t_1^A = S_{\mathrm{L}}(u_1^B - u_1^A) = S_{\mathrm{L}}(u_1^B - u_1^{\mathrm{in}} - u_1^{\mathrm{sc}}) \tag{67a}$$

$$t_2^A = S_{\mathrm{T}}(u_2^B - u_2^A) = S_{\mathrm{T}}(u_2^B - u_2^{\mathrm{in}} - u_2^{\mathrm{sc}}) \tag{67b}$$

For the case of $S_{\mathrm{L}} \to \infty$, $S_{\mathrm{T}} \to \infty$, we have

$$u_i^B = u_i^A = u_i^{\mathrm{in}} + u_i^{\mathrm{sc}}, \quad t_i^B = -t_i^A \tag{67c,d}$$

On the surface of the half-plane and on the faces of the failure region, the tractions are assumed to vanish, thus

$$t_j^A(\mathbf{x}) = t_j^{\mathrm{sc}}(\mathbf{x}) + t_j^{\mathrm{in}}(\mathbf{x}) = 0, \quad \mathbf{x} \text{ on } \Gamma_A \text{ and } C_A \tag{68}$$

$$t_j^B(\mathbf{x}) = 0 \qquad ; \quad \mathbf{x} \text{ on } \Gamma_B \text{ and } C_B \tag{69}$$

where t_j^{sc} and t_j^{in} are the tractions related to the back-scattered and incident waves respectively. Since $t_j^{\mathrm{in}}(\mathbf{x}) = 0$, for $\mathbf{x}$ on Γ_A, we can rewrite (69) as

$$t_j^{\mathrm{sc}}(\mathbf{x}) = 0, \quad \mathbf{x} \text{ on } \Gamma_A \tag{70}$$

$$t_j^{\mathrm{sc}}(\mathbf{x}) = -f_j^{\mathrm{in}}(\mathbf{x}), \quad \mathbf{x} \text{ on } C_A \tag{71}$$

Since the displacements are bounded, eqns (66)-(67a,b) imply that for this interphase model the stresses remain bounded at the tips of failure regions for finite S_{L} and S_{T}.

The multi-region method will be used to solve the problem defined by eqns (57)-(71). The integral representation for the back-scattered field in region A can be written as

$$\epsilon_A(\mathbf{x}_p)u_k^{\mathrm{sc}}(\mathbf{x}) = \int_{\Gamma_A+C_A+D_A} [u_{ik}^{\mathrm{G}A}(\mathbf{y}-\mathbf{x})t_i^{\mathrm{sc}}(\mathbf{y}) - \sigma_{ijk}^{\mathrm{G}A}(\mathbf{y}-\mathbf{x})u_i^{\mathrm{sc}}(\mathbf{y})n_j^A(\mathbf{y})]\mathrm{d}s(\mathbf{y}) \tag{72}$$

where

$$\epsilon_A(\mathbf{x}) = \begin{cases} 1 \ , & \mathbf{x} \in A \\ 0 \ , & \mathbf{x} \notin A \end{cases}$$

and the contours Γ_A and C_A and D_A are indicated in Fig. 6. In (72), $u_{ik}^{\mathrm{G}A}$ and $\sigma_{ijk}^{\mathrm{G}A}$ denote the full-plane displacement and stress Green's functions for the material in region A. The analogous integral representation for the displacement in region B is

$$\epsilon_B(\mathbf{x})u_k^B(\mathbf{x}) = \int_{C_B+D_B+\Gamma_B} [u_{ik}^{\mathrm{G}B}(\mathbf{y}-\mathbf{x})t_i^B(\mathbf{y}) - \sigma_{ijk}^{\mathrm{G}B}(\mathbf{y}-\mathbf{x})u_i^B(\mathbf{y})n_j^B(\mathbf{y})]\mathrm{d}s(\mathbf{y}) \tag{73}$$

where

$$\epsilon_B(\mathbf{x}) = \begin{cases} 1 & , \quad \mathbf{x} \in B \\ 0 & , \quad \mathbf{x} \notin B \end{cases}$$

Again C_B, D_B and Γ_B are indicated in Fig. 6, while u_{ik}^{GB} and σ_{ijk}^{GB} denote the full-plane displacement and stress Green's functions for material B. For $\mathbf{x} \notin A$, substitution of (62) into (72) yields by the use of (66), (67), (70) and (71),

$$\int_{\Gamma_A + C_A} \sigma_{ijk}^{GA} u_i^{sc} n_j^A \mathrm{d}s(\mathbf{y}) + \int_{D_A} [\sigma_{ijk}^{GA} u_i^B n_j^A + (u_{1k}^{GA} + \sigma_{1jk}^{GA} n_j^A / S_{\mathrm{L}}) t_1^B +$$

$$(u_{2k}^{GA} + \sigma_{2jk}^{GA} n_j^A / S_{\mathrm{T}}) t_2^B] \mathrm{d}s(\mathbf{y})$$

$$= - \int_{C_A} u_{ik}^{GA} t_i^{in} \mathrm{d}s(\mathbf{y}) + \int_{D_A} (\sigma_{ijk}^{GA} u_i^{in} n_j^A - u_{ik}^{GA} t_i^{in}) \mathrm{d}s(\mathbf{y}), \ \mathbf{x} \notin A \qquad (74)$$

For $\mathbf{x} \notin B$, (73) becomes by the use of (69)

$$\int_{\Gamma_B + C_B + D_B} \sigma_{ijk}^{GB} u_i^B n_j^B \mathrm{d}s(\mathbf{y}) - \int_{D_B} u_{ik}^{GB} t_i^B \mathrm{d}s(\mathbf{y}) = 0, \mathbf{x} \notin B \qquad (75)$$

Taking the limits as the point $\mathbf{x} \notin A$ approaches the boundary of A, and the point $\mathbf{x} \notin B$ approaches the boundary of B, eqn (74) and (75) become a set of boundary integral equations for the following unknowns:

$$u_i^{sc}(\mathbf{x}) \ , \quad \mathbf{x} \text{ on } \Gamma_A \text{ and } C_A$$

$$u_i^B(\mathbf{x}) \ , \quad \mathbf{x} \text{ on } D_A$$

$$t_i^B(\mathbf{x}) \ , \quad \mathbf{x} \text{ on } D_B$$

$$u_i^B(\mathbf{x}) \ , \quad \mathbf{x} \text{ on } \Gamma_B \text{ and } C_B \qquad (76)$$

After they have been solved, the displacement fields of the back-scattered and transmitted waves at any point in the half space can be calculated from (72) and (73).

By considering an integral representation of the scattered displacement field in terms of the half-plane Green's functions, and by expanding the half-plane Green's functions as $|\mathbf{x}| \to \infty$, Neerhoff[16] concluded that the scattered far-field is composed of body wave parts and Rayleigh surface-wave parts. The body wave parts decay, but the surface wave parts do not decay with increasing $|x_1|$. Hence, when $|\mathbf{x}| \to \infty$, we can write

$$u_k^{sc}(\mathbf{x}) \approx R\, u_k^{SR}(\mathbf{x}) \ , \quad x < 0, |\mathbf{x}| \to \infty \qquad (77)$$

$$u_k^B(\mathbf{x}) \approx T\, u_k^{BR}(\mathbf{x}) \ , \quad x_1 > 0, |\mathbf{x}| \to \infty \qquad (78)$$

where R and T are defined as reflection and transmission coefficients, and $u_k^{SR}(\mathbf{x})$ and $u_k^{BR}(\mathbf{x})$ denote Rayleigh surface waves propagating in the negative and positive x_1 directions, respectively:

$$u_1^{SR}(\mathbf{x}) = U \frac{k_{\mathrm{R}}^A}{k_{\mathrm{L}}^A} \left[\mathrm{e}^{-\gamma_{\mathrm{L}}^A x_2} + \frac{2\gamma_{\mathrm{L}}^A \gamma_{\mathrm{T}}^A}{(k_{\mathrm{T}}^B)^2 - 2(k_{\mathrm{R}}^A)^2} \mathrm{e}^{-\gamma_{\mathrm{T}}^A x_2} \right] \mathrm{e}^{-ik_{\mathrm{R}}^A x_1} \qquad (79a)$$

$$u_2^{SR}(\mathbf{x}) = -iU\frac{\gamma_{\mathrm{L}}^A}{k_{\mathrm{L}}^A}\left[\mathrm{e}^{-\gamma_{\mathrm{L}}^A x_2} + \frac{2(k_{\mathrm{R}}^A)^2}{(k_{\mathrm{T}}^A)^2 - 2(k_{\mathrm{R}}^A)^2}\mathrm{e}^{-\gamma_{\mathrm{T}}^A x_2}\right]\mathrm{e}^{-ik_{\mathrm{R}}^A x_1} \tag{79b}$$

$$u_1^{BR}(\mathbf{x}) = U\frac{k_{\mathrm{R}}^A}{k_{\mathrm{L}}^B}\left[\mathrm{e}^{-\gamma_{\mathrm{L}}^B x_2} + \frac{2\gamma_{\mathrm{L}}^B\gamma_{\mathrm{T}}^B}{(k_{\mathrm{T}}^B)^2 - 2(k_{\mathrm{R}}^B)^2}\mathrm{e}^{-\gamma_{\mathrm{T}}^A x_2}\right]\mathrm{e}^{ik_{\mathrm{R}}^B x_1} \tag{80a}$$

$$u_2^{BR}(\mathbf{x}) = iU\frac{\gamma_{\mathrm{L}}^B}{k_{\mathrm{L}}^B}\left[\mathrm{e}^{-\gamma_{\mathrm{L}}^B x_2} + \frac{2(k_{\mathrm{R}}^B)^2}{(k_{\mathrm{T}}^B}\,)^2 - 2(k_{\mathrm{R}}^B)^2\mathrm{e}^{-\gamma_{\mathrm{T}}^A x_2}\right]\mathrm{e}^{ik_{\mathrm{R}}^B x_1} \tag{80b}$$

The integral expressions for the reflection and transmission coefficients were obtained by Li & Achenbach.[14] In the context of the present paper they can be written in the form

$$R = \int_{C_A + D_A} [Q_i^A(\mathbf{x})\sigma_{ij}^{\mathrm{sc}}(\mathbf{x}) - S_{ij}^A(\mathbf{x})u_i^{\mathrm{sc}}(\mathbf{x})]n_j^A(\mathbf{x})ds(\mathbf{x}) \tag{81}$$

$$T = \int_{C_A + D_B} [Q_i^B(\mathbf{x})\sigma_{ij}^B(\mathbf{x}) - S_{ij}^B(\mathbf{x})u_i^B(\mathbf{x})]n_j^B(\mathbf{x})ds(\mathbf{x}) \tag{82}$$

Explicit expressions for Q_i^A, S_{ij}^B, Q_i^B and S_{ij}^B are listed in Appendix B.

The boundary integral eqns (74) and (75) will now be solved by the use of the boundary element method. In (74) and (75) the integral paths along Γ_A, D_A, D_B and Γ_B extend however, to infinity, which is not suitable for the numerical procedure. These integral paths can be reduced to finite length by the following considerations. The wave system generated by interaction of the incident surface wave with the interface containing a failure region consists of reflected and transmitted surface waves along the free surface, diffracted body waves and possibly interface waves along D_A and D_B. The surface waves and the body waves decay as x_2 increases. If interface waves along D_A and D_B do not exist or decay, then the integral paths D_A and D_B may be truncated without loss of accuracy. The integrals along D_A and D_B can then be replaced by integrals along finite paths, say D_{A1} and D_{B1}, where

$$D_{A1}:\ x_1 = 0,\ x_2 < x_{2A},\ \mathbf{x} \notin C_A$$

$$D_{B1}:\ x_1 = 0,\ x_2 < x_{2A},\ \mathbf{x} \notin C_B$$

Here x_{2A} is a large enough number. For the integrals along Γ_A and Γ_B, on the other hand, the reflected and transmitted waves do not decay as $|x_1| \to \infty$, since the main parts of these waves are outgoing Rayleigh surface waves. If $|x_{1A}|$ $(x_{1A} < 0)$ is large enough so that for $x_1 < x_{1A}$ the back-scattered displacement fields $u_k^{\mathrm{sc}}(\mathbf{x})$, can be expressed approximately by (77), then it is useful to split Γ_A at $x_1 = x_{1A}$ in Γ_{AO} and Γ_{A1}, where

$$\Gamma_{AO}:\ x_2 = 0, x_1 < x_{1A}$$

$$\Gamma_{A1}:\ x_2 = 0, x_{1A} < x_1 < 0$$

If t_i^{SR} denote the traction components corresponding to u_i^{SR}, we can write according to eqn (72) for $\mathbf{x} \notin A$,

$$\int_{\Gamma_{AO}+\Gamma_{A1}+C_A+D_{A1}} \left[u_{ik}^{\mathrm{G}A}(\mathbf{y}-\mathbf{x})t_i^{SR}(\mathbf{y}) - \sigma_{ijk}^{\mathrm{G}A}(\mathbf{y}-\mathbf{x})u_i^{SR}(\mathbf{y})n_j^A(\mathbf{y})\right] ds(\mathbf{y}) = 0 \tag{83}$$

By the use of the traction-free conditions on Γ_A, eqn (83) becomes

$$\int_{\Gamma_{AO}} \sigma_{ijk}^{GA} u_i^{SR} n_j^A ds(\mathbf{y}) = -\int_{\Gamma_{A1}} \sigma_{ijk}^{GA} u_i^{SR} n_j^A ds(\mathbf{y}) + \int_{C_A+D_{A1}} (u_{ik}^{GA} t_i^{SR} - \sigma_{ijk}^{GA} u_i^{SR} n_j^A) ds(\mathbf{y}) \tag{84}$$

Now, substituting (77) into (74) for $\mathbf{x}$ on Γ_{A0} and using (84), we obtain

$$\begin{aligned}
&\int_{\Gamma_{A1}+C_A} \sigma_{ijk}^{GA} u_i^{sc} n_j^A ds(\mathbf{y}) \\
&+ \int_{D_{A1}} [\sigma_{ijk}^{GA} u_i^B n_j^A + (u_{1k}^{GA} + \sigma_{1jk}^{GA} n_j^A / S_L) t_1^B + (u_{2k}^{GA} + \sigma_{2jk}^{GA} n_j^A / S_T) t_2^B] \, ds(\mathbf{y}) \\
&- R\left(\int_{\Gamma_{A1}+C_A+D_{A1}} \sigma_{ijk}^{GA} u_i^{SR} n_j^A ds(\mathbf{y}) - \int_{C_A+D_{A1}} u_{ik}^{GA} t_i^{SR} ds(\mathbf{y})\right) \\
&= \int_{D_{A1}} \sigma_{ijk}^{GA} u_i^{in} n_j^A ds(\mathbf{y}) - \int_{C_A+D_{A1}} u_{ik}^{GA} t_i^{in} ds(\mathbf{y})
\end{aligned} \tag{85}$$

Using a similar approach for the transmitted waves, we have from (75),

$$\begin{aligned}
&\int_{\Gamma_{B1}+C_B+D_{B1}} \sigma_{ijk}^{GB} u_i^B n_j^B ds(\mathbf{y}) - \int_{D_{B1}} u_{ik}^{GB} t_i^B ds(\mathbf{y}) \\
&- T\left(\Gamma_{B1}+C_B+D_{B1} \sigma_{ijk}^{GB} u_i^{BR} n_j^B ds(\mathbf{y}) - \int_{C_B+D_{B1}} u_{ik}^{GB} t_i^{BR} ds(\mathbf{y}\right) = 0
\end{aligned} \tag{86}$$

where Γ_{B1} is the part of the boundary $x_2 = 0$ defined by $0 < x_1 < x_{1B}$, and x_{1B} is large enough such that for $x_1 > x_{1B}$, the transmitted waves can be expressed approximately by (78). Also t_i^{BR} denotes the tractions corresponding to u_i^{BR} . The terms u_i^{SR} and u_i^{BR} in (84) and (86) are defined by (79) and (80), respectively.

Equations (85) and (86) are boundary integral equations which can be solved in the usual manner by the boundary element method. Results can be found in the paper by Li & Achenbach.[14]

8 Concluding comment

For elastic bodies containing interior, surface-breaking and near-surface cracks, we have reviewed various ways to derive sets of boundary integral equations for solution by the boundary element method. Techniques to deal with higher order singularities have been discussed in some detail. Matters of formulation are now well understood, and a wide range of static and dynamic problems for cracked bodies is now amenable to solution by the boundary element method.

Acknowledgment

This work was carried out in the course of research supported by the Department of Energy, Basic Energy Sciences Engineering Research Program for Contract DEFG02-86-ER13484.

References

1. Achenbach, J.D. *Wave Propagation in Elastic Solids*, North-Holland, Amsterdam/New York, 1973.

2. Banerjee, P.K. & Butterfield, R. *Boundary Element Methods in Engineering Science*, McGraw-Hill, London, 1981.

3. Beskos, D.E. Boundary element methods in dynamic analysis, *ASME Applied Mechanics Reviews*, 1987, **40**, 1-23.

4. Budiansky, B. & Rice, J.R. An integral equation for dynamic elastic response of an isolated 3-D crack, *Wave Motion*, 1979, **1**, 187-192.

5. Budreck, D.E. & Achenbach, J.D. Scattering from three-dimensional planar cracks by the boundary integral equation method, *ASME Journal of Applied Mechanics*, 1988, **55**, 405-412.

6. Cruse, T.A. Fracture mechanics, in *Boundary Element Methods in Mechanics*, B.D. Beskos (ed), North-Holland, 1987, pp. 333-365.

7. Erdogan, F. Stress intensity factors, *ASME Journal of Applied Mechanics*, 1983, **50**, 992-1002.

8. Guiggiani, M., Krishnaswamy, G. Rudolphi, T.J. & Rizzo, F.J. A general algorithm for the numerical solution of hypersingular boundary integral equations, *ASME Journal of Applied Mechanics*, 1992, **50**, 604-614.

9. Hirose, S. & Achenbach, J.D. Time-domain boundary element analysis of elastic wave interaction with a crack, *International Journal for Numerical Methods in Engineering*, 1989, **28**, 629-644.

10. Kitahara, M., Hirose, S. & Achenbach, J.D. Transient elastodynamic fracture mechanics of three-dimensional solids, in *Advanced Dynamic Analysis by Boundary Element Methods (Developments in Boundary Element Methods - 7)*, P.K. Banerjee & S. Kobayashi (ed), Elsevier Science Publishers, New York, 1992, 253-281.

11. Kobayashi, S. Some problems of the boundary integral equation method in elastodynamics, in *Boundary Elements: Proceedings 5th International Conference*, Springer-Verlag, Berlin, 1983, pp. 775-784.

12. Kobayashi, S. Elastodynamics, in *Boundary Element Methods in Mechanics*, D.E. Beskos (ed), North-Holland, 1987, pp. 191-225.

13. Krishnaswamy, G., Schmerr, L.W., Rudolphi, T.J. & Rizzo, F.J. Hypersingular boundary integral equations: Some application in acoustic and elastic wave scattering, *ASME Journal of Applied Mechanics*, 1990, **51**, 404-414.

14. Li, Z.L. & Achenbach, J.D. Interaction of a Rayleigh wave with a disbond in a material interphase normal to a free surface, *Ultrasonics*, 1990, **29**(1), 45-52.

15. Lin, W. & Keer, L.M. Scattering by a planar three-dimensional crack, *J. Acoust. Soc. Am.*, 1987, **82**(4), 1442-1448.

16. Neerhoff, F.L. Reciprocity and power-flow theorem for the scattering of plane elastic waves in a half-space, *Wave Motion*, 1980, **2**, 99-113.

17. Nishimura, N. & Kobayashi, S. An improved boundary integral equation method for crack problems, in *Proc. IUTAM Symposium on Advanced Boundary Element Methods*, T.A. Cruse (ed), Springer-Verlag, 1987.

18. Rice, J.R. Elastic fracture mechanics concepts for interfacial cracks, *ASME Journal of Applied Mechanics*, 1988, **55**, 98-103.

19. Schmerr, L.W. The scattering of elastic waves by isolated cracks using a new integral equation method, in *Review of Progress in Quantitative Nondestructive Evaluation*, Vol. 1, D.O. Thompson & D.E. Chimenti (ed), Plenum Press, New York, 1982, pp. 511-515.

20. Sih, C.F. & Asaro, R.J. Elastic-plastic analysis of cracks on bimaterial interfaces: Part 1 - Small Scale Yielding, *ASME Journal of Applied Mechanics*, 1988, **55**, 299-316.

21. Sládek, V. & Sládek, J. Transient elastodynamic three-dimensional problems in cracked bodies, *Appl. Math. Modelling*, 1984, **8**, 2-10.

22. Tan, T.H. Diffraction theory for time-harmonic elastic waves, Doctoral Dissertation, Delft University of Technology, 1975.

23. Zhang, C. & Achenbach, J.D. Scattering by multiple crack configurations, *ASME Journal of Applied Mechanics*, 1988, **55**, 104-110.

24. Zhang, C. & Achenbach, J.D. Numerical analysis of surface-wave scattering by the boundary element method, *Wave Motion*, 1988, **10**, 365-374.

25. Zhang, C. & Achenbach, J.D. A new boundary integral equation formulation for elastodynamic and elastostatic crack analysis, *ASME Journal of Applied Mechanics*, 1989, **56**, 284-290.

Appendix A: Time-harmonic fundamental elastodynamic solutions

The expressions for the displacement are

$$u_{ik}^{\mathrm{G}}(\mathbf{y}-\mathbf{x}) = A\left[U_1(r)\delta_{ik} - U_2(r)\frac{\partial r}{\partial y_i}\frac{\partial r}{\partial y_k}\right] \tag{A1}$$

where $r = |\mathbf{x} - \mathbf{x}|$ and

(i) for the three-dimensional case:

$$A = \frac{1}{4\pi\mu} \tag{A2}$$

$$U_1 = \left[1 + \frac{i}{k_\mathrm{T} r} - \frac{1}{(k_\mathrm{T} r)^2}\right] \frac{\mathrm{e}^{ik_\mathrm{T} r}}{r} - \left(\frac{k_\mathrm{L}}{k_\mathrm{T}}\right)^2 \left[\frac{i}{k_\mathrm{L} r} - \frac{1}{(k_\mathrm{L} r)^2}\right] \frac{\mathrm{e}^{ik_\mathrm{L} r}}{r} \tag{A3}$$

$$U_2 = \left[1 + \frac{3i}{k_\mathrm{T} r} - \frac{3}{(k_\mathrm{T} r)^2}\right] \frac{\mathrm{e}^{ik_\mathrm{T} r}}{r} - \left(\frac{k_\mathrm{L}}{k_\mathrm{T}}\right)^2 \left[1 + \frac{3i}{k_\mathrm{L} r} - \frac{3}{(k_\mathrm{L} r)^2}\right] \frac{\mathrm{e}^{ik_\mathrm{L} r}}{r} \tag{A4}$$

(ii) for the two-dimensional case:

$$A = \frac{i}{4\mu} \tag{A5}$$

$$U_1 = H_o^{(1)}(k_\mathrm{T} r) - \frac{1}{k_\mathrm{T} r}\left[H_1^{(1)}(k_\mathrm{T} r) - \frac{k_\mathrm{L}}{k_r} H_1^{(1)}(k_\mathrm{L} r)\right] \tag{A6}$$

$$U_2 = H_o^{(1)}(k_\mathrm{T} r) - \left(\frac{k_\mathrm{L}}{k_\mathrm{T}}\right)^2 H_o^{(1)}(k_\mathrm{T} r) - \frac{2}{k_\mathrm{T} r}\left[H_1^{(1)}(k_\mathrm{T} r) - \frac{k_\mathrm{L}}{k_\mathrm{T}} H_1^{(1)}(k_\mathrm{L} r)\right] \tag{A7}$$

Here $H_n^{(1)}$ is the Hankel function of the nth order and the first kind, and k_L and k_T are the longitudinal and transverse wave numbers, respectively. The corresponding stresses follow from Hooke's law:

$$\sigma_{ijk}^\mathrm{G}(\mathbf{y} - \mathbf{x}) = \lambda \frac{\partial u_{mk}^\mathrm{G}}{\partial y_m} \delta_{ij} + \mu \left(\frac{\partial u_{ik}^\mathrm{G}}{\partial y_j} + \frac{\partial u_{ik}^\mathrm{G}}{\partial y_i}\right) \tag{A8}$$

Appendix B

$$Q_i^A(\mathbf{x}) = \frac{i}{2(\lambda^A + 2\mu^A)\gamma_\mathrm{L}^A D^A}\left[N_\mathrm{LL}^A k_i^{\mathrm{L}A} \mathrm{e}^{ik_\mathrm{L}^A k_j^{\mathrm{L}A} x_j} + N_\mathrm{TL}^A \epsilon_{ki} k_i^{\mathrm{T}A} \mathrm{e}^{ik_\mathrm{T}^A k_j^{\mathrm{T}A} x_j}\right] \tag{B1}$$

$$S_{ij}^A(\mathbf{x}) = \frac{-1}{2(\lambda^A + 2\mu^A)\gamma_\mathrm{L}^A D^A}\left[N_\mathrm{LL}^A k_\mathrm{L}^A (\lambda^A \delta_{ij} + 2\mu^A k_i^{\mathrm{L}A} k_j^{\mathrm{L}A})\, \mathrm{e}^{ik_\mathrm{L}^A k_\ell^{\mathrm{L}A} x_\ell}\right.$$
$$\left. + N_\mathrm{TL}^A k_\mathrm{T}^A \mu^A (\epsilon_{\ell j} k_i^{\mathrm{T}A} + \epsilon_{\ell i} k_j^{\mathrm{T}A}) k_\ell^{\mathrm{T}A} \mathrm{e}^{ik_\mathrm{T}^A k_\ell^{\mathrm{T}A} x_\ell}\right] \tag{B2}$$

In these expressions, ϵ_{ij} is the two-dimensional permutation tensor, and

$$N_\mathrm{LL}^A = -2\left[(k_\mathrm{T}^A)^2 - 2(k_\mathrm{R}^A)^2\right]^2 \tag{B3}$$

$$N_\mathrm{TL}^A = \tfrac{4i}{k_\mathrm{L}^A} k_\mathrm{T}^A k_\mathrm{R}^A \gamma_\mathrm{L}^A \left[(k_\mathrm{T}^A)^2 - 2(k_\mathrm{R}^A)^2\right] \tag{B4}$$

$$D^A = 8k_\mathrm{R}^A \gamma_\mathrm{L}^A \gamma_\mathrm{T}^A + 4(k_\mathrm{R}^A)^3(\gamma_\mathrm{L}^A/\gamma_\mathrm{T}^A + \gamma_\mathrm{T}^A/\gamma_\mathrm{L}^A) + 8k_\mathrm{R}^A\left[(k_\mathrm{T}^A)^2 - 2(k_\mathrm{R}^A)^2\right] \tag{B5}$$

$$k_j^{\mathrm{L}A} = \tfrac{1}{k_\mathrm{L}^A}(-k_\mathrm{R}^A, i\gamma_\mathrm{L}^A) \tag{B6}$$

$$k_j^{\mathrm{T}A} = \tfrac{1}{k_\mathrm{T}^A}(-k_\mathrm{R}^A, i\gamma_\mathrm{T}^A) \tag{B7}$$

Replacing all A's by B's in eqns (B1)-(B3) and (B5), and changing (B4), (B6) and (B7) to

$$N_{\mathrm{TL}}^{B} = -4i\frac{k_{\mathrm{T}}^{B}}{k_{\mathrm{L}}^{B}}k_{\mathrm{R}}^{B}\gamma_{\mathrm{L}}^{B}\left[(k_{\mathrm{T}}^{B})^{2} - 2(k_{\mathrm{R}}^{B})^{2}\right] \tag{B8}$$

$$k_{j}^{\mathrm{L}B} = \frac{1}{k_{\mathrm{L}}^{B}}(k_{\mathrm{R}}^{B}, i\gamma_{\mathrm{L}}^{B}) \tag{B9}$$

$$k_{j}^{\mathrm{T}B} = \frac{1}{k_{\mathrm{T}}^{B}}(k_{\mathrm{R}}^{B}, i\gamma_{\mathrm{T}}^{B}) \tag{B10}$$

respectively, we obtain the expressions for Q_{i}^{B} and S_{ij}^{B}.

Chapter 3

The boundary element method as a tool in the design of composite laminates - theory and application

D. Klingbeil

Federal Institute for Materials Research and Testing (BAM), D-1000 Berlin 45, Germany

Abstract

A particular failure mode of composite laminates is the delamination of layers from the free-edge. For non-delaminated cases the stress concentrations near the free-edge are calculated. The numerical treatment of the delaminated structures is carried out by introducing crack tip elements and the use of the energy release rate. All numerical investigations are done by the boundary element method (BEM). Several examples are presented compared with alternative solutions.

1 Introduction

Engineers are interested in the calculation of layered materials in order to ensure the reliability of the constructions where these materials are used. Well-known examples of layered materials are composite laminates, such as glass-epoxy or graphite-epoxy laminates, but also plywood. Two main analyses are required during the design of a layered structure, i.e. the global calculation of the whole structure and the calculation of local effects near free boundaries, holes, notches or cracks. Because layered structures are usually thin in one direction the overall calculation can be done by using the theory of plates and shells. Local effects can only be solved by the continuum theory because the stresses are high and increase rapidly near free boundaries and cracks where three-dimensional stress states occur.

One layer consists of unidirectional fibres and a matrix so that the layer is very stiff in the direction of the fibres and flexible perpendicular to the fibres. In order to obtain a stiff construction the layers are packed with different fibre orientations and

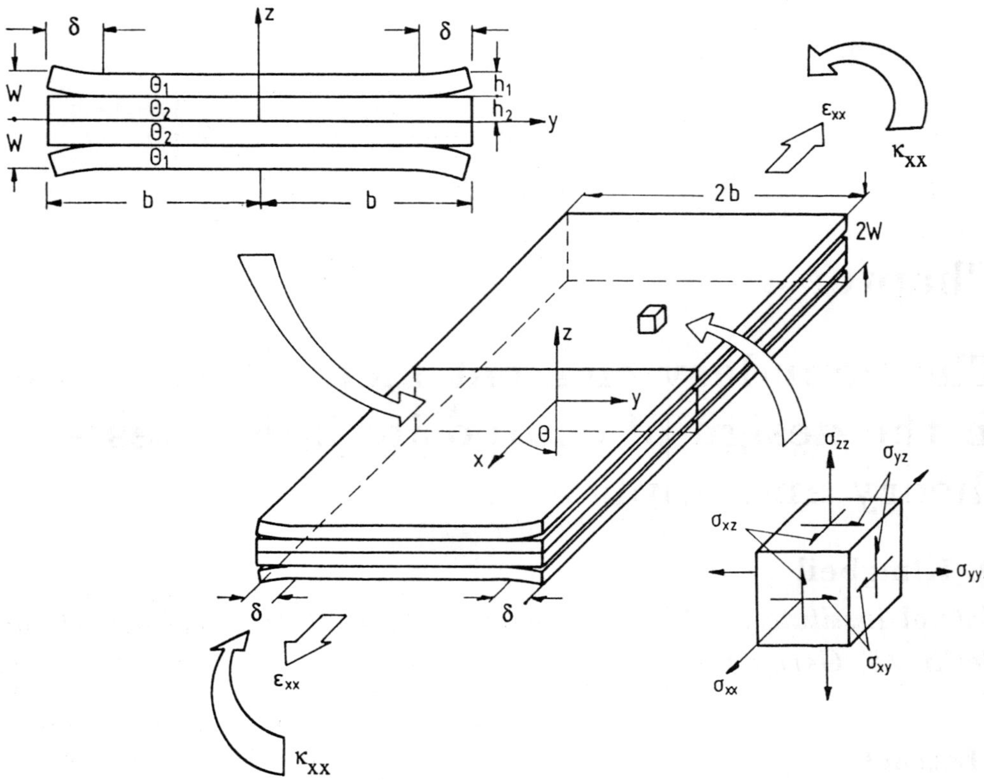

Figure 1: Geometry and loading of a symmetrical unidirectional four layer laminate $[\Theta_1/\Theta_2/\Theta_2/\Theta_1]$ under uniaxial strain ε_{xx} and constant curvature κ_{xx}

form the composite laminate. These different orientations lead to a stiffness varying from layer to layer. The deformation of one layer under loading is restricted by its neighbouring layers which results in high or, within a linear theory, infinite stresses near the free boundaries and cracks. These high stresses can lead to a delamination between the layers which may cause a catastrophic failure of the laminate. This particular failure mode can occur, even if the limit load of a laminate calculated by the theory of plates and shells has not been reached. The problem of the stress concentration is solved here for both the non-delaminated and delaminated case.

For the delaminated case, fracture mechanics offers the concept of stress-intensity factors and energy release rates describing the crack tip behaviour. Both methods are applied here to the delamination of laminates from the free boundary. All calculations regarding local effects for the non-delaminated, as well as for the delaminated, cases are carried out by the boundary element method.

The numerous works listed in the References section are related to one of the following subjects, which are concerned with the work presented here: boundary element method, global and local behaviour of laminates, fracture mechanics and special numerical aspects.

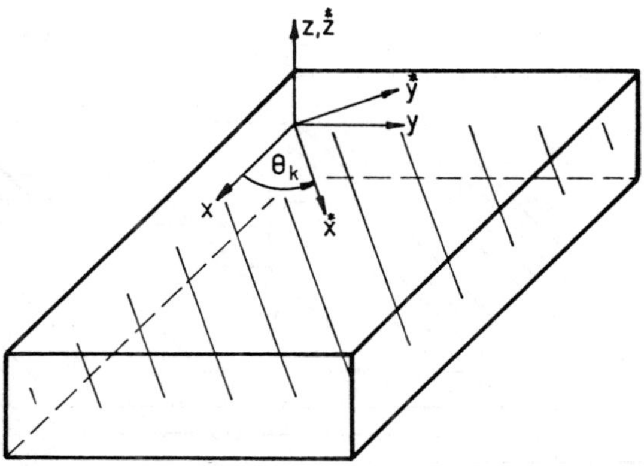

Figure 2: One unidirectional (UD) layer of a composite laminate

2 Mechanics of composite laminates

The total displacement and stress state in laminated composites is decomposed into a basic stress state related to theory of plates and shells and an appropriate stress state related to free boundaries or cracks. Because the cross-section has to keep its shape within the plate and shell theory, stresses occur on the free boundaries. These stresses are called basic stresses and have to be applied to the laminate with the opposite sign in order to satisfy the boundary traction free condition. These stresses are applied to the free boundary and lead to stress concentrations inside the laminate. Furthermore, it is described how the basic stress state is calculated and which known analytical solutions are available for the stress concentration problem.

2.1 Basic stress state of a composite laminate

A unidirectional (UD) layer consists of parallel fibres and a matrix where the fibres are orientated parallel to the x^*-axis (Fig. 2). One unidirectional layer k is considered to be homogeneous with the constitutive equation for one layer k

$$\begin{bmatrix} \sigma_{\overset{*}{x}\overset{*}{x}} \\ \sigma_{\overset{*}{y}\overset{*}{y}} \\ \sigma_{\overset{*}{x}\overset{*}{y}} \end{bmatrix}_k = \begin{bmatrix} \overset{*}{S}_{11} & \overset{*}{S}_{12} & 0 \\ \overset{*}{S}_{21} & \overset{*}{S}_{22} & 0 \\ 0 & 0 & \overset{*}{S}_{16} \end{bmatrix}_k \begin{bmatrix} \varepsilon_{\overset{*}{x}\overset{*}{x}} \\ \varepsilon_{\overset{*}{y}\overset{*}{y}} \\ \varepsilon_{\overset{*}{x}\overset{*}{y}} \end{bmatrix} \quad \text{or} \quad \overset{*}{\boldsymbol{\sigma}}_k = \overset{*}{\mathbf{S}}_k \overset{*}{\boldsymbol{\varepsilon}} \tag{1}$$

where Voigt's notation is used.[24] Here, shear and normal stresses are uncoupled in the x^*-, y^*-coordinate system. The matrix

$$\mathbf{T}_k = \begin{bmatrix} \cos^2\theta_k & \sin^2\theta_k & 2\sin\theta_k\cos\theta_k \\ \sin^2\theta_k & \cos^2\theta_k & -2\sin\theta_k\cos\theta_k \\ -\sin\theta_k\cos\theta_k & \sin\theta_k\cos\theta_k & \cos^2\theta_k - \sin^2\theta_k \end{bmatrix} \tag{2}$$

transforms eqn (1) from the local to the global coordinate system:

$$\boldsymbol{\sigma}_k = \mathbf{T}_k^{-1}\, \overset{*}{\boldsymbol{\sigma}}_k = \mathbf{T}_k^{-1}\, \overset{*}{\mathbf{S}}_k\, \mathbf{T}_k^{-T} \boldsymbol{\varepsilon}_k = \mathbf{S}_k \boldsymbol{\varepsilon}_k \tag{3}$$

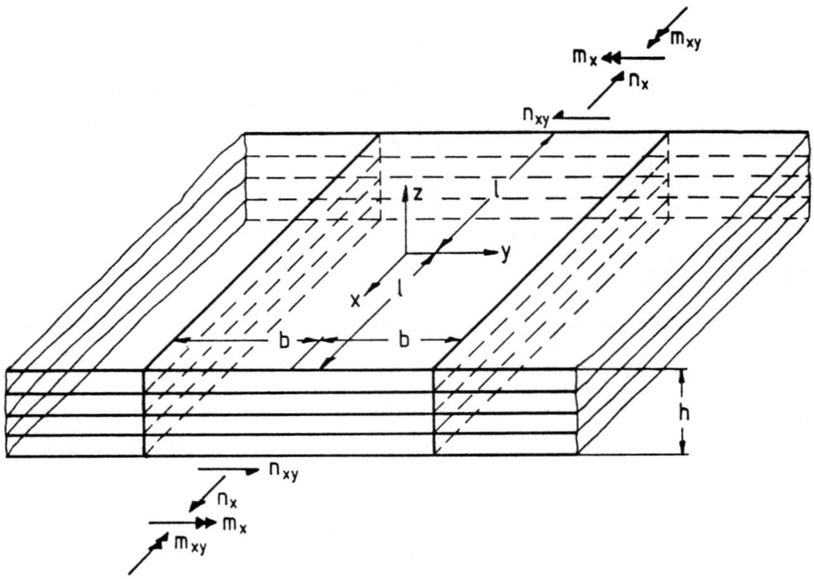

Figure 3: Composite laminate with statical loads for the basic stress state

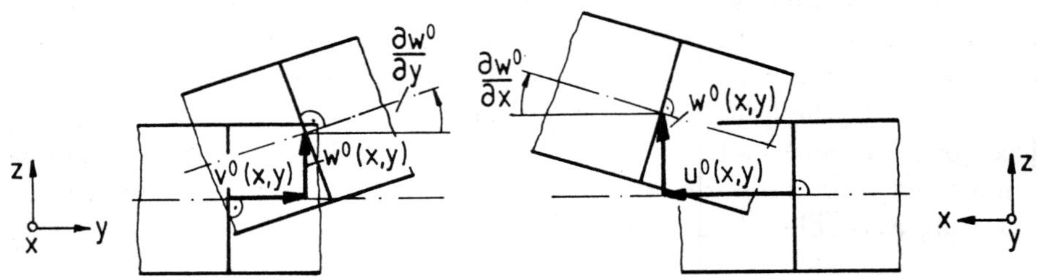

Figure 4: Geometrical relations of a cross-section

Generally, this constitutive equation is anisotropic. If Θ_k is equal to 90° or an integral multiple of 90°, the terms coupling shear and normal stresses vanish. A composite laminate consists of numerous layers whose orientation angles vary but it only has a small extension in the z-direction (Fig. 2). Therefore, the plate and shell theory is used which is written here for plane composite laminates.
A laminate is considered which is unbounded in y-direction and has the length $2l$ in x-direction (Fig. 3). This composite laminate is loaded by the normal and shear forces, n_x and n_{xy}, and by the bending and torsion moments, m_x and m_{xy}. Geometrical boundary conditions can be applied as well as statical boundary conditions. The stresses do not vary once inside the composite laminate at a certain distance away from the free boundaries where the loads are applied. This stress state will be calculated in the following. Therefore, the geometrical relations are used from Fig. 4. Hence, it follows

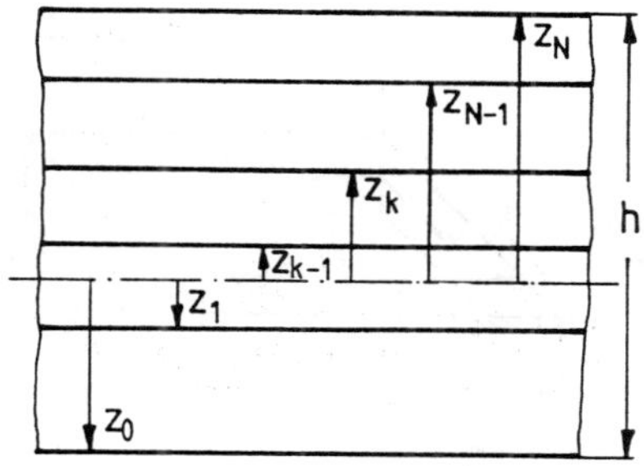

Figure 5: Cross-section of a composite laminate

$$\begin{aligned} u(x,y,z) &= u^0(x,y) - z\frac{\partial w^0(x,y)}{\partial x} \\ v(x,y,z) &= v^0(x,y) - z\frac{\partial w^0(x,y)}{\partial x} \\ w(x,y,z) &= w^0(x,y) \end{aligned} \tag{4}$$

where $u^0(x,y)$, $v^0(x,y)$ are the in-plane displacements, $w^0(x,y)$ is the out-of-plane displacement of the neutral plane. The strains are derived by the derivatives of the displacements as follows

$$\begin{aligned} \varepsilon_{xx}(x,y,z) &= \frac{\partial u}{\partial x} = \frac{\partial u^0(x,y)}{\partial x} - z\frac{\partial^2 w^0(x,y)}{\partial x^2} &&= \varepsilon^0_{xx} + z\kappa^0_{xx} \\ \varepsilon_{yy}(x,y,z) &= \frac{\partial v}{\partial y} = \frac{\partial v^0(x,y)}{\partial y} - z\frac{\partial^2 w^0(x,y)}{\partial y^2} &&= \varepsilon^0_{yy} + z\kappa^0_{yy} \\ \gamma_{xy}(x,y,z) &= \frac{\partial u}{\partial y} + \frac{\partial v}{\partial x} = \frac{\partial u^0(x,y)}{\partial y} + \frac{\partial v^0(x,y)}{\partial x} - 2z\frac{\partial^2 w^0(x,y)}{\partial x \partial y} &&= \gamma^0_{xy} + z\kappa^0_{xy} \end{aligned} \tag{5}$$

with

$$\frac{\partial^2 w^0(x,y)}{\partial x^2} = -\kappa^0_{xx} \ , \quad \frac{\partial^2 w^0(x,y)}{\partial y^2} = -\kappa^0_{yy} \text{ and } 2\frac{\partial^2 w^0(x,y)}{\partial x \partial y} = -\kappa^0_{xy} \tag{6}$$

where ε^0_{xx}, ε^0_{yy} and ε^0_{xy} are the strains and κ^0_{xx}, κ^0_{yy} and κ_{xy} are the curvatures of the neutral plane. Introducing the vectors

$$\boldsymbol{\varepsilon} = \begin{bmatrix} \varepsilon_{xx} \\ \varepsilon_{yy} \\ \gamma_{xy} \end{bmatrix}, \boldsymbol{\varepsilon}^0 = \begin{bmatrix} \varepsilon^0_{xx} \\ \varepsilon^0_{yy} \\ \gamma^0_{xy} \end{bmatrix}, \boldsymbol{\kappa}^0 = \begin{bmatrix} \kappa^0_{xx} \\ \kappa^0_{yy} \\ \kappa^0_{xy} \end{bmatrix} \tag{7}$$

eqns (5) are rewritten:

$$\boldsymbol{\varepsilon} = \boldsymbol{\varepsilon}^0 + z\boldsymbol{\kappa}^0 \tag{8}$$

The forces and moments are derived by integrating the stresses of a typical cross-section (Fig. 5)

$$n_x = \int_h \sigma_{xx} \mathrm{d}z \quad n_y = \int_h \sigma_{yy} \mathrm{d}z \quad n_{xy} = \int_h \sigma_{xy} \mathrm{d}z \quad n_{yx} = \int_h \sigma_{yx} \mathrm{d}z \tag{9}$$

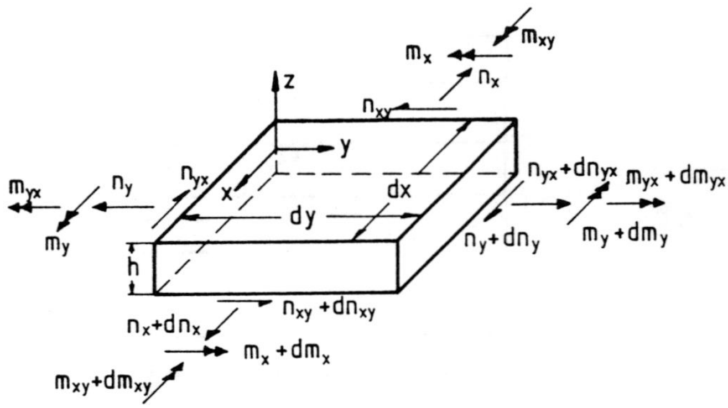

Figure 6: Infinitesimal element of a plate

$$m_x = \int_h \sigma_{xx} z \mathrm{d}z \quad m_y = \int_h \sigma_{yy} z \mathrm{d}z \quad m_{xy} = \int_h \sigma_{xy} z \mathrm{d}z \quad m_{yx} = \int_h \sigma_{yx} z \mathrm{d}z \tag{10}$$

and they are acting on an infinitesimal element (Fig. 6).

By introducing the vectors

$$\mathbf{n} = \begin{bmatrix} n_x \\ n_y \\ n_{xy} \end{bmatrix}, \ \mathbf{m} = \begin{bmatrix} m_x \\ m_y \\ m_{xy} \end{bmatrix}, \ \boldsymbol{\sigma} = \begin{bmatrix} \sigma_x \\ \sigma_y \\ \sigma_{xy} \end{bmatrix} \tag{11}$$

eqns (9, 10) are rearranged:

$$\mathbf{n} = \int_h \boldsymbol{\sigma} \mathrm{d}z \ , \ \mathbf{m} = \int_h \boldsymbol{\sigma} z \mathrm{d}z \tag{12}$$

The integration over the cross-section is done by summarizing all layers and introducing eqns (3)-(8) into eqns (12)

$$\begin{aligned} \mathbf{n} &= \sum_{k=1}^{N} \int_{z_{k-1}}^{z_k} \boldsymbol{\sigma}_k \mathrm{d}z = \sum_{k=1}^{N} \int_{z_{k-1}}^{z_k} \mathbf{S}_k \boldsymbol{\varepsilon} \mathrm{d}z \\ &= \boldsymbol{\varepsilon}^0 \sum_{k=1}^{N} \int_{z_{k-1}}^{z_k} \mathbf{S}_k \mathrm{d}z + \boldsymbol{\kappa}^0 \sum_{k=1}^{N} \int_{z_{k-1}}^{z_k} \mathbf{S}_k z \mathrm{d}z \\ \mathbf{m} &= \sum_{k=1}^{N} \int_{z_{k-1}}^{z_k} \boldsymbol{\sigma}_k z \mathrm{d}z = \sum_{k=1}^{N} \int_{z_{k-1}}^{z_k} \mathbf{S}_k \boldsymbol{\varepsilon} z \mathrm{d}z \\ &= \boldsymbol{\varepsilon}^0 \sum_{k=1}^{N} \int_{z_{k-1}}^{z_k} \mathbf{S}_k z \mathrm{d}z + \boldsymbol{\kappa}^0 \sum_{k=1}^{N} \int_{z_{k-1}}^{z_k} \mathbf{S}_k z^2 \mathrm{d}z \end{aligned} \tag{13}$$

Because the stiffness is constant for one layer, the integrations in eqns (13) can easily be carried out:

$$\begin{aligned} \mathbf{n} &= \boldsymbol{\varepsilon}^0 \sum_{k=1}^{N} \mathbf{S}_k (z_k - z_{k-1}) \mathrm{d}z + \boldsymbol{\kappa}^0 \frac{1}{2} \sum_{k=1}^{N} \mathbf{S}_k (z_k^2 - z_{k-1}^2) \\ \mathbf{m} &= \boldsymbol{\varepsilon}^0 \frac{1}{2} \sum_{k=1}^{N} \mathbf{S}_k (z_k^2 - z_{k-1}^2) + \boldsymbol{\kappa}^0 \frac{1}{3} \sum_{k=1}^{N} \mathbf{S}_k (z_k^3 - z_{k-1}^3) \end{aligned} \tag{14}$$

Introducing the matrices

$$\mathbf{A} = \sum_{k=1}^{N} \mathbf{S}_k (z_k - z_{k-1}) \qquad \mathbf{B} = \frac{1}{2} \sum_{k=1}^{N} \mathbf{S}_k (z_k^2 - z_{k-1}^2)$$
$$\mathbf{D} = \frac{1}{3} \sum_{k=1}^{N} \mathbf{S}_k (z_k^3 - z_{k-1}^3) \tag{15}$$

the eqns (14) are written as

$$\mathbf{n} = \mathbf{A}\boldsymbol{\varepsilon}^0 + \mathbf{B}\boldsymbol{\kappa}^0 \qquad \mathbf{m} = \mathbf{B}\boldsymbol{\varepsilon}^0 + \mathbf{D}\boldsymbol{\kappa}^0 \tag{16}$$

which is the global constitutive equation for a composite laminate.

In order to calculate the basic stresses in a composite laminate, the following steps have to be carried out:

- The stiffness in local coordinates of all layers k is calculated or measured.
- The orientation angles Θ_k are determined and all local matrices $\mathbf{S}_k^*$ are transformed by using eqn (3). The results are the matrices $\mathbf{S}_k$ in the global coordinate system.
- The matrices $\mathbf{A}$, $\mathbf{B}$ and $\mathbf{D}$ are calculated. If the composite laminate is symmetrical to its neutral plane, the matrix $\mathbf{B}$ vanishes. In this case the forces $\mathbf{n}$ and moments $\mathbf{m}$ are uncoupled in the constitutive eqn (16).
- If a geometrical load $\boldsymbol{\varepsilon}^0$ or $\boldsymbol{\kappa}^0$ is applied, the strain $\boldsymbol{\varepsilon}$ is calculated for every layer by using eqn (8). The basic stresses are obtained by eqn (3).
- If a static load $\mathbf{n}$ or $\mathbf{m}$ is applied, eqns (16) have to be inversed and with the resulting strains $\boldsymbol{\varepsilon}^0$ and curvatures $\boldsymbol{\kappa}^0$ the basic stresses are calculated by using eqns (8), (3) again.

2.2 Decomposition of the total stress state of a composite laminate

If the plate theory is applied to a laminate, stresses remain on the free boundaries in order to satisfy the conditions that the cross-section remains plane and perpendicular to the neutral plane. Here, these remaining stresses are applied to the free boundary with the opposite sign so that the boundary traction free condition is satisfied. The cross-section of the laminate is no longer plane and perpendicular to its neutral plane.

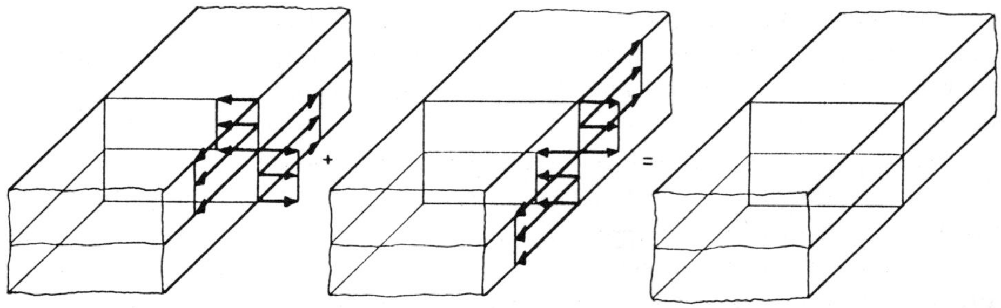

Figure 7: Decomposition of the solution

The interlaminar stresses inside the laminate near the free boundary are high and decrease rapidly inside the laminate. A typical distance wherein the stresses tend to zero from the free boundary inside the laminate is the thickness of one layer.
The following equations show the decomposition into the part which is due to the plate theory referring to eqn (4) and the parts $U(y,z)$, $V(y,z)$ and $W(y,z)$ resulting from applying the boundary stresses with the opposite sign:

$$\begin{aligned} u(x,y,z) &= u^0(x,y) - z\frac{\partial w^0(x,y)}{\partial x} + U(y,z) \\ v(x,y,z) &= v^0(x,y) - z\frac{\partial w^0(x,y)}{\partial x} + V(y,z) \\ w(x,y,z) &= w^0(x,y) \qquad\qquad\quad + W(y,z) \end{aligned} \tag{17}$$

This decomposition is valid inside the composite laminate a certain distance away from the boundaries $x = \pm l$ loaded (Fig. 3) by external forces, moments, strains or curvatures. Here, the stress concentration problem is solved by the BEM.

2.3 Known solutions

Literature offers an arbitrary number of publications concerned with the calculation of the free-edge and crack stress states in composite laminates. Therefore, only a small part of this work can be mentioned here. In order to get a comparison of the BEM calculations with other methods the publications are classified as to whether they are concerned with the non-delaminated or delaminated case.

2.3.1 Known solutions for the non-delaminated case

Generally, it is impossible to solve the stress concentration problem analytically. Analytical solutions are only given for special orientations of the layers, material constants and boundary conditions. Giencke[20] showed, that an unlimited number of [0°/90°] layers leads to finite but high interlaminar stress near the free-edge. Furthermore, Giencke[20] calculated a logarithmic singularity of the stresses for an unlimited number of [+45°/-45°] layers.

Pipes & Pagano[38] gave some further analytical results for certain [0°/90°] layers. Wang & Choi[59,60] showed for a [-45°/+45°] laminate that the stresses become singular and they calculated the power of the singularity and its strength. Delale[16] offered a more general approach assuming that the stresses behave like $Ar^{-\alpha}$ near the free-edge, wherein α means the power and A the strength. The power α was given for different material combinations.

Some problems were solved by numerical methods, i.e. finite differences,[39] finite elements[40,58] or boundary elements.[25,26] Raju & Crews[40] determined the power of the stress singularities by using regular elements near the free-edge and neglecting the results in some points closest to the crack tip. An extrapolation to the crack tip is used to determine the type of singularity. So far, the BEM[25,26] was used with regular elements for the non-delaminated case.

2.3.2 Known solutions for the delaminated case

In contrast to the non-delaminated case, it is possible to give a clear classification of known solutions for the delaminated case. The simplest case is a crack on the interface between two different isotropic materials. At first, this problem was solved by England[17] (Fig. 8(a)), who found that the basic solution of the stresses is

$$\sigma_{zz,yz,xz} = r^{-0.5}\left(A_{zz,yz,xz}\cos(C_{zz,yz,xz}\ln r) + B_{zz,yz,xz}\sin(C_{zz,yz,xz}\ln r)\right) \qquad (18)$$

with the assumptions of linear elasticity where r is the distance from the crack tip. The constants $A_{zz,yz,xz}$, $B_{zz,yz,xz}$ and $C_{zz,yz,xz}$ depend on the material constants of the two domains and are different for the normal and shear stresses. The crack length is the only geometrical parameter affecting the result. Equation (18) means that the solution oscillates infinitely before the crack front which contradicts to the physical reality of the problem. This wild oscillation can be avoided by closing the crack a little[8,9,10,60] or by introducing a non-linear constitutive equation.[28]

Wang[61,62] and Wang & Choi[60] gave some solutions for the stresses near the crack tip for an unbounded laminate (Fig. 8(c)) and for a four layer symmetrical laminate with a $[+\Theta/-\Theta]_S$-orientation (Fig. 8(b)) where the index S denotes symmetry to the neutral plane. The basic solution for the stresses is the same as in eqn (18). The constants $A_{zz,yz,xz}$, $B_{zz,yz,xz}$ and $C_{zz,yz,xz}$ depend on the anisotropic material constants and the oscillation still occurs. All mentioned examples are particular cases because the global constitutive eqn (16) is still orthotropic. Figure 8(d) shows an example where the two unbounded layers have different orientation angles and therefore the global constitutive equation becomes anisotropic. Delale[16] solved this case and showed that the solution only oscillates, if the orientation of the layers is $+\Theta/-\Theta$. In all other cases, i.e. the orientation angles are Θ/Ψ, the solution takes the power form $\sigma \sim Ar^{-\alpha}$ with $0 < \alpha < 0,5$.

These general statements can be done by analysing the eigenvalues of the problem which occur in the stress function. If an eigenvalue has an imaginary part, the solution is singular and oscillates. Otherwise, the solution is singular without any oscillation.[16] Table 1 shows the different authors and the solved problems.

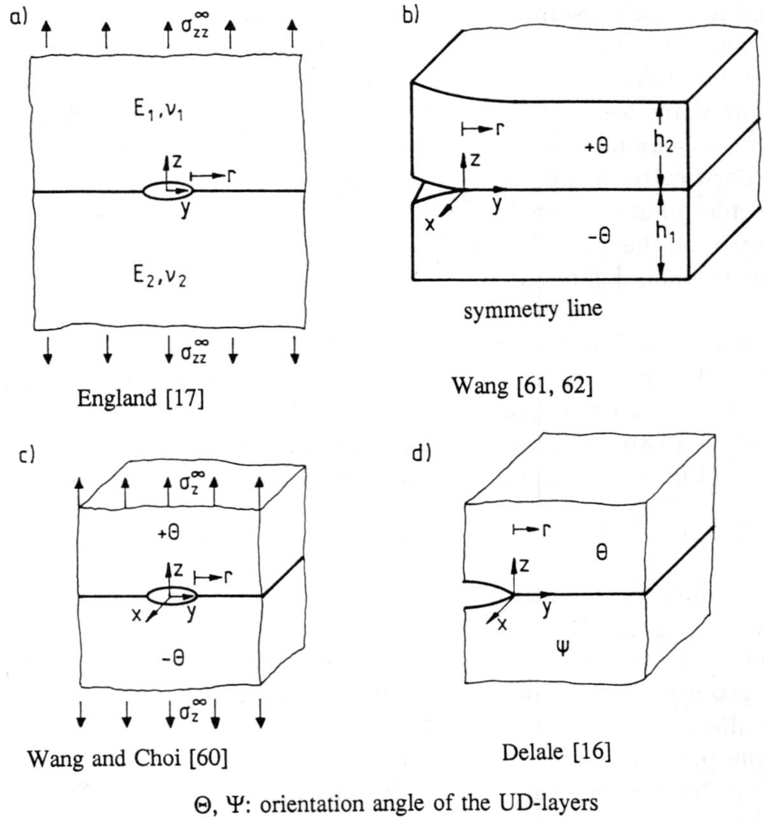

Figure 8: Examples from Refs 17, 60, 61, 62, and 16

Table 1: Relation between the works of different authors and the solved problems

Non-delaminates composite		Delaminated composite	
Regular solutions	Singular solutions	Only singular solutions	
		[0°/90°], [+Θ/-Θ]	All other orientations
Analytical: - Giencke[20] - Pipes & Pagano[38] Numerical: - Wang & Crossman[57] - Pipes & Pagano[39] - Raju & Crews[40] - Klingbeil[25,26,27]	Analytical: - Delale[16] - Wang & Choi[58,59] - Giencke[20] - Zwiers *et al.*[65] Numerical: - Raju & Crews[40]	Analytical: - Wang & Choi[60] - Wang[61,62] Numerical: - Klingbeil[25,26]	Analytical: - Delale[16] Numerical: - Not known to the author's knowledge

2.4 Definitions of the stress-intensity factors and energy release rates in anisotropic elasticity

In isotropic elasticity the stress-intensity factors are defined by

$$K_{I,II,III} = \lim_{r\to 0} \sqrt{2\pi r} \quad \sigma_{zz,yz,xz}(r) \tag{19}$$

where index I refers to in-plane normal stress, index II to in-plane shear stress and index III to out-of-plane shear stress.

Because the stresses have an inverse square root singularity the terms with $\sqrt{r}$ cancel each other and it remains a fixed quantity.

This simple definition is no longer valid for anisotropic elasticity, because the stresses are not only singular with a power of r but oscillate shortly in front of the crack tip. The solution is a sum whose oscillating and singular parts may be different so that there is no common factor which could be canceled. If any load in a direction is applied, all other directions are affected. Therefore, Rice & Sih[41] and Wang[62] proposed different definitions of stress-intensity factors. The definition of Wang[62]

$$K_{I,II,III} = \lim_{r\to 0} \sqrt{2\pi} \quad r^{-\delta_j}\sigma_{zz,yz,xz}(r,\delta_j) \tag{20}$$

is used where δ_j is a complex number and j is the summation index for the different parts of the solution. The character of the solution depends on δ_j. If the imaginary part of δ_j vanishes, the stresses are singular with a power less than the inverse square root singularity[16] and without any oscillation. If the arrangement of a four layer composite is $[+\Theta/-\Theta]$s, it holds for the real part $\mathrm{Re}(\delta_j) = -0,5$ and the imaginary part of δ_j causes an oscillation so that the stresses have the general behaviour[62]

$$\sigma_{zz,xz,yz} = r^{-0.5}(\cos(C_{zz,xz,yz}\ln r) + B_{zz,xz,yz}\sin(C_{zz,xz,yz}\ln r)) \tag{21}$$

which is equal to eqn (18) and where $A_{zz,yz,xz}$, $B_{zz,yz,xz}$ and $C_{zz,yz,xz}$ are constants depending on the material properties and the orientation angle Θ. If eqn (21) is introduced into the definition of the stress-intensity factors (20), the oscillation is canceled and well defined K-factors are available. This procedure does not apply any more in general numerical procedure because none of the parts of any solution are available. Therefore, a modified definition for the stress-intensity factor is introduced

$$K_{I,II,III} = \sqrt{\Delta r}\sigma_{zz,xz,yz}(\Delta r) \tag{22}$$

where Δr is a small but finite distance from the crack tip. It is important to note that Δr should be taken outside the domain where the wild oscillation occurs in order to achieve a stress-intensity factor which is stable for different distances Δr. The domain of the oscillation is about 10^{-8} ... 10^{-4} of the crack length.[8,9,10,17,60,62]

The energy release rate or crack extension force is defined as the work which has to be done when a crack is closed by a virtual crack length δ and δ takes the limiting value of zero

$$G_I = \lim_{r\to 0}\frac{1}{\delta}\int_0^{\delta} \sigma_{zz}(y)W(y-\delta)\mathrm{d}y \tag{23}$$

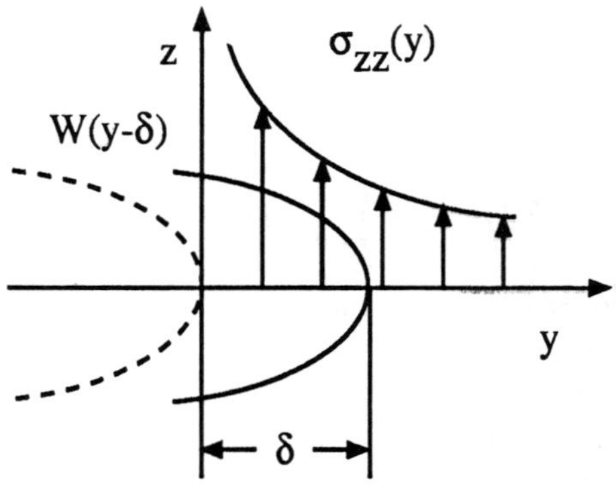

Figure 9: Basic principle of the energy release rate

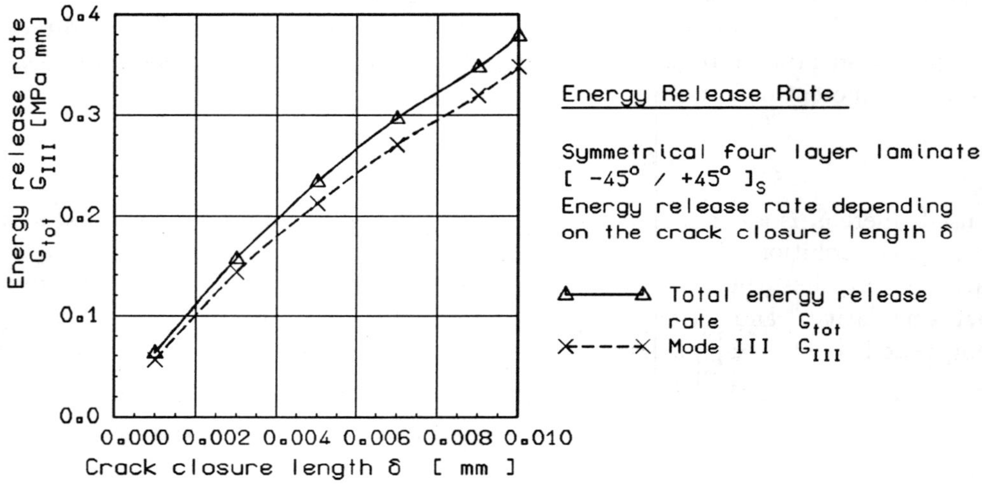

Figure 10: Influence of the virtual crack closure length δ on the energy release rate

where $\sigma_{zz}(y)$ is the stress of the closed crack and $W(y-\delta)$ is the displacement for the opened crack, where all notations refer to Fig. 9.
It is significant that in linear isotropic elasticity the energy release rate is always a fixed, but nonzero quantity, because the stresses behave like $1/\sqrt{r}$ and the displacements behave like $\sqrt{r}$ so that the terms with $\sqrt{r}$ cancel each other in the integrand of eqn (23).

This conventional concept no longer holds in anisotropic elasticity,[4,5,27,62] because both crack fronts contribute different parts to the energy release rate and because the stress states in the non-delaminated and delaminated case are totally different.[16] In this case, not only the strength but also the power of the stresses varies from the non-delaminated to the delaminated case. Because the singularity of the non-delaminated case is always weaker than that of the delaminated case[16] the energy release tends to zero for the virtual crack closure length δ approaching zero (Fig. 10).
The total energy release rate is defined by

$$G_{\text{tot}} = G_I + G_{II} + G_{III} \tag{24}$$

and its parts G_I, G_{II} and G_{III} are introduced as[4,5,27] (Fig. 1)

$$\begin{aligned}
G_I &= \frac{1}{\delta}\left(\frac{1}{2}\int_{y=-b}^{-b+\delta} \sigma_{zz}(y)W(y-\delta)\mathrm{d}y + \frac{1}{2}\int_{y=-b}^{-b+\delta} \sigma_{zz}(y)W(y-\delta)\mathrm{d}y\right) \\
G_{II} &= \frac{1}{\delta}\left(\frac{1}{2}\int_{y=-b}^{-b+\delta} \sigma_{yz}(y)V(y-\delta)\mathrm{d}y + \frac{1}{2}\int_{y=-b}^{-b+\delta} \sigma_{yz}(y)V(y-\delta)\mathrm{d}y\right) \qquad (25) \\
G_{IIII} &= \frac{1}{\delta}\left(\frac{1}{2}\int_{y=-b}^{-b+\delta} \sigma_{zx}(y)U(y-\delta)\mathrm{d}y + \frac{1}{2}\int_{y=-b}^{-b+\delta} \sigma_{zx}(y)U(y-\delta)\mathrm{d}y\right)
\end{aligned}$$

Furthermore, it is assumed that the crack is between the layers I and $I+1$ where all notation refer to Fig. 1. Two calculations are always necessary to obtain one energy release rate: one calculation for the non-delaminated case and the other one for the delaminated case.

3 Basic mechanical equations for the stress concentration problem

The mechanical equations are introduced to solve the stress concentration problem for a cross-section. Because the equations are written with indices, the notation of the coordinates is changed so that x_1 refers to y, x_2 to z and x_3 to x (Fig. 1).

Furthermore, the notation of the displacements is changed so that $u_1(x_1, x_2)$ refers to $V(y,z)$, $u_2(x_1,x_2)$ to $W(y,z)$ and $u_3(x_1,x_2)$ to $U(x_1,x_2)$ (Fig. 1). The following assumptions are introduced: small displacements, linear constitutive law, homogeneous layers, and every layer is in full contact with its neighbouring layers. The equations describing the problem are those of equilibrium, geometrical compatibility and the general anisotropic constitutive law for the three-dimensional case:

$$\sigma_{ji,j} + p_i = 0 \qquad (26)$$

$$\varepsilon_{ij} = \frac{1}{2}(u_{i,j} + u_{j,i}) \qquad (27)$$

$$\sigma_{ij} = E_{ijkl}\varepsilon_{kl} = E_{ijkl}u_{k,l} \qquad (28)$$

Latin subscripts can take the values 1 to 3. Referring to Section 2.2, the assumption is now introduced where the structure is long in x_3-direction so that any variation of all mechanical quantities are negligible in this direction. Therefore, all derivatives of u_i and σ_{ij} with respect to x_3 vanish and the eqns (26)-(28) are rewritten:

$$\sigma_{\alpha i,\alpha} + p_i = 0 \qquad (29)$$

$$\varepsilon_{ij} = \frac{1}{2}(u_{i,j} + u_{j,i}) \qquad (30)$$

$$\sigma_{i\beta} = E_{ijk\beta}u_{k,\beta} \qquad (31)$$

where greek subscripts can take the values 1 and 2. It follows that only the (x_1, x_2)-plane of the laminate has to be taken into consideration. In general, the constitutive law is anisotropic.

In the next step, the geometrical and statical compatibility has to be postulated on those parts of the boundaries of each layer, where layers or subdomains contact each other

$$u_i(\chi_\alpha)_{\textbf{boundary } I} = u_i(\chi_\alpha)_{\textbf{boundary } I+1} \quad (32)$$

$$t_i(\chi_\alpha)_{\textbf{boundary } I} = -t_i(\chi_\alpha)_{\textbf{boundary } I+1} \quad \text{for } I = 1, ..., N-1 \quad (33)$$

where N is the total number of layers or subdomains, $t_i = \sigma_{i\alpha} n_\alpha$ is the stress vector and n_α is the outward normal vector of the boundary. The problem is formulated completely, when the geometrical and/or statical conditions on the remaining parts of the boundaries are prescribed.

4 Boundary integral equation

Somigliana's well-known identity[2,3]

$$u_i(x'_\alpha) = \oint_\Gamma T_{ij}(x_\alpha, x'_\alpha) u_j(x_\alpha) \mathrm{d}\Gamma + \oint_\Gamma U_{ij}(x_{alpha}, x'_\alpha) t_j(x_\alpha) \mathrm{d}\Gamma \quad (34)$$

gives the relation between the displacements $u_i(x'_\alpha)$ of an interior or exterior point x'_α of a domain Ω and the displacements $u_j(x_\alpha)$ and tractions $t_j(x_\alpha)$ on its boundary Γ. Body forces are omitted, because they do not appear in this application. The domain Ω is two-dimensional, but the displacements and tractions have 3 components where u_3 and t_3 are the out-of-plane parts. U_{ij} is the fundamental solution and represents the displacement in direction x_i related to a unit line force F_j. According to the assumptions made above, F_j is a constant unit line force in the x_3-direction and in contrast to the full three-dimensional case it has no point load, because all mechanical quantities do not vary in x_3-direction. By introducing this assumption it is possible to calculate the functions U_{ij} and T_{ij} analytically,[18,33,25] while the full three-dimensional solution can only be obtained by an expensive numerical calculation.[33,64] The equations for the analytical calculation of U_{ij} and T_{ij} are given in Section 5.

The boundary integral equation follows, when the load point x'_α lies on the boundary Γ. Because this procedure causes singular integrands in both integrals of eqn (34), Cauchy's principle values have to be taken

$$C_{ij} u_j(x'_\alpha) + \oint_\Gamma T_{ij}(x_\alpha, x'_\alpha) u_j(x_\alpha) \mathrm{d}\Gamma = \oint_\Gamma U_{ij}(x_\alpha, x'_\alpha) t_j(x_\alpha) \mathrm{d}\Gamma \quad (35)$$

The matrix C_{ij} represents the integral free terms. This boundary integral equation describes the behaviour of one subdomain or layer.

The discretization of this boundary integral equation is done by introducing elements on the boundary wherein the boundary Γ itself, the displacements u_i and the tractions t_i are interpolated by polynomials. The subdomains or layers are coupled by postulating the geometrical and statical compatibility on the parts of the boundaries which are in contact with their neighbouring layers according to eqns (32)-(33). This procedure is well-known and described by different authors.[2,3,25]

Nevertheless, the derivation of crack tip elements is given in the next but one section.

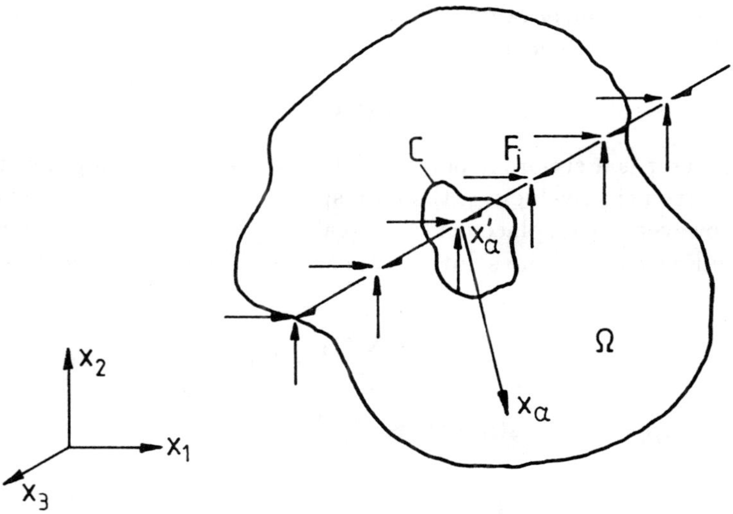

Figure 11: Line forces F_j used for the fundamental solutions

5 Derivation of the fundamental solutions U_{ij} and T_{ij}

An infinite homogeneous continuum is considered with an arbitrary anisotropic constitutive law. In contrast to the full three-dimensional case, where point forces are applied in the load point, line forces are used here which do not vary in x_3-direction (Fig. 11). Here we are following the work of Mura,[33] but Eshelby *et al.*[18] introduced the equations for the first time.

The differential equation for the displacements

$$E_{i\beta k\alpha}u_{k,\alpha\beta} = 0 \tag{36}$$

is valid for all points in the domain Ω with the exception of the load point x'_α. Furthermore, the constitutive equation (28) and the geometrical compatibility (27) are valid. The solution for the displacements is given by an arbitrary function[18]

$$u_k = A_k f(z) \tag{37}$$

where the variable z is a linear combination of x_1 and x_2:

$$z = x_1 + px_2 \tag{38}$$

Equation (37) is differentiated twice and introduced in eqn (36), which leads to

$$\left(E_{i1k1} + p(E_{i1k2} + E_{i2k1}) + p^2 E_{i2k2}\right) A_k \frac{\partial^2 f}{\partial z^2} = 0 \tag{39}$$

Introducing the abbreviation

$$\phi_{ik} = E_{i1k1} + p(E_{i1k2} + E_{i2k1}) + p^2 E_{i2k2} \tag{40}$$

eqn (39) leads to an eigenvalue problem whose eigenvalues $p_{(l)}$ are obtained, if the determinant of the coefficients vanishes:

$$\det(\phi_{ik}) = 0 \tag{41}$$

This determinant is a six-order polynomial, whose complex roots occur in conjugate pairs. For a given eigenvalue $p_{(l)}$, the corresponding eigenvector $A_{(l)}$ is determined by eqn (39). However, the displacements are real, which requires that the imaginary parts of corresponding pairs of the solution cancel out. The displacements are represented by

$$u_k = \mathrm{Re}\left(\sum_{l=1}^{3} A_{(l)k} f_{(l)}(z_{(l)})\right) \tag{42}$$

with $z_{(l)} = x_1 + p_{(l)}x_2$. The stresses are given by

$$\sigma_{ij} = \mathrm{Re}\left(\sum_{l=1}^{3}(E_{ijk1} + p_{(l)}E_{ijk2})A_{(l)k}f'_{(l)}(z_{(l)})\right) \tag{43}$$

A line integral along a closed curve C (Fig. 11) around the load point x'_α gives the line force per unit length

$$\oint_C \sigma_{i\beta}n_\beta \mathrm{d}s = F_i \tag{44}$$

whereas a similar integral over a dislocation

$$\oint_C \Delta u_i \mathrm{d}s = 0 \tag{45}$$

vanishes in order to obtain the fundamental solution. A stress function Ψ_i is introduced following the relations

$$\sigma_{i1} = -\frac{\partial \Psi_i}{\partial x_2} \quad ; \quad \sigma_{i2} = \frac{\partial \Psi_i}{\partial x_1} \tag{46}$$

so that eqn (44) can be rewritten as

$$-\oint_C \mathrm{d}\Psi_i = F_i \tag{47}$$

It follows with eqns (43), (46)

$$\Psi_i = \mathrm{Re}\left(\sum_{l=1}^{3}(E_{i2k1} + p_{(l)}E_{i2k2})A_{(l)k}f_{(l)}(z_{(l)})\right) \tag{48}$$

Eshelby *et al.*[18] showed that the functions $f_{(l)}$ are given by

$$f_{(l)}(z_{(l)}) = \mathrm{sign}(\mathrm{Im}(p_{(l)}))\frac{1}{2\pi i}D_{(l)}\log(z_{(l)} - z'_{(l)}) \tag{49}$$

with

$$z'_{(l)} = x'_1 + p_{(l)} x'_2 \tag{50}$$

Then, eqns (42), (43) become

$$\mathrm{Re}\left(\sum_{l=1}^{3} A_{(l)k} D_{(l)}\right) = 0 \tag{51}$$

and

$$\mathrm{Re}\left(\sum_{l=1}^{3} (E_{i2kl} + p_{(l)} E_{i2k2}) A_{(l)k} D_{(l)}\right) = F_i \tag{52}$$

and they represent a system of six linear equations for the real and imaginary parts of $D_{(l)}$. The different components of F_i are alternately equal the unit line force $F_i = 1_i$ and all other components of F_i vanish. The results are called $\mathrm{d}_{(l)m}$ and represent a linear combination

$$D_{(l)} = \mathrm{d}_{(l)m} F_m \tag{53}$$

The displacements can be expressed now by introducing eqns (49), (53) in (42)

$$u_k(x_i - x'_i) = \mathrm{Re}\left(\sum_{l=1}^{3} A_{(l)k}\, \mathrm{sign}(\mathrm{Im}(p_{(l)})) \frac{\mathrm{d}_{(l)m} F_m}{2\pi i} \log(z_{(l)} - z'_{(l)})\right) \tag{54}$$

The fundamental solution U_{ij} follows, when $F_m = 1_m$ and the index m is used to indicate the direction of the unit line force

$$U_{km}(x_i - x'_i) = \mathrm{Re}\left(\sum_{l=1}^{3} A_{(l)k}\, \mathrm{sign}(\mathrm{Im}(p_{(l)})) \frac{\mathrm{d}_{(l)m}}{2\pi i} \log(z_{(l)} - z'_{(l)})\right) \tag{55}$$

In order to achieve a formulation for the fundamental solution for T_{ij}, the stress vector t_i is expressed as a function of the line force F_m

$$\begin{aligned} t_i &= \sigma_{i\beta} n_\beta \\ &= E_{i\beta kl} \varepsilon_{kl} n_\beta \\ &= E_{i\beta k\alpha} u_{k\alpha} n_\beta \\ &= E_{i\beta k\alpha} U_{km,\alpha} n_\beta F_m \end{aligned} \tag{56}$$

from which T_{ij} follows by a comparison with its definition

$$t_i = T_{im} F_m \tag{57}$$

i.e.

$$T_{im} = E_{i\beta k\alpha} U_{km,\alpha} n_\beta \tag{58}$$

The derivations of U_{ij} have to be carried out

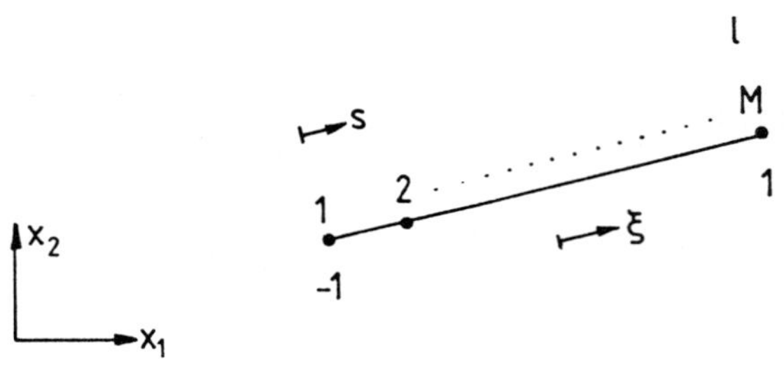

Figure 12: Typical straight boundary element

$$
\begin{aligned}
\frac{\partial U_{km}}{\partial x_1} &= \mathrm{Re}\left(\sum_{l=1}^{3} A_{(l)k}\,\mathrm{sign}(\mathrm{Im}(p_{(l)}))\frac{\mathrm{d}_{(l)m}}{2\pi i}\frac{1}{z_{(l)}-z'_{(l)}}\right)\\
\frac{\partial U_{km}}{\partial x_2} &= \mathrm{Re}\left(\sum_{l=1}^{3} A_{(l)k}\,\mathrm{sign}(\mathrm{Im}(p_{(l)}))\frac{\mathrm{d}_{(l)m}}{2\pi i}\frac{p_{(l)}}{z_{(l)}-z'_{(l)}}\right)\\
\frac{\partial U_{km}}{\partial x_3} &= 0
\end{aligned}
\tag{59}
$$

6 Crack tip elements

Consider a two-dimensional element on the boundary Γ with the global coordinate s and the local coordinate ξ, where ξ can take values from -1 to 1. The displacements and tractions are interpolated by Lagrange polynomials

$$\phi_{(m)}(\xi) = \prod_{\substack{n=1\\ n\neq m}}^{M} \frac{\xi-\xi_n}{\xi^*_m-\xi_n} \tag{60}$$

with an arbitrary order M:

$$u_j(\xi) = \sum_{m=1}^{M} \phi_{(m)}(\xi)u_{j(m)} \tag{61}$$

$$t_j(\xi) = \sum_{m=1}^{M} \phi_{(m)}(\xi)t_{j(m)} \tag{62}$$

The coordinates $x_{\alpha(m)}$ are interpolated by the same polynomials $\phi_{(m)}(\xi)$ and the same order M

$$x_\alpha(\xi) = \sum_{n=1}^{M} \phi_{(m)}(\xi)x_{\alpha(m)} \tag{63}$$

where $x_{\alpha(m)}$ are the coordinates at the point ξ^*_m. It can be shown[19,23] that this interpolation follows the exponential series expansions for the displacements

$$u_j(s) = \sum_{n=1}^{M-1} A_{j(n)}s^{n-1} \tag{64}$$

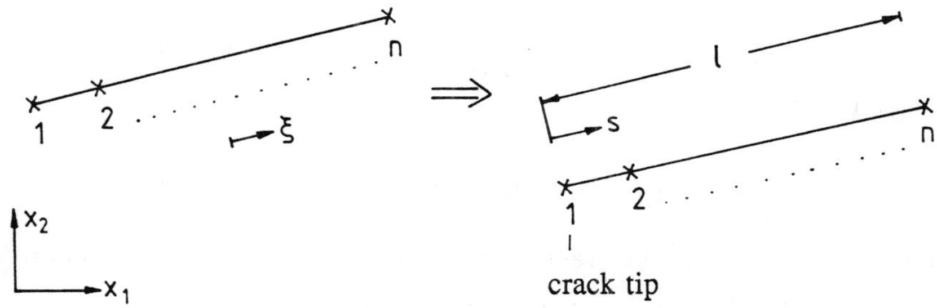

Figure 13: Transformation from the local coordinate ξ to the global coordinate s

and for the tractions

$$t_j(s) = \sum_{n=1}^{M-1} B_{j(n)} s^{n-1} \tag{65}$$

in the global coordinate s. It is seen that these exponential series contain only regular terms. If the solution is singular, these series would not be able to produce an appropriate result.

Therefore, crack tip elements are introduced which are created by shifting the midpoint of an element towards the crack tip. This type of element was created first by Henshel & Shaw[23] in a finite element analysis. This procedure was adopted by different authors[1,19,25,26] for boundary elements. Here, this procedure is extended to anisotropic elasticity where others, rather than square root singularities, occur.
The mapping

$$x_\alpha(\xi) = \left(\frac{1+\xi}{2}\right)^{\frac{1}{\lambda}} (x_{\alpha(M)} - x_{\alpha(1)}) + x_{\alpha(1)} \tag{66}$$

is introduced between the global coordinate s and the local coordinate ξ in an element. It follows from eqn (66)

$$\xi = 2\left(\frac{s}{l}\right)^{\lambda} - 1 \tag{67}$$

where l is the length of the element. The following equation is introduced for the displacements

$$u_j(\xi) = \left(\frac{1+\xi}{2}\right)^{\frac{1}{\lambda}-1} \left(\phi_1(\xi) u_{j(1)} + \sum_{m=2}^{M} \phi_{(m)}(\xi) \left(\frac{2}{1+\xi_m^*}\right)^{\frac{1}{\lambda}-1} u_{j(m)}\right) \tag{68}$$

where the factor $(\frac{1+\xi}{2})^{\frac{1}{\lambda}-1}$ reproduces the $s^{1-\lambda}$-behaviour of the displacements and the tractions are modeled by

$$t_j(\xi) = \left(\frac{2}{1+\xi}\right) \left(\phi_1(\bar{\xi}) t_{j(1)} + \sum_{m=2}^{M} \phi_{(m)}(\bar{\xi}) \; \frac{1+\xi_m^*}{2} \; t_{j(m)}\right) \tag{69}$$

where the factor $\frac{2}{1+\xi}$ reproduces their $s^{-\lambda}$ behaviour. The modified local coordinate

$$\bar{\xi} = 2\left(\frac{1+\xi}{2}\right)^{\frac{1}{\gamma}} - 1 \, , \; \gamma = \text{ constant} \tag{70}$$

is introduced. A series for the displacements is derived by substituting eqn (67) into eqn (68)

$$u_j(s) = A_{j(1)}s^{1-\lambda} + A_{j(2)}s + \cdots + A_{j(M)}s^{1+(M-2)\lambda} \tag{71}$$

It follows for $\lambda = \frac{1}{2}$

$$u_j(s) = A_{j(1)}\sqrt{s} + A_{j(2)}s + \cdots + A_{j(M)}s^{1+\frac{(M-2)}{2}} \tag{72}$$

A similar series is found for the tractions by substituting eqn (67) into eqn (69)

$$t_j(s) = B_{j(1)}s^{-\lambda} + B_{j(2)}s^{\frac{\lambda}{\gamma}-\lambda} + \cdots + B_{j(M)}s^{(M-1)\frac{\lambda}{\gamma}-\lambda} \tag{73}$$

which is specified by fixed numbers for λ and γ. It follows for $\lambda = \frac{1}{2}$ and $\gamma = 1$

$$t_j(s) = B_{j(1)}\frac{1}{\sqrt{s}} + B_{j(2)} + B_{j(3)}\sqrt{s} + \cdots + B_{j(M)}s^{\frac{(M-2)}{2}} \tag{74}$$

which is equal to the single mapping and for $\lambda = \frac{1}{2}$ and $\gamma = \frac{1}{2}$ it follows

$$t_j(s) = B_{j(1)}\frac{1}{\sqrt{s}} + B_{j(2)}\sqrt{s} + B_{j(3)}s\sqrt{s} + \cdots + B_{j(M)}s^{M-\frac{3}{2}} \tag{75}$$

which is equal to double mapping. It is seen that the $\frac{1}{\sqrt{s}}$ singularity always appears, but the following terms in the series (74), (75) depend on the choice of γ.

If the crack is not at the beginning but at the end of the element, then eqns (76)-(79) are valid.

$$x_\alpha(\xi) = \left(\frac{1-\xi}{2}\right)^{\frac{1}{\lambda}} (x_{\alpha(1)} - x_{\alpha(M)}) + x_{\alpha(M)} \tag{76}$$

$$u_j(\xi) = \left(\frac{1-\xi}{2}\right)^{\frac{1}{\lambda}-1} \left(\phi_M(\xi)u_{j(M)} + \sum_{m=1}^{M-1} \phi_{(m)}(\xi)\left(\frac{2}{1-\xi_m^*}\right)^{\frac{1}{\lambda}-1} u_{j(m)}\right) \tag{77}$$

$$t_j(\xi) = \left(\frac{2}{1-\xi}\right)\left(\phi_M(\bar{\xi})t + j(M) + \sum_{m=1}^{M-1} \phi_{(m)}(\bar{\xi})\,\frac{1-\xi_m^*}{2}\,t_{j(m)}\right) \tag{78}$$

$$\bar{\xi} = -2\left(\frac{1-\xi}{2}\right)^{\frac{1}{\lambda}} + 1\,,\ \gamma = \text{constant} \tag{79}$$

The stress-intensity factors are identical $t_{j(1)}$ for elements where the crack tip is at the beginning of the element (eqn (69)) and they are identical $t_{j(M)}$ for elements where the crack tip is at the end of the element (eqn (78)).

All numerical investigations are done with quadratic polynomials $\phi_{(m)}(\xi)$ so that the series (72), (74), (75) consist of three terms. The stress-intensity or K-factors are direct unknowns at the crack tip in the linear system of equations. All other unknowns are displacements or tractions.

7 Examples

A graphite-epoxy laminate is considered for all examples, which has been established in literature by several authors, whose material properties for one layer are

$$\begin{array}{llll} E_{xx} = 138.0 & E_{yy} = 14.5 & E_{xx} = 14.5 & [GPa] \\ G_{xy} = 5.86 & G_{yz} = 5.86 & G_{zx} = 5.86 & [GPa] \\ \nu_{xy} = 0.21 & \nu_{yz} = 0.21 & \nu_{zx} = 0.21 & [1] \end{array}$$

from which the stiffness matrix $\mathbf{S}^*$ is calculated in the x^*, y^*, z^*-system:[24]

$$\text{with } \mathbf{S}^* = \begin{bmatrix} 139.5 & 3.89 & 3.89 & 0.0 & 0.0 & 0.0 \\ & 15.26 & 3.29 & 0.0 & 0.0 & 0.0 \\ & & 15.26 & 0.0 & 0.0 & 0.0 \\ & & & 5.86 & 0.0 & 0.0 \\ & \text{sym.} & & & 5.86 & 0.0 \\ & & & & & 5.86 \end{bmatrix} [GPa] \tag{80}$$

The constitutive equation is:

$$\boldsymbol{\sigma} = \mathbf{S}\boldsymbol{\varepsilon} \quad \text{with} \quad \boldsymbol{\sigma}^T = [\sigma_{xx}, \sigma_{yy}, \sigma_{zz}, \sigma_{yz}, \sigma_{zx}, \sigma_{xy}] \tag{81}$$
$$\text{and} \quad \boldsymbol{\varepsilon}^T = [\varepsilon_{xx}, \varepsilon_{yy}, \varepsilon_{zz}, \varepsilon_{yz}, \varepsilon_{zx}, \varepsilon_{xy}]$$

from which the x and y components are used for the calculation of the overall behaviour of one layer according to eqn (1) and the full matrix $\mathbf{S}$ is received by transforming the local matrix $\mathbf{S}^*$ into global coordinates which is needed for the three-dimensional constitutive law in eqn (31).[24,25]

7.1 Non-delaminated composite laminate

The first example is a [+45°/-45°]$_S$-laminate under uniaxial strain $\varepsilon_{xx} = 1$. Figure 14 shows the geometry used and Fig. 15 shows the elementation for the boundary elements wherein double symmetry of the cross-section is considered. The ratio of h to b is 1/8. Linear elements are used for the interpolation of the boundary as well as for the displacements and tractions. The number of unknowns in the linear system of equations is approximately 550.

Figure 16 shows the normal stress σ_{zz} on the free boundary at $y = 0$. It is seen that this stress has its maximum at the interface between the layers. The interlaminar shear stress σ_{xz} is plotted in Fig. 17, which also has a maximum on the free boundary. Both stresses would not occur in the plate theory, but they are significantly high and can lead to a delamination.

More comprehensive examples are the [45°/-45°/0°/90°]$_S$- and [90°/-0°/-45°/45°]$_S$-laminates which are symmetrical and consist of eight layers. Figure 18 shows the geometry and Fig. 19 shows the elementation used. It is made use of the double symmetry of the cross-section in all calculations. The Figs 20(a),(b) demonstrate visually how the cross-sections are deformed. The +45°/-45° layers can easily be recognized in both figures, because they show a significant curvature on the free-edge. Figure 21 shows the interlaminar stress σ_{zz} for the [45°/-45°/0°/90°]$_S$-laminate at different

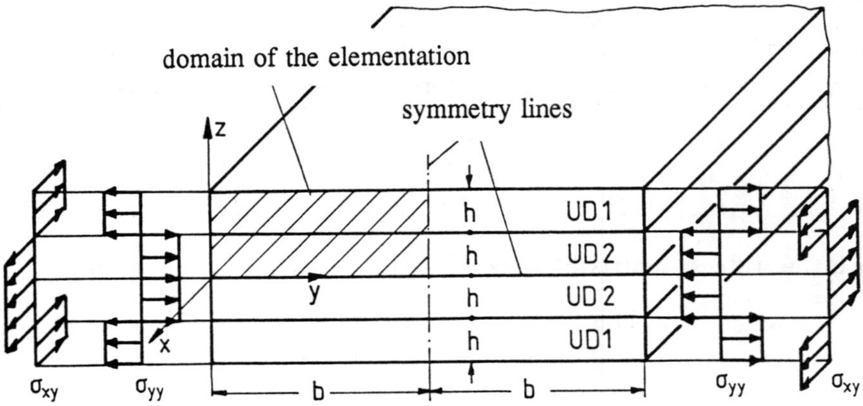

Figure 14: Geometry used for the [+45°/-45°]S-laminate

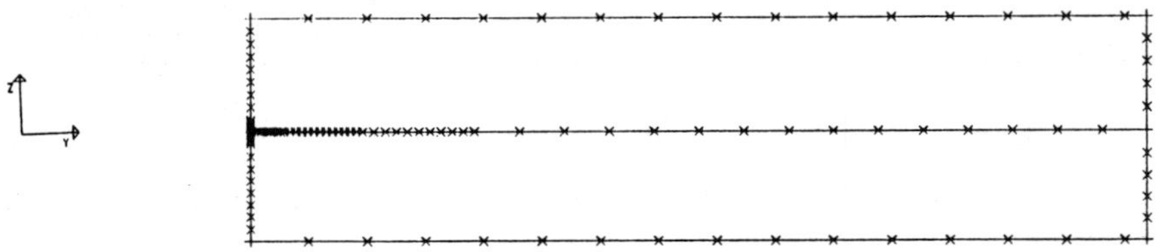

Figure 15: Discretization used for the [+45°/-45°]S-laminate

interfaces which reaches its maximum at or near the free boundary at $y = 0$. The calculations of Wang & Crossman[57] are done by the FEM and compared to the BEM solution. It is seen that both methods lead to similar results. The result in Ref. 57 do not show such significant high stress concentrations as the BEM calculation. This difference is due to the fact that in Ref. 57 the basic stresses on the free boundary are smoothed at the interfaces between the layers. All solutions (Figs 21, 22 and 23) received by the BEM agree very well with the results from other authors.

7.2 Delaminated case

7.2.1 Stress-intensity factor concept

The first example of a delaminated laminate is an unbounded two layer structure whose load is a normal stress σ_{zz}^{∞} acting at infinity (Fig. 24). The crack is at the centre between the two layers. This example is very suitable for the study of the basic behaviour of boundary elements, because no boundary influences the results near the crack tip and because only the orientation angle Θ affects the results. Wang & Choi[60] calculated an analytical solution so that a comparison can easily be done. Figure 25 shows the used discretization which simulates infinity and Fig. 26 presents the K-factors from the analytical solution[60] and the boundary element calculation where different ratios $\Delta r/a$ are used, all plotted versus the orientation angle Θ. The length of the crack tip element is Δr. The stress intensity factors from the boundary element

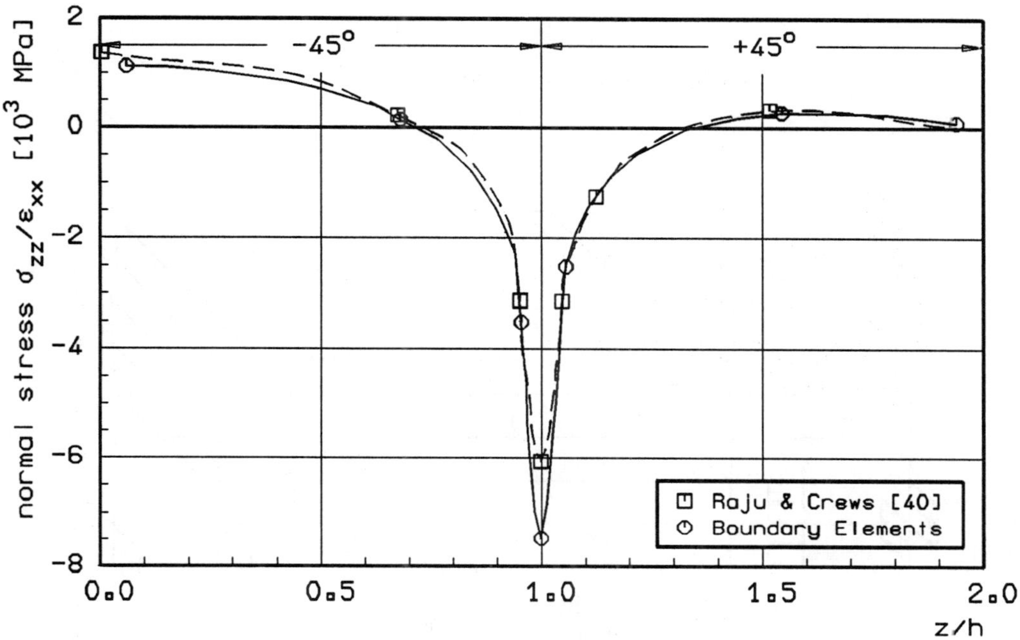

Figure 16: Symmetrical four layer [+45°/-45°]$_S$-laminate: normal stress σ_{zz} on the free boundary at $y = 0$

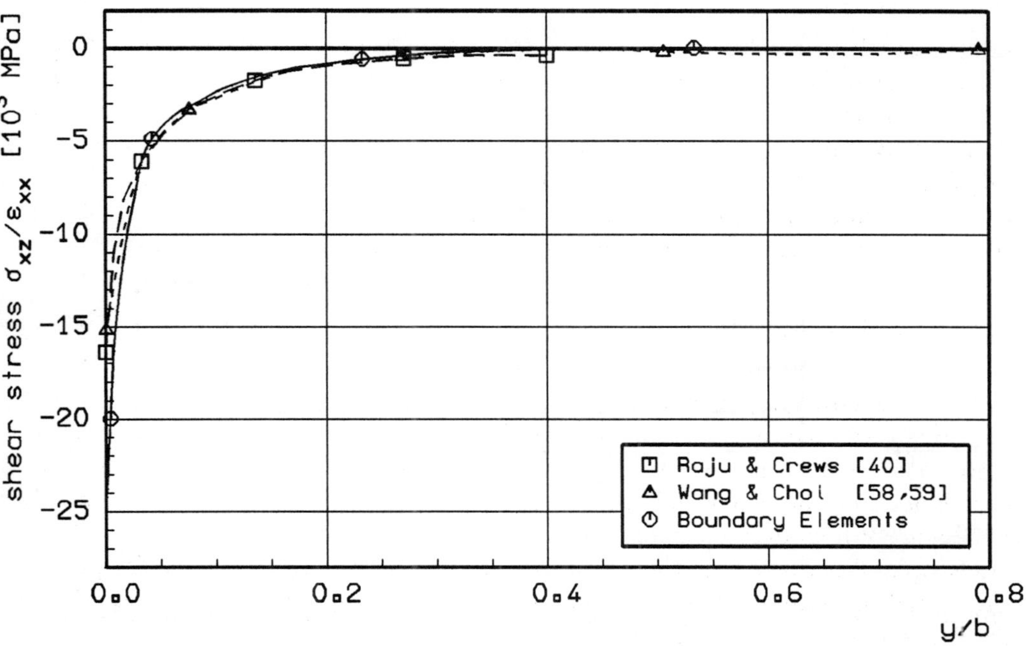

Figure 17: Symmetrical four layer [+45°/-45°]$_S$-laminate: out-of-plane shear stress σ_{xz} on the interface between the layers at $z = h$

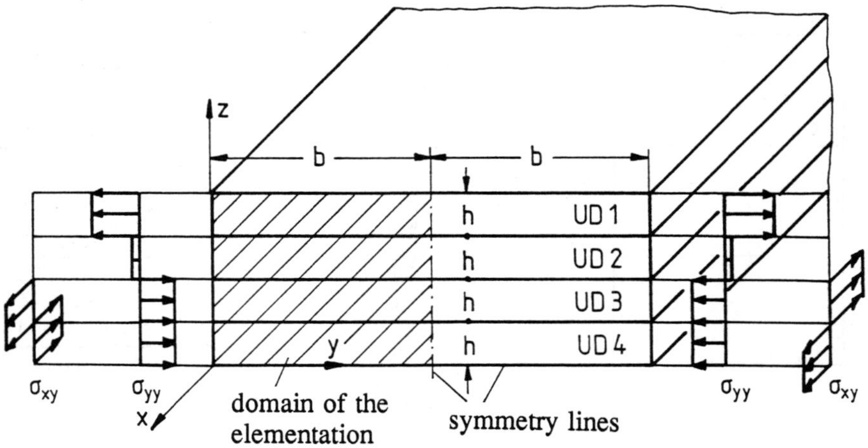

Figure 18: Geometry used for the [90°/0°/-45°/45°]$_S$- and [45°/-45°/0°/90°]$_S$-laminates

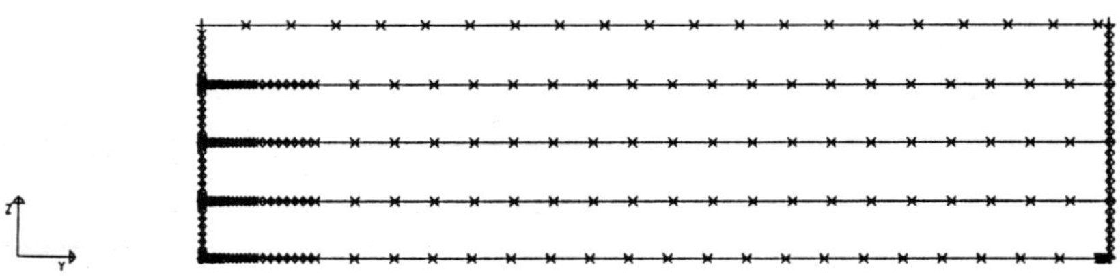

Figure 19: Discretization used for the [90°/-0°/-45°/45°]$_S$- and [45°/-45°/0°/90°]$_S$-laminates

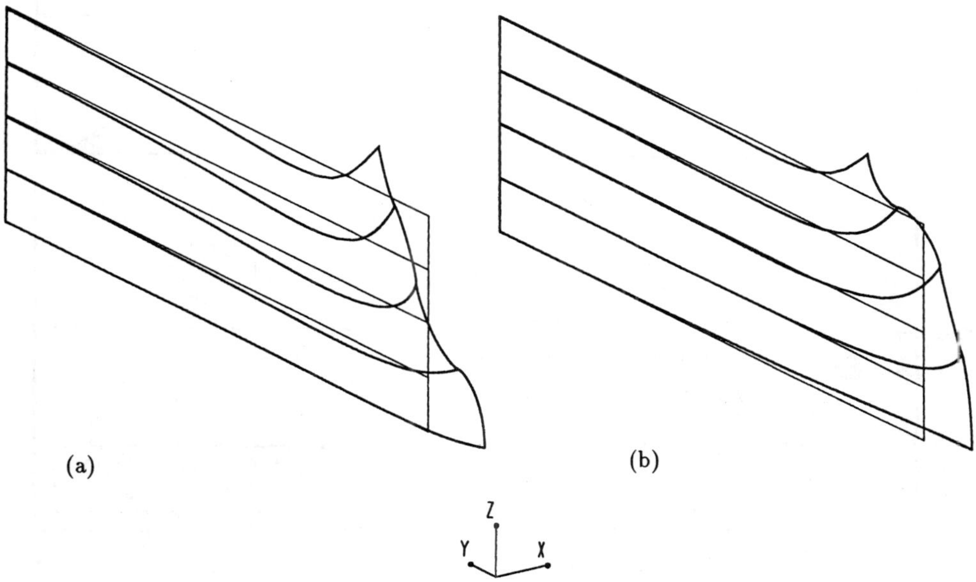

Figure 20: (a) Displacements of the [90°/0°/-45°/45°]$_S$-laminate; (b) Displacements of the [45°/-45°/0°/90°]$_S$-laminate

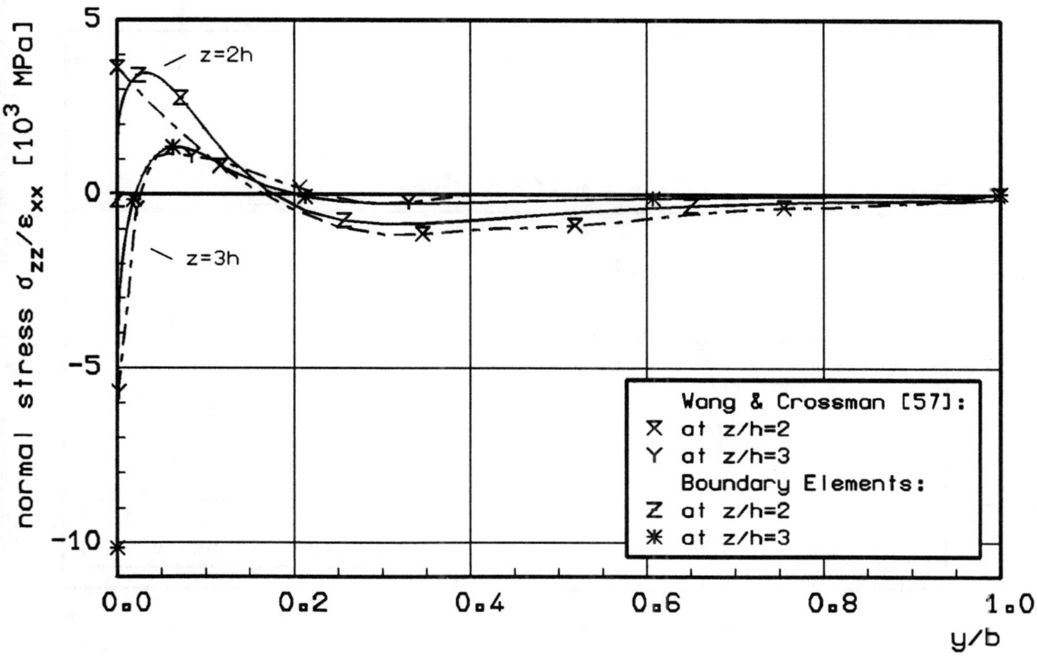

Figure 21: Symmetrical eight layer [45°/-45°/0°/90°]$_S$-laminate: normal stress σ_{zz} on different interfaces between the layers

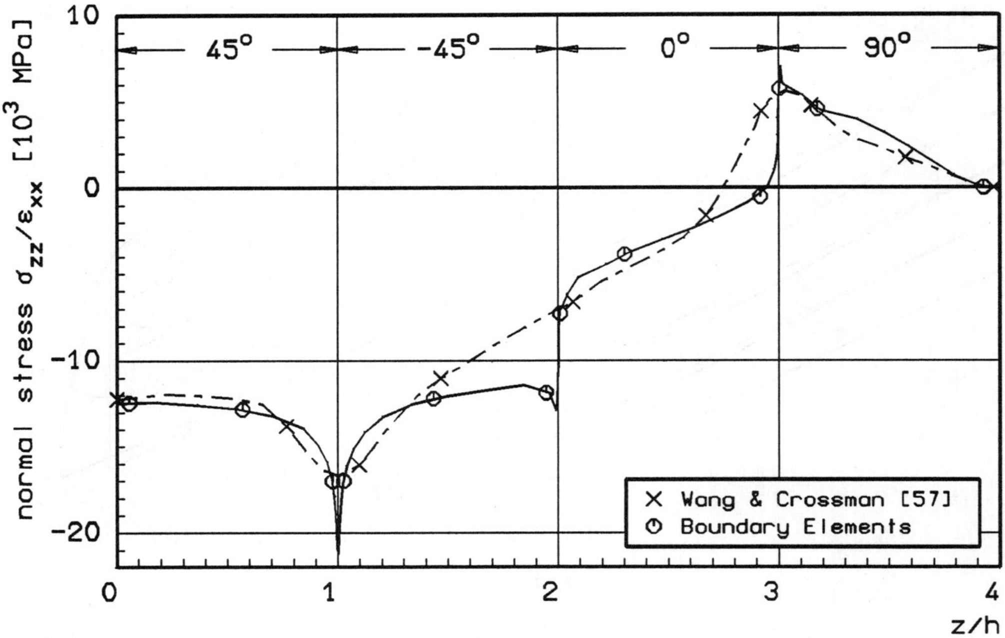

Figure 22: Symmetrical eight layer [90°/0°/-45°/45°]$_S$-laminate: normal stress σ_{zz} on the free boundary at $y = 0$

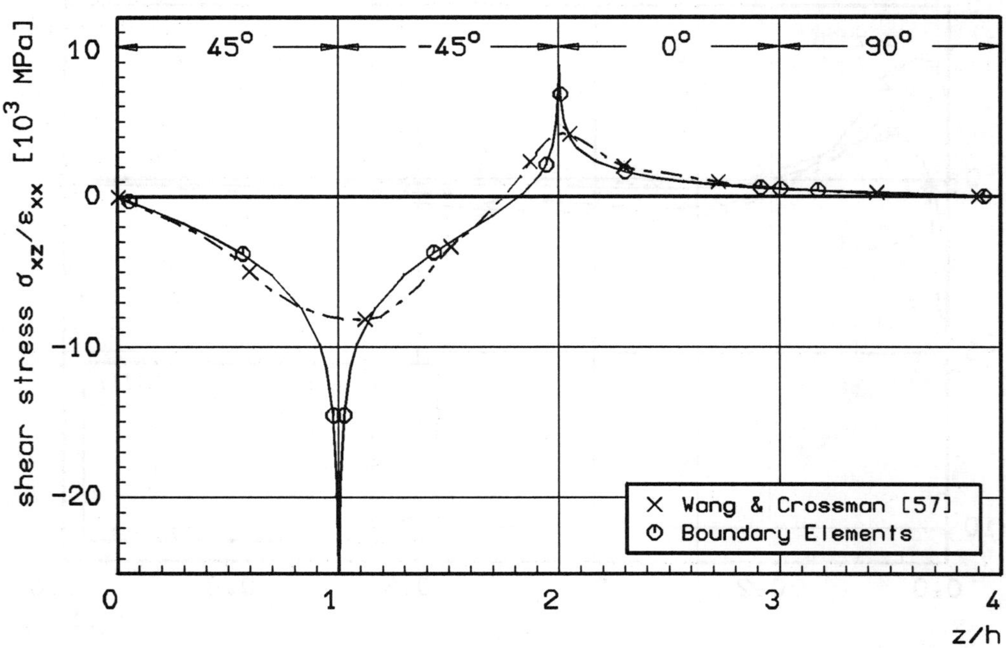

Figure 23: Symmetrical eight layer [90°/0°/-45°/45°]$_S$-laminate: out-of-plane shear stress σ_{xz} on the free boundary at $y = 0$

calculation are those from the linear system of equations. No extrapolation or interpolation is done. All three solutions (Fig. 26) agree very well but it is recognized that the element size Δr affects the result for the stress intensity factor K_I. Furthermore, it should be noticed that the element sizes are much more smaller in the case of an interface crack[25,26,50,51,52] than in homogeneous elasticity and that certain attempts are necessary to obtain a satisfactory result. Nevertheless, the results show that the method is very well suited for this kind of problem.

The basic geometry of a cracked four layer laminate is shown in Fig. 1. The size of a laminate is assumed to be $h_1 = h_2 = h, b/h = 4$ and $\delta = h$. The orientation of the fibres is $[+\Theta/\text{-}\Theta]_S$. This orientation leads on the free-edge to the shear stresses σ_{yx} which alternates from the $+\Theta$ to the $-\Theta$ layer. Therefore, the stresses are a pure mode III load. Figure 27 shows the stress-intensity factors depending on the orientation angle Θ. The K_{III}-factor is apparently much greater than the other ones. One reason is that the applied uniaxial strain $\varepsilon_{xx} = 1$ leads to high boundary stresses σ_{yx} at $\Theta \approx 18°$. The other reason is that only mode III shear stresses are applied and therefore the K_{III}-factor is dominant. Because of the anisotropy of the problem, K_I- and K_{II}-factors occur. Small differences occur between the analytical solution of Wang & Choi[61] and the BEM solution. The reason is the different definition of the stress-intensity factors in the analytical solution in eqn (20) and the requirement of a modified definition in any numerical procedure given in eqn (22). For instance, if an oscillating term in the complete solution has a significant influence (see eqn (21)), the numerical solution would always reproduce a part of the oscillation. This fact may lead to different results from the analytical and numerical solution. Noticeably, this effect is seen for the stress-intensity factors K_I and K_{II}, where oscillating parts occur in the solution.[61] As shown in the previous example, the result is affected by the length of the crack tip element, which is not discussed here in detail.

Another analytical solution is available for a $[+45°/\text{-}45°]_S$-laminate whose ratio h_1/h_2 of the layer thicknesses is varied (Figs 1 and 28). The geometry is $W = h_1 + h_2, W/b = 1, \delta = W/2$ and the ratio h_1/h_2 varies from 0.1 to 10. The analytical solution for the stress-intensity factor was given by Wang.[62] Because of the orientation of the layers only the stress σ_{yx} occurs on the free boundary which leads to a dominance of the K_{III}-factor. The K_I and K_{II} stress-intensity factors occur because of the anisotropy of the problem. The results from Ref. 63 and from the BEM calculations are compared in Fig. 28. Both solutions agree for the K_{III}-factor while differences occur for the in-plane stress intensity factor K_{II}, especially for extreme ratios h_1/h_2. Again, these differences are due to different definitions of the stress-intensity factors for the analytical solution[63] and the numerical solution as mentioned above.

7.2.2 Energy release rate concept

The designer of a laminate has to take care of three main features: the laminate must be stiff, its global limit load has to hold a certain value and the danger of a delamination before reaching the limit load must be minimized. The first two points are solved according to Section 2.1 while the numerical treatment of a delamination is obtained here by using the energy release rate. In contrast to the examples mentioned above it is possible to get important information on the design of a laminate by using the energy release rate concept. A $[\Theta/\Theta\text{-}90°]_S$- and a $[\Theta/\Theta\text{-}90°/\text{-}45°/+45°]_S$-laminate are

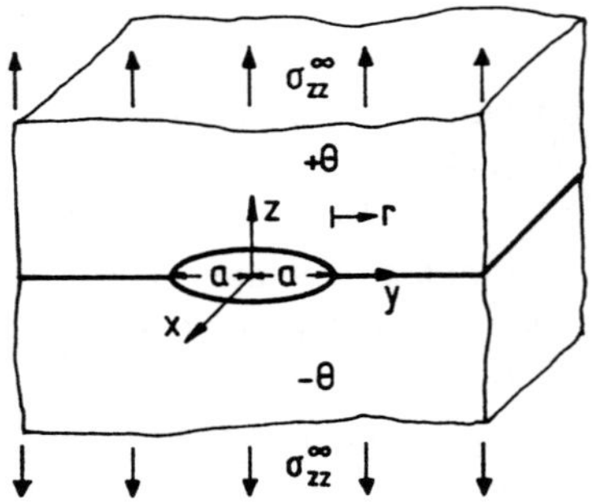

Figure 24: Geometry of the unbounded [+Θ/-Θ]S-laminate

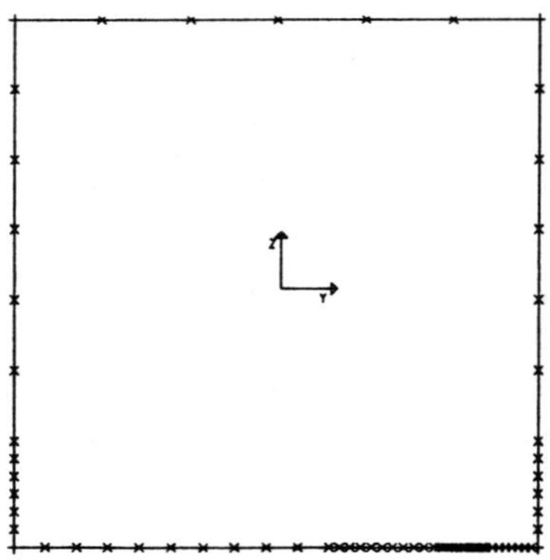

Figure 25: Discretization used for the [+Θ/-Θ]S-laminate

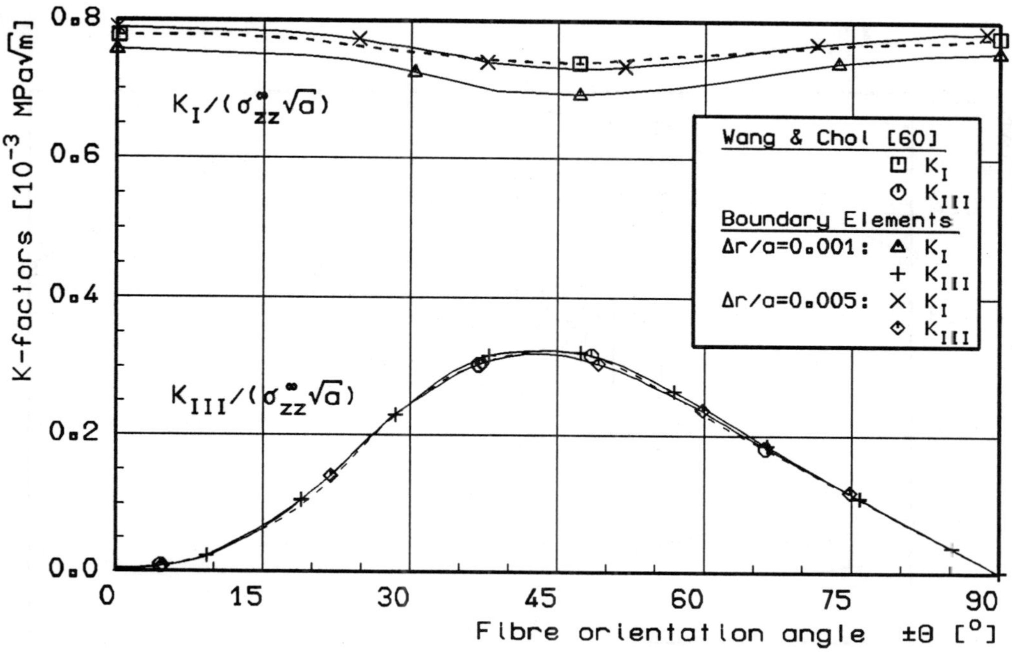

Figure 26: Symmetrical four layer $[+\Theta/\text{-}\Theta]_S$-laminate: stress-intensity factors at the crack front depending on the fibre orientation angle Θ

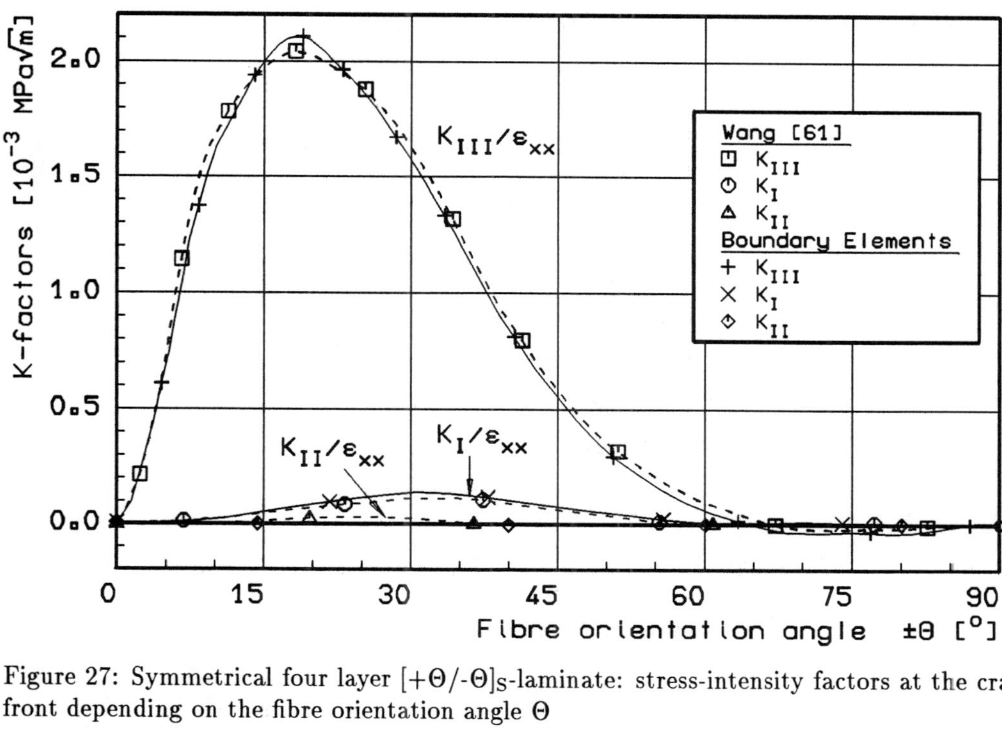

Figure 27: Symmetrical four layer $[+\Theta/\text{-}\Theta]_S$-laminate: stress-intensity factors at the crack front depending on the fibre orientation angle Θ

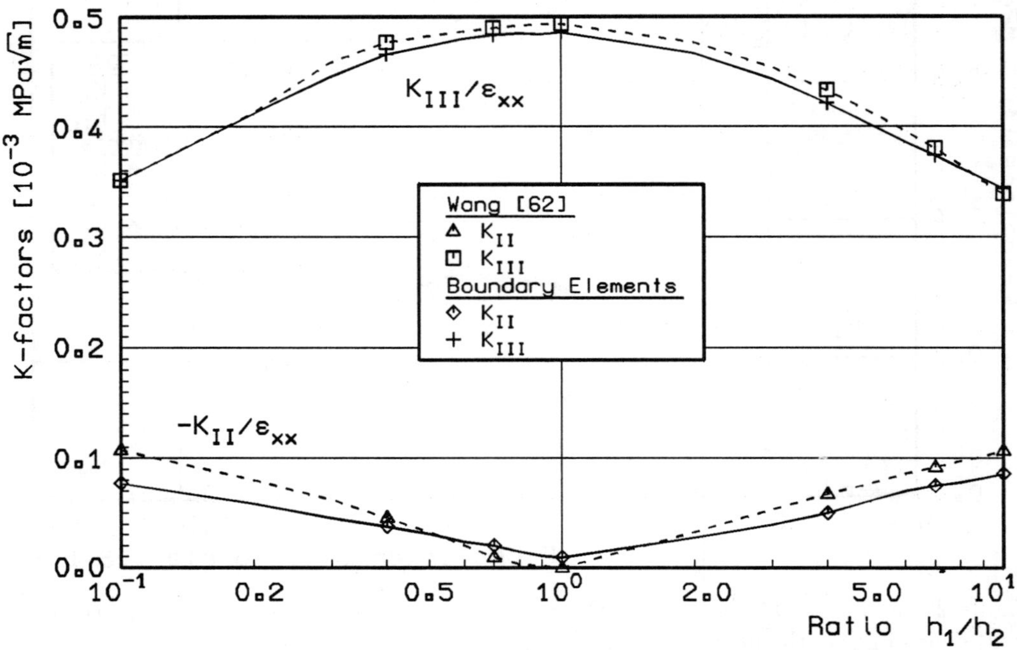

Figure 28: Symmetrical four layer [+45°/-45°]$_S$-laminate: stress-intensity factors depending on the ratio h_1/h_2 of the thicknesses of the layers

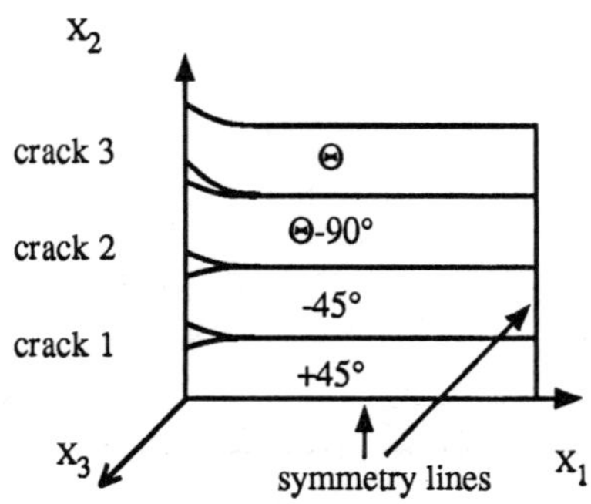

Figure 29: Numbering of the cracks

presented to demonstrate the method. The orientation angle Θ of the fibres is varied and the energy release rates are calculated under uniaxial strain or constant curvature. The virtual crack closure length δ is chosen to 0.01 mm for all calculations and the ratio b/h is set 1/5. As mentioned in Section 2.4, two calculations have to be carried out in order to get one value for the energy release rate: one for the non-delaminated case and another one for the delaminated case. All calculations are carried out with regular elements because the tractions are not used for the delaminated case in the ligament of the crack according to eqn (25).

Figures 30 and 31 show the results for the $[\Theta/\Theta\text{-}90°]_S$-laminate under uniaxial strain $\varepsilon_{xx} = 1$. It is significant that the energy release rate has marked minimums and maximums depending on the orientation angle Θ. The part G_{III} is dominant because the applied load ε_{xx} causes mainly interlaminar shear stresses σ_{xz}. The smaller parts G_I and G_{II} appear because of the anisotropy of the constitutive law (31). For $\Theta = 0°$ and $\Theta = 90°$ G_{tot} has its absolute minimum. In this case, the constitutive law is only orthotropic and no interlaminar shear stresses σ_{xz} occur. The lack of this orientation is the small shear stiffness of the whole laminate so that it is more convenient to choose Θ in the range from 30° to 60°.

A more realistic case is the following symmetric eight layer $[\Theta/\Theta\text{-}90°/\text{-}45°/+45°]_S$-laminate under uniaxial strain and constant curvature. All results in the Figs 32, 33, 34 and 35 refer to crack number 3 (Fig. 29). The two examples show that the anisotropy has an essential influence on the results because the energy release rates G_I and G_{II} supply significant parts to the total energy release rate G_{tot}. The energy release rates resulting from the constant curvature are much higher than those under uniaxial strain because the distance from the neutral plane of the laminate to crack 3 (Fig. 29) is large which leads to high stresses in the case of bending.

In both cases it is quite simple to choose the ranges that should be avoided and that should be used for the design. In the case of uniaxial strain (Figs 32,33) the range from 10° to 60° and under constant curvature (Figs 34,35) the range from 5° to 30° should not be used. It is mentioned that this result only refers to crack 3 while a complete analysis must include the other cracks, too.

8 Summary and conclusions

It is concluded that any loading of a laminate leads to stress concentrations on the free-edge. These stress concentrations may cause a delamination of the layers from the free-edge so that interlaminar cracks occur. Therefore, two different problems arise, namely the non-delaminated and the delaminated case. These two problems have been solved in parts by many authors using various methods. The given references are classified to get a general view. The basic solutions for the stresses are discussed.

Fracture mechanics offers the stress-intensity factor and energy release rate concept to describe the mechanical behaviour at the crack tip. Both concepts are adopted for anisotropic elasticity.

The basic mechanical equations are given introducing an arbitrary anisotropic linear constitutive law. A boundary integral equation is used to describe the displacement and stress state. A full three-dimensional application of the boundary integral equation requires an expensive numerical procedure for the calculation of the funda-

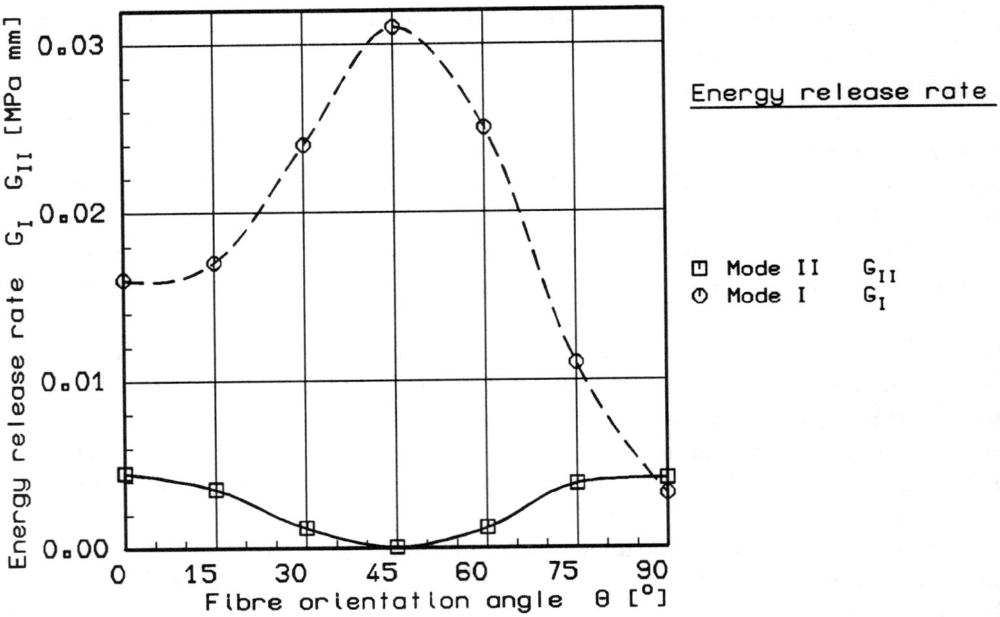

Figure 30: Symmetrical four layer $[\Theta/\Theta\text{-}90°]_S$-laminate: Energy release rate under uniaxial strain $\varepsilon_{xx} = 1$ depending on the fibre orientation angle Θ

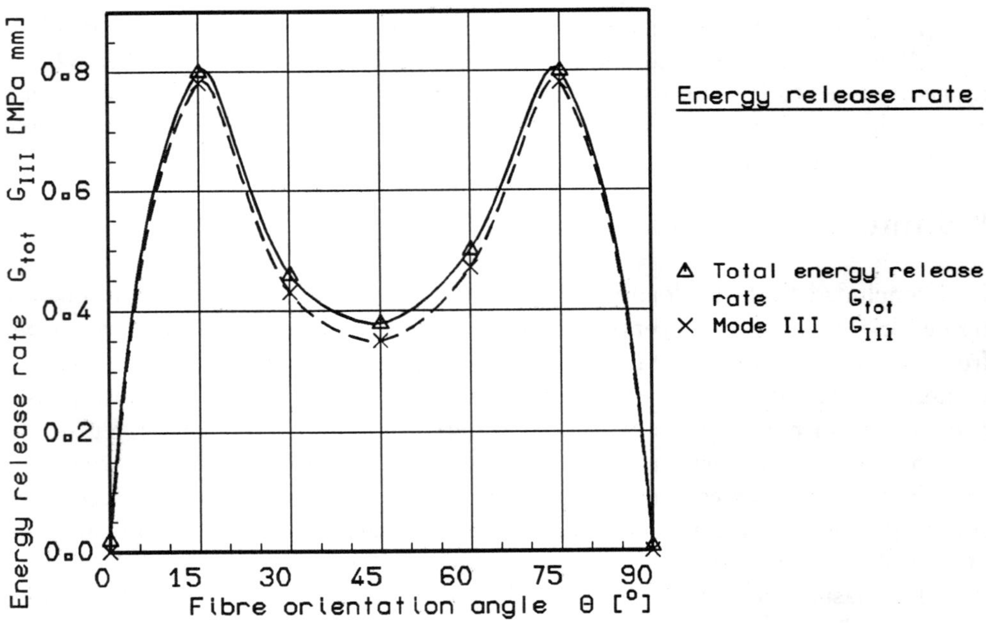

Figure 31: Symmetrical four layer $[\Theta/\Theta\text{-}90°]_S$-laminate: Energy release rate under uniaxial strain $\varepsilon_{xx} = 1$ depending on the fibre orientation angle Θ

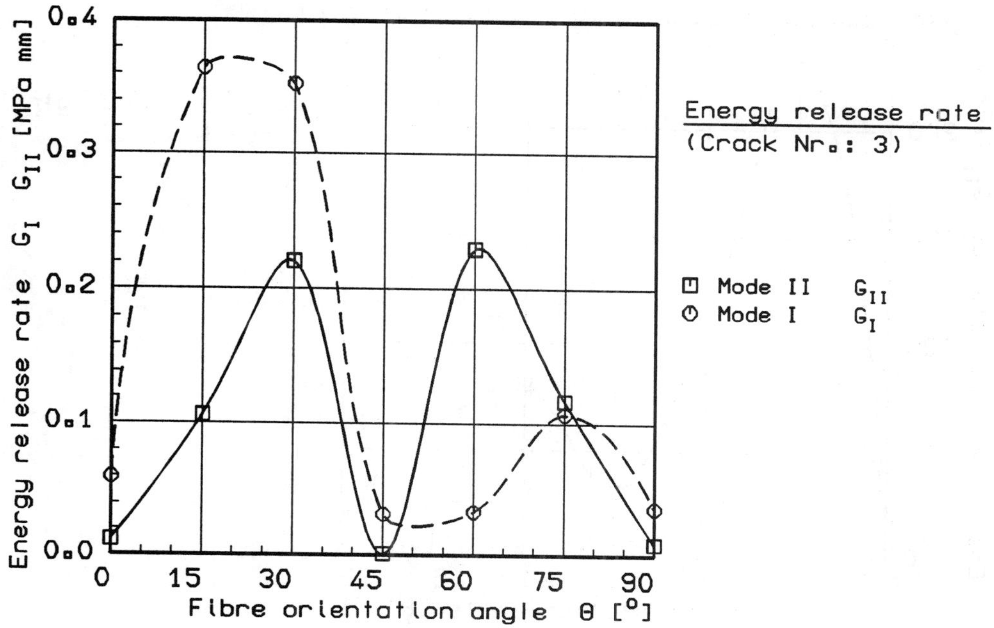

Figure 32: Symmetrical eight layer [Θ/Θ-90°/-45°/ + 45°]s-laminate: Energy release rate under uniaxial strain $\varepsilon_{xx} = 1$ depending on the fibre orientation angle Θ

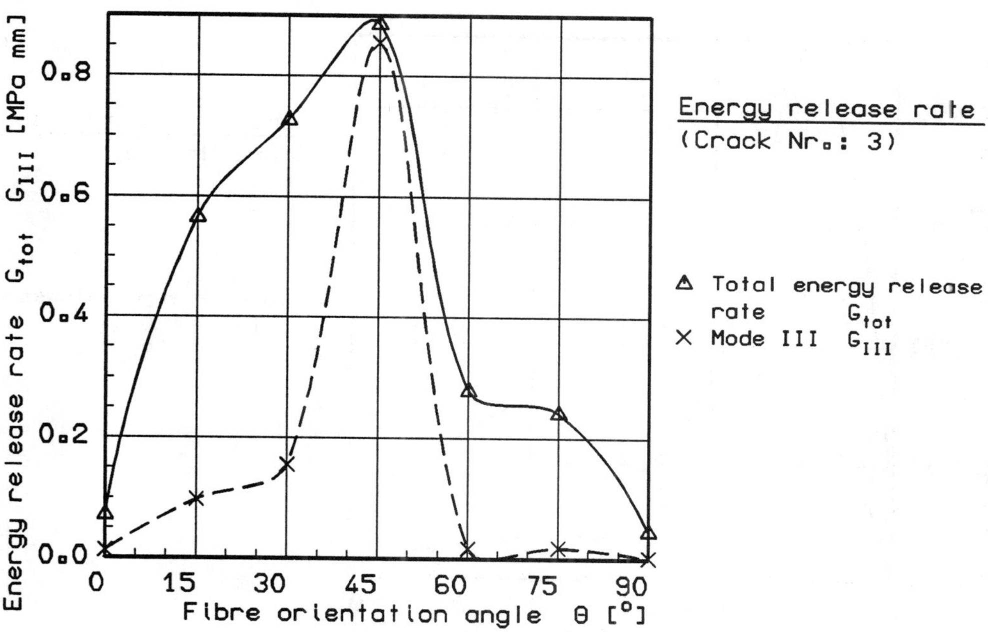

Figure 33: Symmetrical eight layer [Θ/Θ-90°/-45°/ + 45°]s-laminate: Energy release rate under uniaxial strain $\varepsilon_{xx} = 1$ depending on the fibre orientation angle Θ

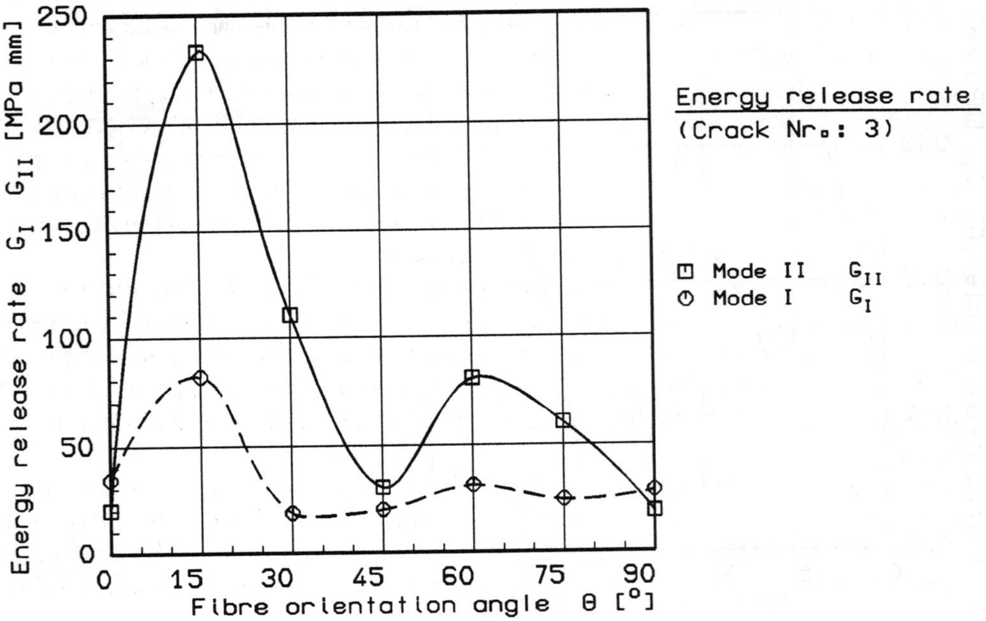

Figure 34: Symmetrical eight layer $[\Theta/\Theta\text{-}90°/\text{-}45°/+45°]_S$-laminate: Energy release rate under constant curvature $\kappa_{xx} = 1$ depending on the fibre orientation angle Θ

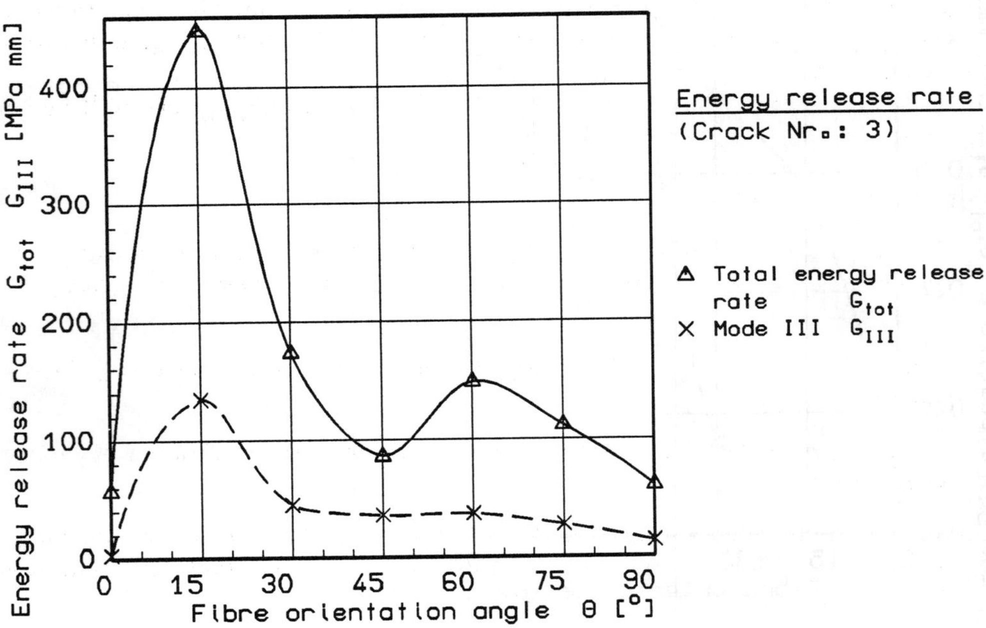

Figure 35: Symmetrical eight layer $[\Theta/\Theta\text{-}90°/\text{-}45°/+45°]_S$-laminate: Energy release rate under constant curvature $\kappa_{xx} = 1$ depending on the fibre orientation angle Θ

mental solutions. Introducing the assumption that all mechanical quantities do not vary in the longitudinal direction of the laminate, the fundamental solutions are calculated analytically which has the great advantage that less computer time is required.

The boundary integral equation is discretized by introducing interpolation functions for the boundary, the displacements and the tractions which leads to the boundary element method (BEM). In the case of the non-delaminated structure, regular elements are used. Crack tip elements are introduced to simulate the singular behaviour of the stresses using a general $r^{-\lambda}$ formulation. The stress-intensity factors are direct unknowns in the linear system of equations.

Several examples for the non-delaminated as well as for the delaminated case are presented in comparison with the results from other authors received by different methods. The BEM turns out to be a very good method for the solution of the stress concentration of non-delaminated free-edges. The rapidly decreasing stresses inside the laminate from the free boundary are reproduced very well both qualitatively and quantitatively by the BEM.

The crack tip elements are applied to delaminated structures where the stress-intensity factors are calculated. The crack tip element lengths have to be chosen very small to simulate the rapidly decreasing stresses near the crack tip. The comparison with analytical solutions shows a good agreement with the results obtained by the BEM.

The application of the energy release rate concept offers a tool to get an optimum design of composite laminates in order to avoid delamination from the free-edge. The BEM turns out to be easy to use because it requires less input for the geometry and less computer time during the solution. Unfortunately, no values for a critical total energy release rate G_{tot} are available from experiments so that all calculations can only give qualitative results. Nevertheless, the results presented here give important information for the design of a composite laminate.

Finally, it is emphasized that the BEM solved all problems very well and no difficulties occurred relating to the method.

References

1. Blandford, G.E., Ingraffea, A.R. & Ligget, J.A. Two-dimensional stress intensity factor computations using the boundary element method, *Int. J. Num. Meth. Eng.*, 1981, **17**, 387-404.

2. Brebbia, C.A. & Walker, S. *Boundary Element Techniques in Engineering*, Newnes-Butterworths, 1980.

3. Brebbia, C.A., Telles, J.C.F. & Wrobel, L.C. *Boundary Element Techniques*, Springer-Verlag, 1984.

4. Buchholz, F.G., Bürger, M., Kumosa, M.& Eggers, H. Mixed mode fracture analysis of orthotropic laminates by local and global energy methods, *Proc. 5th. Int. Conf. Num. Meth. Fract. Mech.*, 1990, 391-402.

5. Buchholz, F.G. Mixed mode fracture analysis of debonding and matrix crack processes by the virtual crack closure method, *Proc. Joint FEFG/ICF Int. Conf. on Frac. Eng. Mat. and Struc.*, 1991, 265-270.

6. Charalanbides, P.G., Lund, J., Evans, A.G. & McMeeking, R.M. A test specimen for determining the fracture resistance of bimaterial interfaces, *J. Appl. Mech.*, 1989, **56**, 77-82.

7. Chaudonneret, M. On the discontinuity of the stress vector in the boundary integral equation method for elastic analysis, *Recent Advances in Boundary Element Methods*, C. A. Brebbia (ed), 1978, pp. 185-194.

8. Comninou, M. The interface crack, *J. Appl. Mech.*, 1977, **44**, 631-636.

9. Comninou, M. The interface crack in a shear field, *J. Appl. Mech.*, 1978, **45**, 287-290.

10. Comninou, M. & Schmueser, D. The interface crack in a combined tension-compression and shear field, *J. Appl. Mech.*, 1979, **46**, 345-348.

11. Cruse, T.A. An improved boundary-integral-equation method of three dimensional elastic analysis, *Comp. & Struct.*, 1974, **4**, 741-754.

12. Cruse, T.A. Boundary integral equation method for three-dimensional elastic fracture mechanics analysis, Air Force Office of Scientific Research TR-75-0813, 1975.

13. Cruse, T.A. Boundary-integral equation fracture mechanics analysis, *Boundary-Integral Equation Method: Computational Applications in Applied Mechanics*, ASME, 1975, AMD-11, 31-46.

14. Cruse, T.A. & Wilson, R.B. Boundary-integral equation method for elastic fracture mechanics analysis, AFOSR-TR-78-0355, 1975.

15. Cruse, T.A. Two dimensional BIE fracture mechanics analysis, *Appl. Math. Modelling*, 1978, **2**, 287-293.

16. Delale, F. Stress singularities in bonded anisotropic materials, *Int. J. Solids & Structures*, 1984, **20**, 31-40.

17. England, A.H. A crack between dissimilar media, *J. Appl. Mech.*, Trans. ASME 87, Series E, 1965, **32**, 400-402.

18. Eshelby, J.D., Read, W.T. & Shockley, W. Anisotropic elasticity with application to dislocation theory, *Acta Metallurgica*, 1953, **1**, 251 - 259.

19. Ezawa, E. & Okamoto, N. Singularity modelling in two-and three-dimensional stress intensity factor computation using boundary elements, *Boundary Elements VII* C. A. Brebbia & M. Tanaka (eds), Proc. of the 7th Int. Conf., 1985, pp. 7-3-7-12.

20. Giencke, E. Einfluss der Anisotropie auf die interlaminaren Randspannungen, DGLR-Bericht 87/02 1987, 41-67 (in German).

21. Gründemann, H. Boundary integral methods in linear elastic fracture mechanics, *Zamm*, 1984, **64**, M452-M453.

22. Gründemann, H. Randelementemethode in der Festkörpermechanik, *Fachbuchverlag Leipzig*, 1991, (in German).

23. Henshell, R.D. & Shaw, K.G. Crack tip finite elements are unnecessary, *Int. J. Num. Meth. Eng.*, 1975, **9**, 495-507.

24. Jones, R.M. *Mechanics of Composite Materials*, Washington D.C., 1975.

25. Klingbeil, D. Zur Berechnung von räumlich beanspruchten, geschichteten anisotropen bauteilen mit starker spannungsvariation in querrichtung mit einem randelementverfahren, *VDI-Bericht Reihe*, 1989, **18**(69), (in German).

26. Klingbeil, D. The calculation of interlaminar stresses and stress intensity factors in non-delaminated composite laminates by a boundary element method (BEM), *Proc. 5th Int. Conf. Num. Meth. Fract. Mech.*, 1990, pp. 367-378.

27. Klingbeil, D. & Karakaya, V. The calculation of energy release rates in composite laminates using the boundary element method, *Proceedings of the 13th Int. Conf. on Boundary Element Methods*, 1991, pp. 623-634.

28. Knowles, J.K. & Sternberg, E. On the singularity induced by certain mixed boundary conditions in linearized and non-linear elastostatics, *Int. J. Solids Structures*, 1973, **11**, 1173-1201.

29. Krenk, S. Stress-concentration around holes in anisotropic sheets, *Proc. 1st Int. Seminar on Recent Advances in Boundary Element Methods*, Southampton University, July 1978, Pentech Press, pp. 215-232.

30. Kuo, M.C. & Bogy, D.B. Plane solutions for the displacement and traction-displacement problem for anisotropic elastic wedges, *J. Appl. Mech.*, 1974, **41**, 197-202.

31. Kuo, M.C. & Bogy, D.B. Plane solutions for the traction problems on orthotropic unsymmetrical wedges and symmetrical twinned wedges, *J. Appl. Mech.*, 1974, **41**, 203-208.

32. Lekhnitskii, S.G. *Theory of Elasticity of an Anisotropic Body*, Holden-Day, San Francisco, 1963.

33. Mura, T. *Micromechanics of Defects in Solids*, Nijhoff Publishers, The Hague, 1982.

34. Nishimura, N. & Kobayashi, S. A boundary integral equation formulation for three dimensional anisotropic elastostatics, *Boundary Elements V*, Proc. of the 5th Int. Conf., 1983, pp. 345-354.

35. Pagano, N.J. On the calculation of interlaminar normal stress in composite laminate, *J. Comp. Mat*, 1974, **8**, 65-81.

36. Piessens, R., de Droncker-Kapenga, E., Überhuber, C.W. & Kahhauer, D.K. *Quadpack - A Subroutine Package for Automatic Integration*, Springer-Verlag, 1983.

37. Pinchas, B.-Y. & Pian, T.H.H. Calculation of interlaminar stress concentration in composite laminates, *J. Comp. Mat.*, 1981, **15**, 225-239.

38. Pipes, R.B. & Pagano, N.J. Interlaminar stresses in composite laminates - an approximate elasticity solution, *J. Appl. Mech.*, Trans. ASME, 1974, **41**, 668-672.

39. Pipes, R.B. & Pagano, N.J. Interlaminar stresses in composite laminates under uniform axial extension, *J. Comp. Mat.*, 1970, **4**, 538-548.

40. Raju, I.S. & Crews, J.H., jr. Interlaminar stress singularities at a straight free edge in composite laminates, NASA TM-81976, 1980.

41. Rice, J.R. & Sih, G.C. Plane problems of cracks in dissimilar media, *J. Appl. Mech.*, Trans. ASME, Series E , 1965, 418-423.

42. Rizzo, F.J. & Shippy, D.J. A formulation and solution procedure for the general non-homogeneous elastic inclusion problem, *Int. J. Solids & Structures*, 1968, **4**, 1161-1179.

43. Rizzo, F.J. & Shippy, D.J. A method for stress determination in plane, anisotropic bodies, *Int. J. Comp. Mat.*, 1970, **4**, 33-61.

44. Sih, G.C. & Chen, E.P. (eds). *Cracks in Composite Materials*, Mechanics of Fracture 6, Nijhoff Publishers, 1980.

45. Sládek, V. & Sládek, J. Three-dimensional crack analysis for an anisotropic body, *Appl. Math. Modelling*, 1982, **6**, 374-380.

46. Sládek, V. & Sládek, J. Boundary element method in fracture mechanics, *Acta Technic*, 1982, SAV 27, 718-732.

47. Sládek, V., Sládek, J. & Balăs, J. Boundary integral formulation of crack problems, *ZAMM* 1983, **66**, 83-94.

48. Snyder, M.D. & Cruse, T.A. Boundary integral equation analysis of cracked anisotropic plates, *Int. J. Frac.*, 1975, **11**, 315-328.

49. Strout, A. H. & Secrest, D. *Gaussian Quadrature Formulas*, 1966.

50. Tan, C.L. & Gao, Y.L. Stress intensity factors for cracks at spherical inclusions by the boundary integral equation method, *Journal of Strain Analysis*, 1990, **25**, 197-206.

51. Tan, C.L. & Gao, Y.L. Treatment of bimaterial interface crack problems using the boundary element method, *Eng. Frac. Mech.*, 1990, **36**, 919-932.

52. Tan, C.L. & Gao, Y.L. Axisymmetric boundary integral equation analysis of interface cracks between dissimilar media, *Comp. Mech.*, 1991, **7**, 381-396.

53. Tauchert, Th.R. On the validity of elementary bending theory for anisotropic elastic slabs, *J. Comp. Mat.*, 1975, **9**, 207-214.

54. Tauchert, Th.R. Stresses in an anisotropic elastic slab due to distributed surface loads and displacements, *Developments in Mechanics*, 1975, **8**, Proceedings of the 14th Midwestern Mechanics Conference, 89-101.

55. Vogel, S.M. & Rizzo, F.J. An integral equation formulation of three-dimensional anisotropic boundary value problems, *J. Elas.*, 1973, **3**, 203-216.

56. Walpole, L.J. Elastic behaviour of composite materials: theoretical foundations, *Advances in Applied Mechanics*, C.-S. Yih (ed), 1978, **21**, 169-242.

57. Wang, A.S.D. & Crossman, F.W. Some new results on edge effects in symmetric composite laminates, *J. Comp. Mat.*, 1977, **11**, 92-106.

58. Wang, S.S. & Choi, I. Boundary layer effects in composite laminates, part I., *J. Appl. Mech.*, 1982, **49**, 541-548.

59. Wang, S.S. & Choi, I. Boundary layer effects in composite laminates , part II, *J. Appl. Mech.*, 1982, **49**, 549-560.

60. Wang, S.S. & Choi, I. The interface crack between dissimilar anisotropic materials, *J. Appl. Mech.*, 1983, **50**, 169-178.

61. Wang, S.S. Fracture mechanics for delamination problems in composite materials, *J. Comp. Mat.*, 1983, **17**, 210-223.

62. Wang, S.S. Edge delamination in angle-ply composite laminates, *AIAA Journal*, 1984, **22**, 256-264.

63. Willis, J.R. Variational and related methods for the overall properties of composites, *Advances in Applied Mechanics*, C.-S. Yih (ed), 1978, **21**, 1-78.

64. Wilson, R.B. & Cruse, T.A. Efficient implementation of anisotropic three dimensional boundary-integral equation stress analysis, *Int. J. Num. Meth. Eng.*, 1978, **12**, 1383-1397.

65. Zwiers, R.J., Ting, T.C.T. & Spilker, R.L. On the logarithmic singularity of free- edge stress in laminated composites under uniform extension, *J. Appl. Mech.*, 1982, **49**, 561-569.

Chapter 4
Body force method

H. Nisitani[a] & D.H. Chen[b]

[a] *Faculty of Engineering, Kyushu University, Fukuoka 812 Japan*
[b] *Faculty of Computer Science and Systems Engineering, Kyusyu Institute of Technology, Iizuka 820, Japan*

1 Introduction

The body force method (BFM) was originally proposed by Nisitani[1] in 1967. In the early stages during the progress of the method, it was mainly applied to the elastostatic problems for calculating the stress concentration factors of notches or the stress intensity factors of cracks.[2–10] Recently, the method has been extended to be applicable to elastic-plastic problems[1–13] and elastodynamic problems.[14]

BFM is a method based on the principle of superposition. The solutions in BFM are obtained by superposing the fundamental solutions. That is, in an infinite body the point forces and the discrepancies* are continuously distributed along an imaginary boundary so that the given boundary conditions are satisfied, and the required elastic field is obtained simply by superposing the fundamental elastic fields due to the point forces and discrepancies applied in an infinite body. Since the basic consideration of BFM is simple and intuitive, various techniques for improving the solution accuracy are easy to understand.

In this chapter, firstly, the principle of BFM is illustrated and the most general formula based on the principle of superposition is proposed. Secondly, the basic techniques for improving the accuracy of numerical results are given and their applications to various problems are shown. Finally, the extension of BFM to the elastic-plastic problem and to the elastodynamic problem is explained.

* A discrepancy corresponds to the 'nuclei of strain' defined in *A Theory on the Mathematical Theory of Elasticity* by Love.

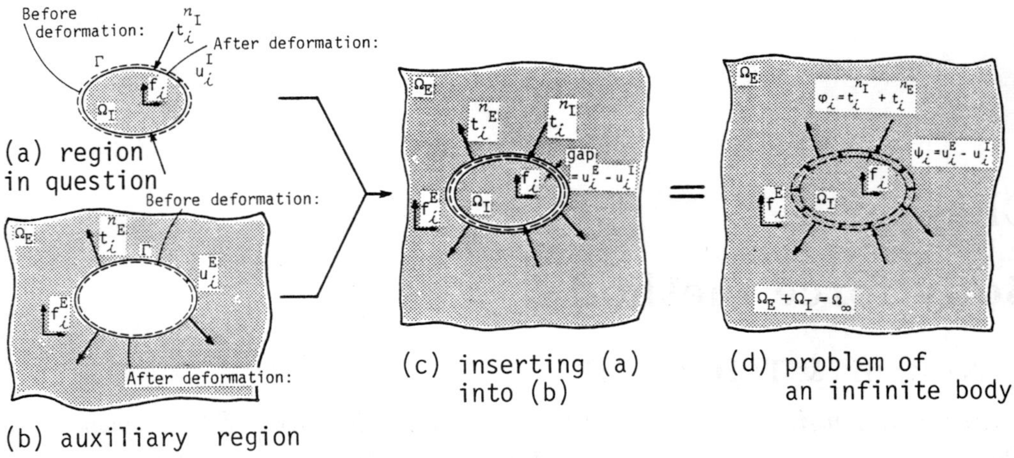

Figure 1: Transformation of the problem of a finite region into the problem of an infinite region

2 Fundamental concept and formulae

In the present section, the method of analysis applicable to elastostatic problems is described and illustrated.

Solving an elastic problem can be considered as a search for a reasonable elastic field which satisfies the given boundary conditions. Since the elastic fields can be superposed, the required elastic field can be obtained by superposing the known elastic fields. BFM is directly based on this principle of superposition. In BFM the specific known elastic fields, which may be written in closed form, are superposed with some unknown weight-magnitudes so that the required elastic field in question can be obtained. That is,

$$\boxed{\begin{array}{l}\text{required}\\ \text{elastic fields}\end{array}} = \sum_i \boxed{\begin{array}{l}\text{specific known}\\ \text{solutions of}\\ \text{elastic field}\end{array}}_i \times \boxed{\text{unknowns}}_i \tag{1}$$

If the given boundary conditions can be satisfied by adjusting the unknowns in eqn (1), then the superposed elastic field is the solution of the given problem, because the solution of the elastic problem is unique. Therefore, the validity of this method is dependent upon the existence of the suitable unknowns in eqn (1) which can satisfy the given boundary conditions. In view of this, in BFM, the fundamental elastostatic fields due to point forces or discrepancies applied in an infinite body are used as the known specific elastic fields, and the unknowns in eqn (1) are the densities of point forces and discrepancies continuously distributed along an imaginary boundary. The imaginary boundary is a surface imagined in the infinite body which stands for the prospective boundary for the given problem. In these cases the existence of the suitable unknowns in eqn (1) can be seen through Fig. 1.

In Fig. 1(a) an arbitrarily shaped elastic body (region Ω_I in question) is subjected to an arbitrarily given surface force $t_i^{n\mathrm{I}}(Q)$; $Q \in \Gamma$ and body force $f_i(Q); Q \in \Omega_I$, where

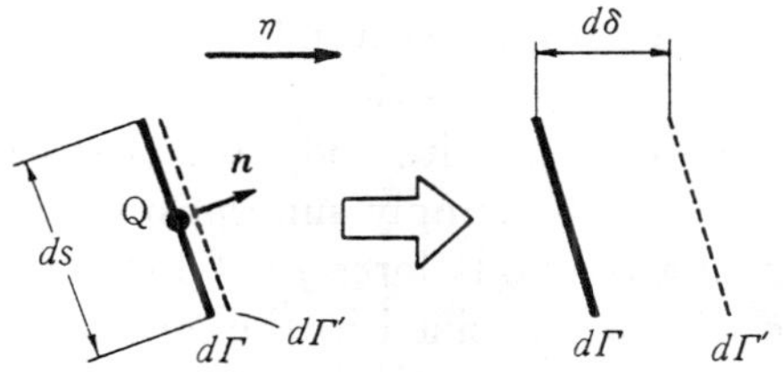

Figure 2: Discrepancy having magnitude of d$s \times$ dδ

Γ is the boundary of the region $\Omega_{\rm I}$ in question, $n_{\rm I}$ is the outward normal vector of Γ and i denotes the coordinate component ($i = 1, 2, 3$ for three-dimensional case and $i = 1, 2$ for two-dimensional case). The shape of the elastic body before deformation is shown by a broken line, and the shape after deformation is shown by a solid line. The surface displacement of $\Omega_{\rm I}$ (the deformation corresponds to the difference between the contour of the broken line and that of the solid line) is denoted by $u_i^{\rm I}(Q); Q \in \Gamma$.

In BFM, the given problem is always replaced by the problem of an infinite body, so that the known fundamental elastic fields in an infinite body can be used. For this purpose, an infinite body with a cavity, which has the same shape as the boundary of the region in question, has been taken into account as an auxiliary region Ω_E, as shown in Fig. 1(b). In the case of Fig. 1 the region in question is a finite body Ω_I and the auxiliary region is an infinite body with a cavity Ω_E. If the region in question is Ω_E, then the auxiliary region becomes Ω_I.

The auxiliary region is subjected to an arbitrary surface force $t_i^{n_E}(Q); Q \in \Gamma$ and an arbitrary body force $f_i^E(Q)$; $Q \in \Omega_E$. The surface displacement is denoted by $u_i^E(Q)$. When the region in question is inserted into the auxiliary region as shown in Fig. 1(c), there are gaps between the region in question and the auxiliary region, because the surface displacements of the two are not coincident. However, these gaps can be regarded as the discrepancies embedded in an infinite body. The discrepancy is defined in Fig. 2. Figure 2 shows a discrepancy in η-direction of facets, whose normal vector is in the n-direction, having a magnitude of d$s \times$ dδ (ds: the area of dΓ).

Thus, as shown in Fig. 1(d), a perfect infinite body is obtained, and in this infinite body the discrepancies are continuously distributed along the surface (we call it an imaginary boundary) which corresponds to the real boundary of region Ω_i. The density $\psi_i(Q)$ of the discrepancies equals the difference between the surface displacements as follows:

$$\psi_i(Q) = u_i^E(Q) - u_i^I(Q) \quad (Q \in \Gamma) \tag{2}$$

In the infinite body of Fig. 1(d), there are still distributed forces, besides the discrepancies, along the imaginary boundary Γ, because in Fig. 1(c) the force $t_i^{n_I}(Q)$ is applied on the surface of Ω_I and the force $t_i^{n_E}(Q)$ is applied on the surface of Ω_E. The density $\phi_i(Q)$ of the forces are

$$\phi_i(Q) = t_i^{n_I}(Q) + t_i^{n_E}(Q) \quad (Q \in \Gamma) \tag{3}$$

It is seen that for all the above processes, from Fig. 1(a) to Fig. 1 (d), the elastic field in the region Ω_I has not been disturbed. In Fig. 1(d) the region Ω_I becomes a part of an infinite body. This means that the given problem of a finite region has been reduced to a problem of an infinite body and that the elastic field of the region in question can be represented by simply superposing all the actions applied in the infinite body (the action of given body force $f_i(q)$ distributed in the calculated region, the action of body force $f_i^E(Q)$ specified in the auxiliary region, and the actions of point forces $\phi_i(Q)$ and discrepancies $\psi_i(Q)$ distributed along the imaginary boundary). That is, for the displacement field $u_i(P)$ and the stress field $\sigma_{ij}(P)$ in the region Ω_I, the following equations are obtained.

$$\begin{aligned} u_i(P) &= \int_\Gamma u_i(P,Q_\eta)\phi_\eta(Q)\mathrm{d}\Gamma(Q) + \int_\Gamma u_i^*(P,Q_\eta^n)\psi_\eta(Q)\mathrm{d}\Gamma(Q) \\ &+ \int_{\Omega_I} u_i(P,Q_\eta)f_i(Q)\mathrm{d}\Omega(Q) + \int_{\Omega_E} u_i(P,Q_\eta)f_\eta^E(Q)\mathrm{d}\Omega(Q) \quad (P\in\Omega_I) \end{aligned} \tag{4}$$

$$\begin{aligned} \sigma_{ij}(P) &= \int_\Gamma \sigma_{ij}(P,Q_\eta)\phi_\eta(Q)\mathrm{d}\Gamma(Q) + \int_\Gamma \sigma_{ij}^*(P,Q_\eta^n)\psi_\eta(Q)\mathrm{d}\Gamma(Q) \\ &+ \int_{\Omega_I} \sigma_{ij}(P,Q_\eta)f_\eta(Q)\mathrm{d}\Omega(Q) + \int_{\Omega_E} \sigma_{ij}(P,Q_\eta)f_\eta^E(Q)\mathrm{d}\Omega(Q) \quad (P\in\Omega_I) \end{aligned} \tag{5}$$

In eqns (4) and (5), $u_i(P,Q_\eta)$ and $\sigma_{ij}(P,Q_\eta)$ are the i-direction displacement and the stress σ_{ij} at a point P due to the η-direction unit point force acting at a point Q, respectively. $u_i^*(P,Q_\eta^n)$ and $\sigma_{ij}^*(P,Q_\eta^n)$ are the i-direction displacement and the stress at a point P due to the η-direction unit concentrated discrepancy of facets, whose normal vector is of the n-direction, acting at a point Q, respectively.

For an isotropic elastic body, $u_i(P,Q_\eta)$ and $\sigma_{ij}(P,Q_\eta)$ are given by Kelvin[15] and are found in many references.[16] $u_i^*(P,Q_\eta^n)$ and $\sigma_{ij}^*(P,Q_\eta^n)$ can be calculated from $u_i(P,Q_\eta)$ and $\sigma_{ij}(P,Q_\eta)$, because a concentrated discrepancy may be replaced by some combination of force doublets

$$u_i^*(P,Q_\eta^n) = u_i(P,Q_{\gamma,\delta})\mathrm{D}_{\gamma\delta\eta\zeta}n_\zeta(Q) \tag{6}$$

$$\sigma_{ij}^*(P,Q_\eta^n) = \sigma_{ij}(P,Q_{\gamma,\delta})\mathrm{D}_{\gamma\delta\eta\zeta}n_\zeta(Q) \tag{7}$$

where $\mathrm{D}_{\gamma\delta\eta\zeta}$ is the elastic stiffness tensor, and a comma before a greek letter denotes partial differentiation with respect to coordinative variables.

Equations (4) and (5), as a concrete form of eqn (1), are the most general equations for expressing the elastostatic field based on the principle of superposition. By determining the values of $\phi_\eta(Q)$ and $\psi_\eta(Q)$ appropriately, all problems can be solved through eqns (4) and (5) in principle.

Since Ω_E is not the region in question, the values of the body force $f_i^E(Q)$ and the surface force $t_i^{nE}(Q)$ or the surface displacement $u_i^E(Q)$ may be specified arbitrarily. In Fig. 1(d), the elastic field in Ω_I is all the same independent of the loading conditions of the auxiliary region Ω_E. However, along the imaginary boundary the distributed forces and the incompatibility of surface displacements are dependent of the loading conditions of the auxiliary region. It means that the elastic field of region in question can be represented by various combinations of point forces, discrepancies and body force $f_i^E(Q)$. Therefore, in eqns (4) and (5) one of the two unknowns, $\phi_\eta(Q)$ and $\psi_\eta(Q)$, can be specified arbitrarily and the remaining one is determined from the boundary condition equations, which are obtained from eqn (4) or eqn (5) by moving the calculated point P to the boundary.

By selecting the various special combinations of point forces and discrepancies, the various methods of boundary analysis can be deduced from eqns (4) and (5). That is, in numerical analysis, according to the characteristics of a given problem, we can select the most appropriate combination of $\phi_\eta(Q), \psi_\eta(Q)$ and $f_\eta^E(Q)$ for obtaining an accurate solution. The details about these are found in Ref. 17. Generally, $f_\eta^E(Q)$ is set at zero and the surface force or the surface displacement of an auxiliary region are specified so that the following three kinds of equations of boundary analysis are obtained.

(1) By inserting the region Ω_I in question into such an auxiliary region Ω_E, which is subjected to the surface forces $t_i^E(Q)$ necessary to give the displacements $u_i^E(Q)$ on their boundary coincident with those of the region in question $u_i^I(Q)$, the method, in which the unknowns are only the densities of point forces $\phi_\eta(Q)$, is deduced.

In this case, it is seen from the specified condition $u_\eta^E(Q) = u_\eta^I(Q); Q \in \Gamma$ that $\psi_\eta(Q) = 0$. So eqns (4) and (5) become

$$u_i(P) = \int_\Gamma u_i(P, Q_\eta)\phi_\eta(Q)\mathrm{d}\Gamma(Q) + \int_{\Omega_I} u_i(P, Q_\eta)f_\eta(Q)\mathrm{d}\Omega(Q) \quad (P \in \Omega_I) \tag{8}$$

$$\sigma_{ij}(P) = \int_\Gamma \sigma_{ij}(P, Q_\eta)\phi_\eta(Q)\mathrm{d}\Gamma(Q) + \int_{\Omega_I} \sigma_{ij}(P, Q_\eta)f_\eta(Q)\mathrm{d}\Omega(Q) \quad (P \in \Omega_I) \tag{9}$$

In eqns (8) and (9) the unknowns are only the densities of point forces $\phi_\eta(Q)$, which can be determined from the boundary condition equation for displacement obtained from eqn (8) (in the case of displacement boundary-value problems) or from the boundary condition equation for surface force obtained from eqn (9) (in the case of stress boundary-value problems). If the density $\phi_\eta(Q)$ is obtained, the displacement and the stress at an arbitrary point $P; P \in \Omega_I$ can be calculated from eqns (8) and (9).

(2) By inserting the region Ω_I in question into such an auxiliary region Ω_E, which is subjected to the surface forces $t_i^{nE}(Q)$ equal in magnitude and opposite in direction to those $t_\eta^{nI}(Q)$ acting at the boundary of the region in question, the method, in which the unknowns are only the densities of discrepancies $\psi_\eta(Q)$, is deduced.

In this case, it is seen from the specified condition $t_\eta^{nE}(Q) = -t_\eta^{nI}(Q); Q \in \Gamma$ that $\phi_\eta(Q) = 0$. So eqns (4) and (5) become as follows

$$u_i(P) = \int_\Gamma u_i^*(P, Q_\eta^n)\psi_\eta(Q)\mathrm{d}\Gamma(Q) + \int_{\Omega_I} u_i(P, Q_\eta)f_\eta(Q)\mathrm{d}\Omega(Q) \quad (P \in \Omega_I) \tag{10}$$

$$\sigma_{ij}(P) = \int_\Gamma \sigma_{ij}^*(P, Q_\eta^n)\psi_\eta(Q)\mathrm{d}\Gamma(Q) + \int_{\Omega_I} \sigma_{ij}(P, Q_\eta)f_\eta(Q)\mathrm{d}\Omega(Q) \quad (P \in \Omega_I) \tag{11}$$

In eqns (10) and (11), the unknowns are only the densities of discrepancies $\psi_\eta(Q)$, which can be determined from the boundary condition equation for displacement or surface force, similarly as in the method (1).

(3) By inserting the region Ω_I in question into a stress-free auxiliary region Ω_E, the method, in which the surface force $t_\eta^{n_I}(Q)$ or the surface displacement $u_\eta^I(Q)$ are directly used as unknowns, is deduced.

In this case, it is seen from the specified conditions $u_\eta^E(Q) = 0$ and $t_\eta^{n_E}(Q) = 0$; $Q \in \Gamma$ that the densities of point forces and discrepancies $\phi_\eta(Q), \psi_\eta(Q)$ are equal to the surface force and the surface displacement respectively.

$$\begin{aligned}\phi_\eta(Q) &= t_\eta^{n_I}(Q) \\ \psi_\eta(Q) &= -u_\eta^I(Q)\end{aligned} \qquad (Q \in \Gamma) \tag{12}$$

So eqns (4) and (5) become

$$\begin{aligned} u_i(P) &= \int_\Gamma u_i(P,Q_\eta) t_\eta^{n_I}(Q) \mathrm{d}\Gamma(Q) + \int_\Gamma u_i^*(P,Q_\eta^n) \left\{-u_\eta^I(Q)\right\} \mathrm{d}\Gamma(Q) \\ &+ \int_{\Omega_I} u_i(P,Q_\eta) f_\eta(Q) \mathrm{d}\Omega(Q) \qquad (P \in \Omega_I) \end{aligned} \tag{13}$$

$$\begin{aligned} \sigma_{ij}(P) &= \int_\Gamma \sigma_{ij}(P,Q_\eta) t_\eta^{n_I}(Q) \mathrm{d}\Gamma(Q) + \int_\Gamma \sigma_{ij}^*(P,Q_\eta^n) \left\{-u_\eta^I(Q)\right\} \mathrm{d}\Gamma(Q) \\ &+ \int_{\Omega_I} \sigma_{ij}(P,Q_\eta) f_\eta(Q) \mathrm{d}\Omega(Q) \qquad (P \in \Omega_I) \end{aligned} \tag{14}$$

In the case of displacement boundary-value problems the unknowns in eqns (13) and (14) are the densities of point forces $\phi_\eta(Q)$ (equal to $t_\eta^{n_I}(Q)$), which can be determined from the boundary condition equation for displacement obtained from eqn (13). In the case of stress boundary-value problems, the unknowns are the densities of discrepancies $\psi_\eta(Q)$ (equal to $-u_\eta^I(Q)$), which can be determined from the boundary condition equation for surface force obtained from eqn (14). However, it is found in the method (3) that even in the case of stress boundary-value problems the unknowns $u_\eta^I(Q)$ can also be determined from the displacement boundary equation obtained from eqn (13). This is a characteristic of the method (3) which is not seen in the methods (1) and (2). That is, in the method (3), the unknowns are usually determined from the displacement boundary equation in both the cases of displacement boundary-value problems and of stress boundary-value problems. It can be understood as follows.

By moving the calculated point P to the boundary, eqn (13) becomes

$$\begin{aligned} u_i(P^-) &= \int_\Gamma u_i(P^-,Q_\eta) t_\eta^{n_I}(Q) \mathrm{d}\Gamma(Q) + \int_\Gamma u_i^*(P^-,Q_\eta^n) \left\{-u_\eta^I(Q)\right\} \mathrm{d}\Gamma(Q) \\ &+ \int_{\Omega_I} u_i(P^-,Q_\eta) f_\eta(Q) \mathrm{d}\Omega(Q) \qquad (P \in \Gamma) \end{aligned} \tag{15}$$

For the points P^+ and P^- on both sides of Γ ($P \in \Gamma, P^+$ and P^- are infinitely close to P but $P^+ \in \Omega_E, P^- \in \Omega_I$), we have

$$\int_\Gamma u_i(P^+,Q_\eta)\phi_\eta(Q)\mathrm{d}\Gamma(Q) - \int_\Gamma u_i(P^-,Q_\eta)\phi_\eta(Q)\mathrm{d}\Gamma(Q) = 0 \qquad (P \in \Gamma) \tag{16}$$

$$\int_\Gamma u_i^*(P^+, Q_\eta^n)\psi_\eta(Q)\mathrm{d}\Gamma(Q) - \int_\Gamma u_i^*(P^-, Q_\eta^n)\psi_\eta(Q)\mathrm{d}\Gamma(Q) = \psi_i(P) \qquad (P \in \Gamma) \quad (17)$$

Thus, eqn (15) can be written,

$$\int_\Gamma u_i(P^+, Q_\eta)t_\eta^n(Q)\mathrm{d}\Gamma(Q) + \int_\Gamma u_i^*(P^+, Q_\eta^n)\left\{-u_\eta(Q)\right\}\mathrm{d}\Gamma(Q)$$

$$+ \int_{\Omega_I} u_i(P^+, Q_\eta)f_\eta(Q)\mathrm{d}\Omega(Q) = u_i(P^+) = 0 \qquad (P \in \Gamma) \qquad (18)$$

$u_i(P^+)$ is the boundary displacement of the auxiliary region. Therefore, it is seen that in solving the stress boundary-value problems by eqn (15), we actually use as a boundary condition the fact that the boundary displacement of the auxiliary region is always zero.

Equation (13) is essentially correspondent with the basic integral formula of the direct method of BEM based on Betti's reciprocal relation, because by noting the following identities,

$$u_i(P, Q_\eta) = u_\eta(Q, P_i)$$

and

$$u_i^*(P, Q_\eta^n) = t_\eta^n(Q, P_i)$$

Equation (13) can be rewritten as follows

$$\begin{aligned} u_i(P) &= \int_\Gamma u_\eta(Q, P_i)t_\eta^{n_I}(Q)\mathrm{d}\Gamma(Q) - \int_\Gamma t_\eta^n(Q, P_i)u_\eta^I(Q)\mathrm{d}\Gamma(Q) \qquad (19) \\ &+ \int_{\Omega_I} u_\eta(Q, P_i)f_\eta(Q)\mathrm{d}\Omega(Q) \qquad (P \in \Omega_I) \end{aligned}$$

Equation (19) is the same as the usual formula of the direct method.[18] This means that the comprehension of the direct method of BEM can be attained alternatively through applying the principle of superposition to the case when the region in question is inserted into the stress-free auxiliary region.

3 Singular integral

It is seen from the preceding discussion that the analysis of BFM is equivalent to determining the densities of point forces or discrepancies to be distributed along an imaginary boundary. These densities are determined from the boundary condition equation. In order to obtain the boundary condition equation we need to put the calculated point P on the boundary surface. So, a singular integral is involved, because when $Q \to P$ the integrated function such as $\sigma_{ij}(P, Q_\eta)$ becomes infinite.

Actually, we can regard the boundary points of the region in question as the points P^-, which are not directly on the surface Γ but are infinitesimally close to Γ, as shown in Fig. 3. By taking this fact into account, the problem of finding the principal value of the singular integral can be overcome.

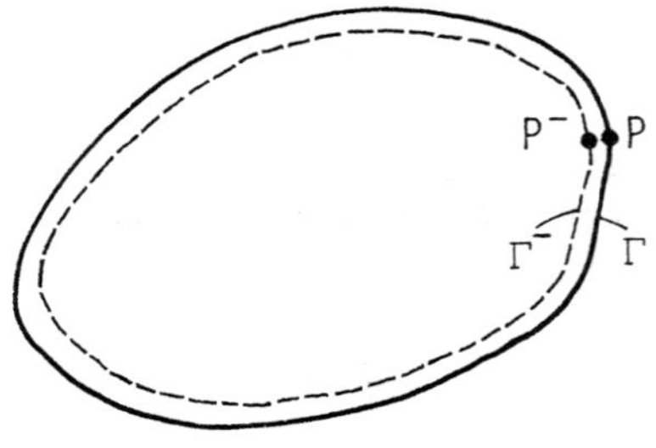

Figure 3: Points P^- where the boundary condition requires to be satisfied

When both the boundary and the density of point forces or discrepancies are smooth, the singular integral is finally reduced to calculating the value in a center point P when the point forces or the discrepancies are uniformly distributed on a circular domain of plate (for three-dimensional cases) or an interval of straight line (for two-dimensional cases). In these cases the principal values of singular integral can be obtained simply.

For example, as shown in Fig. 4, when the distributed point forces having the uniform densities ϕ_1^0, ϕ_2^0 act on an interval of 2δ of the x-axis, the stresses at the point $P(0, \varepsilon)$ due to these point forces are obtained from the fundamental solutions as follows (plane strain condition)

$$\mathrm{d}\sigma_{11} = \frac{-1}{4\pi(1-\nu)}\left\{\left[-\frac{(1-2\nu)x_1}{(\varepsilon^2+x_1^2)} - \frac{2x_1^3}{(\varepsilon^2+x_1^2)^2}\right]\phi_1^0\mathrm{d}x_1 + \left[-\frac{(1-2\nu)\varepsilon}{(\varepsilon^2+x_1^2)} + \frac{2\varepsilon x_1^2}{(\varepsilon^2+x_1^2)^2}\right]\phi_2^0\mathrm{d}x_1\right\}$$

$$\mathrm{d}\sigma_{22} = \frac{-1}{4\pi(1-\nu)}\left\{\left[\frac{(1-2\nu)x_1}{(\varepsilon^2+x_1^2)} - \frac{2\varepsilon^2 x_1}{(\varepsilon^2+x_1^2)^2}\right]\phi_1^0\mathrm{d}x_1 + \left[-\frac{(1-2\nu)\varepsilon}{(\varepsilon^2+x_1^2)} + \frac{2\varepsilon^3}{(\varepsilon^2+x_1^2)^2}\right]\phi_2^0\mathrm{d}x_1\right\}$$

$$\mathrm{d}\sigma_{12} = \frac{-1}{4\pi(1-\nu)}\left\{\left[\frac{(1-2\nu)\varepsilon}{(\varepsilon^2+x_1^2)} + \frac{2\varepsilon x_1^2}{(\varepsilon^2+x_1^2)^2}\right]\phi_1^0\mathrm{d}x_1 + \left[-(1-2\nu)x_1^2)^2\right]\phi_2^0\mathrm{d}x_1\right\} \quad (20)$$

Putting $x_1 = \varepsilon\tan\theta$ and integrating eqn (20), we have

$$\sigma_{11} = \frac{-1}{4\pi(1-\nu)}\int_{-\theta_0}^{\theta_0}\left[2\sin^2\theta - (1-2\nu)\right]\phi_2^0\mathrm{d}\theta$$

$$\sigma_{22} = \frac{-1}{4\pi(1-\nu)}\int_{-\theta_0}^{\theta_0}\left[2\cos^2\theta + (1-2\nu)\right]\phi_2^0\mathrm{d}\theta$$

$$\sigma_{12} = \frac{-1}{4\pi(1-\nu)}\int_{-\theta_0}^{\theta_0}\left[2\sin^2\theta + (1-2\nu)\right]\phi_1^0\mathrm{d}\theta \quad (\theta_0 = \tan^{-1}\frac{\delta}{\varepsilon})$$

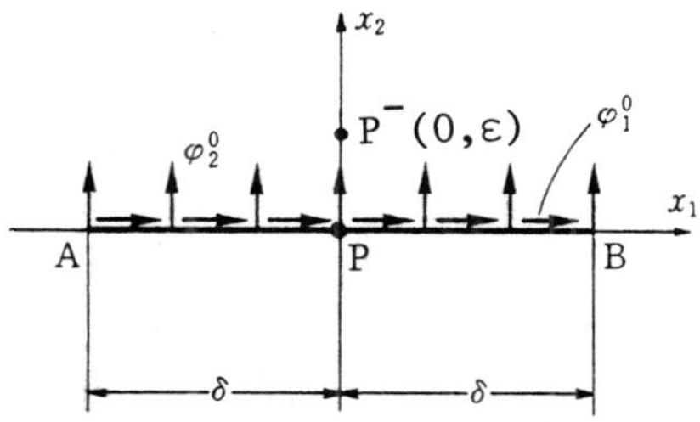

Figure 4: Point forces having uniform densities

The stresses at the point $(0, \varepsilon)$ can be written in an explicit form. By taking the limit $\varepsilon \rightarrow 0$, the stresses at P^- are obtained

$$\begin{aligned} \sigma_{11}(P^-) &= -\frac{\nu}{2(1-\nu)}\phi_2^0 \\ \sigma_{22}(P^-) &= -\frac{1}{2}\phi_2^0 \quad \text{(plane strain)} \\ \sigma_{12}(P^-) &= -\frac{1}{2}\phi_1^0 \end{aligned} \tag{21}$$

In plane stress condition eqn (21) becomes

$$\begin{aligned} \sigma_{11}(P^-) &= -\frac{\nu}{2}\phi_2^0 \\ \sigma_{22}(P^-) &= -\frac{1}{2}\phi_2^0 \quad \text{(plane strain)} \\ \sigma_{12}(P^-) &= -\frac{1}{2}\phi_1^0 \end{aligned} \tag{22}$$

4 Basic density function useful for analyzing stress concentration – techniques of high accuracy and efficiency

In order to determine the densities of distributed point forces or discrepancies from the boundary condition equation, in general, the numerical calculations are necessary. Therefore, the next problem to be considered is how to improve the accuracy of the numerical calculations.

The densities $\phi_\eta(Q)$ and $\psi_\eta(Q)$ are the continuous functions of the point $Q; Q \in \Gamma$. However, in numerical analysis the unknown functions $\phi_\eta(Q)$ and $\psi_\eta(Q)$ are approximately represented by a finite number of discrete values. In the simplest case, we divide the boundary into several elements and assume the densities $\phi_\eta(Q)$ and $\psi_\eta(Q)$ to be constant in each element, that is, we approximate the real density distribution by one parameter within each element, as shown in Fig. 5. Under this approximation, it is impossible to satisfy accurately the boundary conditions at all the points on the boundary. So the boundary conditions are satisfied strictly only at a finite number of points such as at the mid-point of each element. From the finite number

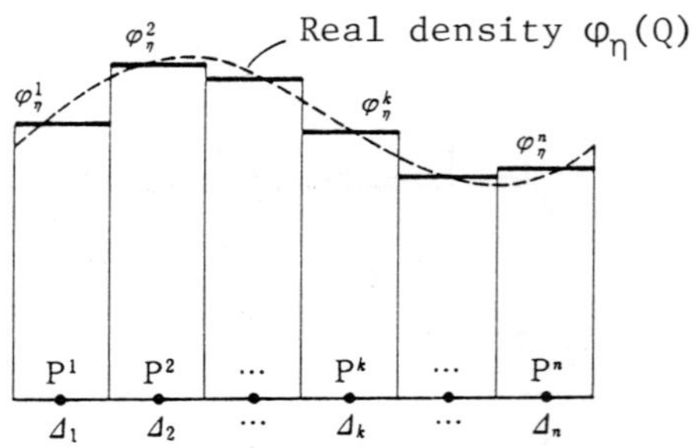

Figure 5: Approximation by stepped distribution

of equations of boundary conditions, the finite number of unknowns, which represent the densities in each element, can be determined easily. The solution obtained by this way is not accurate. The error is due to the fact that we approximate the real density distribution by a finite number of discrete values. Accuracy of calculation can be enhanced by increasing the number of the discrete values, for example by dividing the boundary into smaller elements or by approximating the real density by more number of parameters per element. However, there are more effective techniques which enable us to obtain highly accurate solutions without increasing the number of the discrete values. This is the idea of the basic density functions, shown in the following.

As described in Section 2, the required elastic field are generally expressed as follows.

$$\boxed{\begin{array}{l}\text{required}\\ \text{elastic fields}\end{array}} = \sum_i \boxed{\begin{array}{l}\text{fundamental solutions}\\ \text{due to unit point force}\\ \text{or unit discrepancy}\end{array}}_{\,i} \times \underbrace{\boxed{\begin{array}{l}\text{densities of}\\ \text{point forces}\\ \text{or discrep-}\\ \text{ancies}\end{array}}_{\,i}}_{\text{unknowns}} \tag{23}$$

However, based on the idea of basic density function, eqn (23) is rewritten as follows

$$\boxed{\begin{array}{l}\text{required}\\ \text{elastic fields}\end{array}} = \sum_i \boxed{\begin{array}{l}\text{fundamental solutions}\\ \text{due to unit point force}\\ \text{or unit discrepancy}\end{array}}_{\,i} \times \boxed{\begin{array}{l}\text{basic}\\ \text{density}\\ \text{function}\end{array}}_{\,i} \times \underbrace{\boxed{\text{weight}}_{\,i}}_{\text{unknowns}} \tag{24}$$

If the given problem can be solved accurately, there is no difference between eqn (23) and eqn (24). When the problem is solved approximately by a numerical procedure, the unknowns to be approximated in eqn (23) by discretized values are the density function itself, however, the unknowns to be approximated in eqn (24) by discretized values are the weight function, which has to be multiplied by the basic density functions. By appropriately defining the basic density functions, we can make

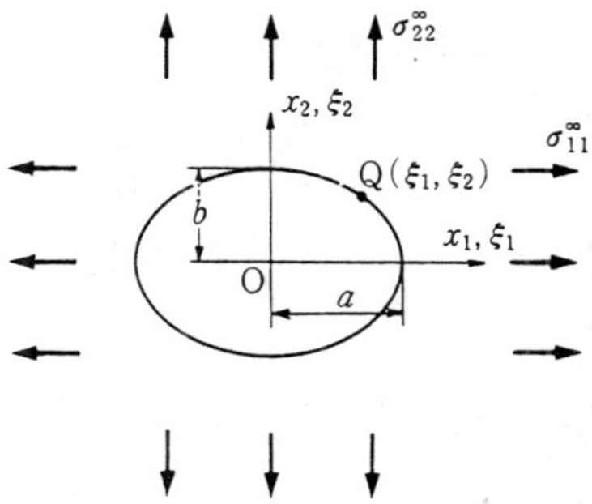

Figure 6: Tension of an infinite plate with an elliptic hole

eqn (24) more effective in getting the accurate solution by the numerical procedure.

Moreover, the availability of the idea of the basic density functions is related to the fact that in the problems of the stress concentration the solutions are not required to have the same accuracy for every point in the calculated region Ω_I. The stress concentration near the point of the calculation is required to be especially accurate. In order to calculate accurately the stresses near the point of stress concentration, the idea of the basic density functions of point forces in notch problems and the basic density functions of force doublets in crack problems was introduced into the body force method in the initial stage of the method.

4.1 Basic density functions of point forces in notch problems

In the notch problems the vicinity of the point where the stress becomes the maximum can generally be approximated effectively by a part of an elliptic hole (or an elliptic cavity). When the problems are solved by BFM, for example, by distributing point forces continuously along an imaginary boundary in an infinite plate, the calculation accuracy of the maximum stress is mainly dependent on the accuracy of the density of the distributed point forces near the point of the stress concentration. Therefore, based on the density of distributed point forces in the problem of an elliptic hole in a uniformly stressed infinite plate, the basic density functions of the distributed point forces in general notch problems can be defined. The density of distributed point forces in the simple problem of an infinite plate with an elliptic hole can be obtained in closed form in the following.

Tension of an infinite plate with an elliptic hole (Fig. 6). First, the plane stress conditions are assumed. In this case the displacements u_1 and u_2 at a point Q on the elliptic hole are expressed as follows:

$$\begin{aligned} u_1(Q) &= \left\{ \frac{\sigma_{11}^\infty}{E}(1 + 2\frac{b}{a}) - \frac{\sigma_{22}^\infty}{E} \right\} \xi_1 \\ u_2(Q) &= \left\{ -\frac{\sigma_{11}^\infty}{E} + \frac{\sigma_{22}^\infty}{E}(1 + 2\frac{a}{b}) \right\} \xi_2 \end{aligned} \quad (25)$$

In this problem the auxiliary region is an elliptic plate having the same size as the elliptic hole. If the displacements at a point Q on the rim of the elliptic plate coincide with those in eqn (25), the strain components $\varepsilon_{11}^0, \varepsilon_{22}^0$ and ε_{12}^0 at an arbitrary point in the plate are constant, and are given by

$$\begin{aligned} \varepsilon_{11}^0 &= \frac{\sigma_{11}^\infty}{E}(1+2\frac{b}{a}) - \frac{\sigma_{22}^\infty}{E} \\ \varepsilon_{22}^0 &= -\frac{\sigma_{11}^\infty}{E} + \frac{\sigma_{22}^\infty}{E}(1+2\frac{a}{b}) \\ \varepsilon_{12}^0 &= 0 \end{aligned} \tag{26}$$

From Hooke's law, the stress $\sigma_{11}^0, \sigma_{22}^0$ and σ_{12}^0 corresponding to the strains in eqn (26) can be expressed as

$$\begin{aligned} \sigma_{11}^0 &= \frac{1}{1-\nu^2}\left[\sigma_{11}^\infty\left\{(1+2\frac{b}{a}) - \nu\right\} - \sigma_{22}^\infty\left\{1-\nu(1+2\frac{a}{b})\right\}\right] \\ \sigma_{22}^0 &= \frac{1}{1-\nu^2}\left[-\sigma_{11}^\infty\left\{1-\nu(1+2\frac{b}{a})\right\} + \sigma_{22}^\infty\left\{(1+2\frac{a}{b}) - \nu\right\}\right] \\ \sigma_{12}^0 &= 0 \end{aligned} \tag{27}$$

So, the elliptic plate has to be subjected to such surface forces as follows,

$$\begin{aligned} t_1^n(Q) &= \frac{1}{1-\nu^2}\left[\sigma_{11}^\infty\left\{(1+2\frac{b}{a}) - \nu\right\} - \sigma_{22}^\infty\left\{1-\nu(1+2\frac{a}{b})\right\}\right] n_1'(Q) \\ t_2^n(Q) &= \frac{1}{1-\nu^2}\left[-\sigma_{11}^\infty\left\{1-\nu(1+2\frac{a}{b})\right\} + \sigma_{22}^\infty\left\{(1+2\frac{a}{b}) - \nu\right\}\right] n_2'(Q) \end{aligned} \tag{28}$$

where $n_1'(Q), n_2'(Q)$ are the components of the outward normal vector on a point Q of the rim of the elliptic plate.

The elliptic plate subjected to the surface forces given by eqn (28) can be inserted into the elliptic hole shown in Fig. 6 without any gap. Therefore, based on eqn (2), the density of the distributed point forces in this case can be obtained

$$\left.\begin{aligned} \phi_1(Q) &= \tfrac{\mathrm{d}F_1}{\mathrm{d}S} = \tfrac{1}{1-\nu^2}\left[\sigma_{11}^\infty\left\{(1+2\tfrac{b}{a}) - \nu\right\} - \sigma_{22}^\infty\left\{1-\nu(1+2\tfrac{a}{b})\right\}\right] n_1'(Q) \\ \phi_2(Q) &= \tfrac{\mathrm{d}F_2}{\mathrm{d}S} = \tfrac{1}{1-\nu^2}\left[-\sigma_{11}^\infty\left\{1-\nu(1+2\tfrac{b}{a})\right\} + \sigma_{22}^\infty\left\{(1+2\tfrac{a}{b}) - \nu\right\}\right] n_2'(Q) \end{aligned}\right\} \tag{29}$$

where S denotes the length of boundary curve measured from a point on the boundary, and $\mathrm{d}F_1$ and $\mathrm{d}F_2$ are respectively x_1- and x_2-components of the point forces acting on the infinitesimal element $\mathrm{d}S = (\mathrm{d}\xi_1^2 + \mathrm{d}\xi_2^2)^{1/2}$.

In the plane strain conditions, eqn (29) becomes

$$\left.\begin{aligned} \phi_1(Q) &= \tfrac{(1-\nu)^2}{(1-2\nu)}\left[\sigma_{11}^\infty\left\{(1+2\tfrac{b}{a}) - \tfrac{\nu}{1-\nu}\right\} - \sigma_{22}^\infty\left\{1-\tfrac{\nu}{1-\nu}(1+2\tfrac{a}{b})\right\}\right] n_1'(Q) \\ \phi_2(Q) &= \tfrac{(1-\nu)^2}{(1-2\nu)}\left[-\sigma_{11}^\infty\left\{1-\tfrac{\nu}{1-\nu}(1+2\tfrac{b}{a})\right\} + \sigma_{22}^\infty\left\{(1+2\tfrac{a}{b}) - \tfrac{\nu}{1-\nu}\right\}\right] n_2'(Q) \end{aligned}\right\} \tag{30}$$

Shear of an infinite plate with an elliptic hole. In this case the density of point forces, which should be distributed along the imaginary ellipse in an infinite plate, are obtained in a similar manner as follows.

In the plane stress conditions:

$$\left.\begin{aligned}\phi_1(Q) &= \frac{\sigma_{12}^{\infty}}{(1+\nu)}\frac{(a+b)^2}{ab}n_2'(Q)\\ \phi_2(Q) &= \frac{\sigma_{12}^{\infty}}{(1+\nu)}\frac{(a+b)^2}{ab}n_1'(Q)\end{aligned}\right\} \tag{31}$$

In the plane strain conditions:

$$\left.\begin{aligned}\phi_1(Q) &= (1-\nu)\sigma_{12}^{\infty}\frac{(a+b)^2}{ab}n_2'(Q)\\ \phi_2(Q) &= (1-\nu)\sigma_{12}^{\infty}\frac{(a+b)^2}{ab}n_1'(Q)\end{aligned}\right\} \tag{32}$$

Basic density functions in notch problems. It is seen from eqns (29) to (32) that in solving the problems of an elliptic hole in a uniformly stressed infinite plate if we assume the expression for the density of point forces as following forms

for tension

$$\left.\begin{aligned}\phi_1(Q) &= W_1(Q)\Phi_1^I(Q)\ ;\ \ \Phi_1^I(Q) = n_1'(Q)\\ \phi_2(Q) &= W_2(Q)\Phi_{21}^I(Q)\ ;\ \ \Phi_2^I(Q) = n_2'(Q)\end{aligned}\right\} \tag{33}$$

for shear

$$\left.\begin{aligned}\phi_1(Q) &= W_1(Q)\Phi_1^{II}(Q)\ ;\ \ \Phi_1^{II}(Q) = n_2'(Q)\\ \phi_2(Q) &= W_2(Q)\Phi_{21}^{II}(Q)\ ;\ \ \Phi_2^{II}(Q) = n_1'(Q)\end{aligned}\right\} \tag{34}$$

then the weight functions $W(Q)$ become constant for every points Q on the imaginary boundary. It means that the accurate solution can be obtained by the numerical calculations, even when the unknown weight functions are simply approximated by such finite discrete values as stepped functions. This idea can be extended to the general notch problems. That is, in solving a general notch problem, if eqn (33) and (34) are used for expressing the densities of distributing point forces, although the weight functions $W(Q)$ are not constant, the variation in weight functions near the point of the stress concentration is then comparatively small, and thus the nearly accurate solutions can be easily obtained.

As an example, the stress concentration factors of semi-infinite plates with a circular notch under tension (Fig. 7) were calculated by using the idea of the basic density functions such as shown in eqn (33). We divide the circular arc into n equal elements and assume the weight functions $W_1(Q), W_2(Q)$ to be constant within the elements, for example $W_1^k, W_2^k; k = 1, 2, ..., n$. When n=48, the obtained stress concentration factors K_t are 3.0612 for the case of a semi-circular notch and 2.4265 for the case of a 120° circular arc notch, which are in good agreement with the reference values, 3.0653 and 2.4240 for the cases of the semi-circular notch and the 120° circular arc notch, respectively, obtained by Isida using the Laurent expansion.[19]

Table 1 shows the values of weight function in the intervals in the case of a semi-circular notch. We find from Table 1 that W_1^k and W_2^k are almost constant near the root of the notch (when $k \simeq 24$). It is this small change in $W_1(Q)$ and $W_2(Q)$ that insures the proposed method to have the high accuracy.

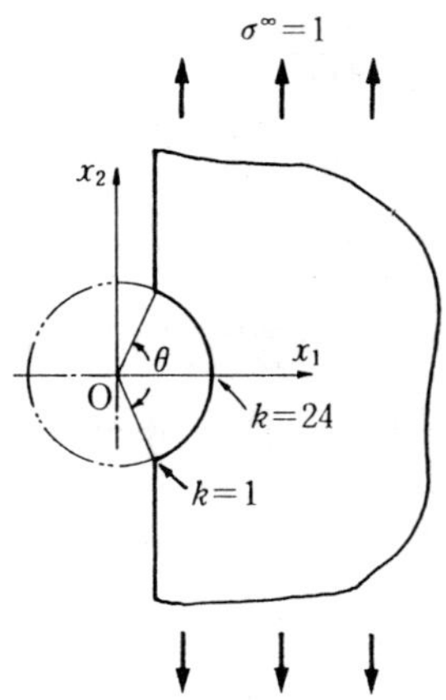

Figure 7: A semi-infinite plate with a circular notch

Table 1: Values of weight function in each intervals, W_1^k, W_2^k

	k	W_2^k	W_1^k	k	W_2^k	W_1^k
Semi-circular notch ($n = 48$)	1	3.06182	1.84118	13	3.26617	-0.01584
	2	3.04641	1.25493	14	3.27610	-0.04670
	3	3.05974	0.98011	15	3.28525	-0.07272
	4	3.08186	0.76507	16	3.29285	-0.09464
	5	3.11279	0.59342	17	3.29934	-0.11301
	6	3.13999	0.45562	18	3.30181	-0.12828
	7	3.16509	0.34411	19	3.30936	-0.14080
	8	3.18768	0.25308	20	3.31304	-0.15086
	9	3.20772	0.17817	21	3.31593	-0.15867
	10	3.22538	0.11604	22	3.31806	-0.16440
	11	3.24086	0.06421	23	3.31946	-0.16815
	12	3.25439	0.02074	24	3.32015	-0.17001

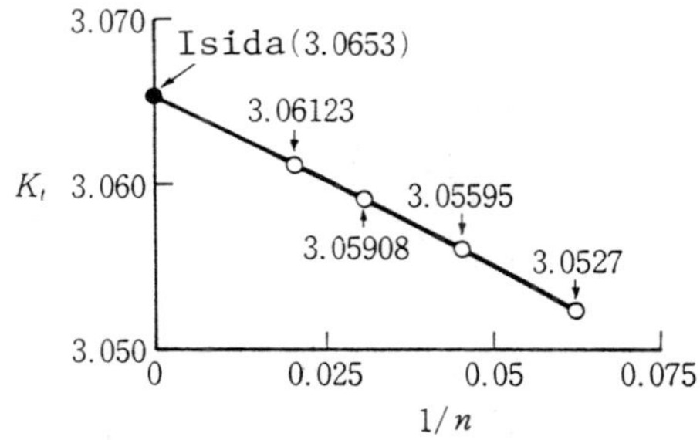

Figure 8: Semi-circular notch

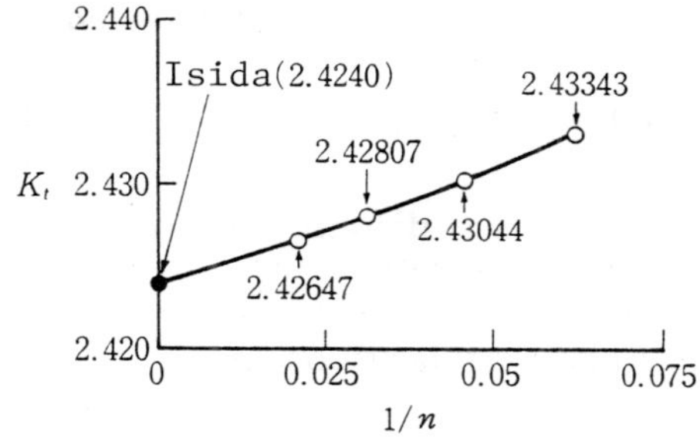

Figure 9: 120° circular arc notch

Moreover, if more accurate solutions are needed, we can simply use the extrapolation procedure, because the numerical results for the systematic increase in the number of discretization n vary systematically in correspondence. For example, the plots of K_t againt $1/n$ are shown in Figs 8 and 9. The discretization errors are nearly proportional to $1/n$. Therefore, based on this characteristic, the value of K_t for the limiting case of $n = \infty$ can be extrapolated. The values of K_t obtained by extrapolation using the values of K_t for the cases of $n = 32$ and 48 are, respectively, 3.0654 and 2.4233 for the cases of the semi-circular notch and the 120° circular arc notch. The errors are 0.003% and 0.03%, respectively.

For three-dimensional problems, for example, when the problem of an infinite body stressed by a uniform stress σ_{33}^{∞} and having an ellipsoidal cavity (which is expressed by the equation $x_1^2/a^2 + x_2^2/b^2 + x_3^2/c^2 = 1$) is analyzed by the present method, the densities of distributed point forces can also be written as the products of the components of the normal vector on the imaginary surface, $n_1'(Q), n_2'(Q), n_3'(Q)$, and the values which are constant on the imaginary surface, A_1, A_2, A_3.

$$\left.\begin{aligned} \phi_1(Q) &= A_1(a,b,c,\nu,\sigma_{33}^{\infty})n_1'(Q) \\ \phi_2(Q) &= A_2(a,b,c,\nu,\sigma_{33}^{\infty})n_2'(Q) \\ \phi_3(Q) &= A_3(a,b,c,\nu,\sigma_{33}^{\infty})n_3'(Q) \end{aligned}\right\} \tag{35}$$

Based on eqn (35), the basic density functions for the three-dimensional problems can be introduced in a similar manner as in the case of the two-dimensional problems.

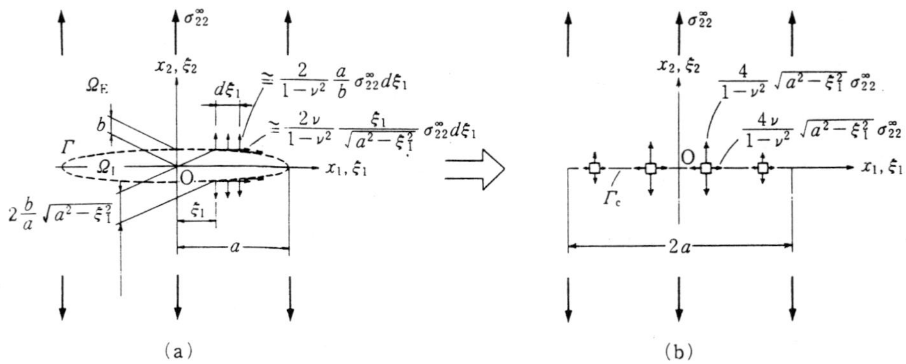

Figure 10: A crack is the limiting shape of a slender elliptic hole

The details about it are found in Refs 5 and 20.

4.2 Basic density functions in two-dimensional crack problems

In crack problems the idea of the basic density functions is especially important because in such problems the stress singularity occurs at the crack tip.

First, consider an infinite plate with a crack subjected to tension. A crack can be considered as an extremely slender elliptic hole ($b/a \to 0$), and the densities of the point forces for the crack can be obtained as a limiting expression of eqn (29) by putting $b/a \to 0$. However, as $b/a \to 0$, the points at which point forces are applied become closer and closer to each other and both $\phi_1(Q)$ and $\phi_2(Q)$ become unbounded. Therefore, the direct introduction of the densities of the distributed point forces to crack problems is inadequate. In order to resolve this difficulty, the point forces acting on the upper and lower parts of the imaginary ellipse are treated as a pair.

For the elliptic hole problem shown in Fig. 10(a) the densities of the point forces are

$$\left.\begin{aligned}\phi_1(Q) &= \tfrac{-\sigma_{22}^{\infty}}{1-\nu^2}\left[1-\nu(1+2\tfrac{a}{b})\right]n_1'(Q) = W_1 n_1'(Q)\\ \phi_2(Q) &= \tfrac{\sigma_{22}^{\infty}}{1-\nu^2}\left[(1+2\tfrac{a}{b})-\nu\right]n_2'(Q) = W_2 n_2'(Q)\end{aligned}\right\} \tag{36}$$

The stress due to the forces distributed along the imaginary boundary Γ with the densities of eqn (36) is given by

$$\sigma_{ij}(P) = \int_{\Gamma}\left[\sigma_{ij}(P,Q_1)W_1 n_1'(Q) + \sigma_{ij}(P,Q_2)W_2 n_2'(Q)\right]\mathrm{d}\Gamma(Q) \qquad (P \in \Omega_E) \tag{37}$$

Based on the divergence theorem, eqn (37) is rewritten as

$$\sigma_{ij}(P) = \int_{\Omega_1}\left\{\frac{\partial}{\partial\xi_1}\left[\sigma_{ij}(P,Q_1)W_1\right] + \frac{\partial}{\partial\xi_2}\left[\sigma_{ij}(P,Q_2)W_2\right]\right\}\mathrm{d}\Omega(Q) \qquad (P \in \Omega_E) \tag{38}$$

As $b/a \to 0$, eqn (38) becomes

$$\lim_{b/a\to 0}\sigma_{ij}(P) = \int_{-a}^{a}\left[\sigma_{ij}(P,Q_{1,1})W_1 + \sigma_{ij}(P,Q_{2,2})W_2\right]2|\xi_2|\mathrm{d}\xi_1 \qquad (P \in \Omega_E) \tag{39}$$

Therefore, by considering

$$\left.\begin{aligned}\lim_{b/a\to 0}[2W_1|\xi_2|] &= \frac{4\nu\sigma_{22}^{\infty}}{1-\nu^2}\sqrt{a^2-\xi_1^2}\\ \lim_{b/a\to 0}[2W_2|\xi_2|] &= \frac{4\nu\sigma_{22}^{\infty}}{1-\nu^2}\sqrt{a^2-\xi_1^2}\end{aligned}\right\} \tag{40}$$

the following equation is obtained

$$\lim_{b/a\to 0}\sigma_{ij}(P) = \int_{-a}^{+a}[\sigma_{ij}(P,Q_{2,2})+\nu\sigma_{ij}(P,Q_{1,1})]\frac{4\sigma_{22}^{\infty}}{1-\nu^2}\sqrt{a^2-\xi_1^2}\mathrm{d}\xi_1 \quad (P\in\Omega_E) \tag{41}$$

where $\sigma_{ij}(P,Q_{2,2})$ and $\sigma_{ij}(P,Q_{1,1})$ are the stresses at a point P due to the x_2x_2-directional and x_1x_1-directional unit force doublets acting at a point Q, respectively.

Therefore, for the crack problems the force doublets, instead of point forces, are distributed along an imaginary crack line in an infinite plate without any crack. For the mode I crack in an infinite plate, the densities of the force doublets are obtained from eqn (41).

$$\left.\begin{aligned}T_{22}(\xi_1) &= \tfrac{4\sigma_{22}^{\infty}}{1-\nu^2}\sqrt{a^2-\xi_1^2}\\ T_{11}(\xi_1) &= \nu\tfrac{4\sigma_{22}^{\infty}}{1-\nu^2}\sqrt{a^2-\xi_1^2}\end{aligned}\right\} \tag{42}$$

In a similar manner, for the mode II crack in an infinite plate, the densities of the force doublets can be deduced from eqn (31).

$$\left.\begin{aligned}T_{12}(\xi_1) &= \tfrac{2\sigma_{12}^{\infty}}{1-\nu}\sqrt{a^2-\xi_1^2}\\ T_{21}(\xi_1) &= \tfrac{2\sigma_{12}^{\infty}}{1-\nu}\sqrt{a^2-\xi_1^2}\end{aligned}\right\} \tag{43}$$

In plane strain conditions the densities of the force doublets will be given by the following, eqns (44) and (45) respectively.

$$\left.\begin{aligned}T_{22}(\xi_1) &= \tfrac{4(1-\nu)^2\sigma_{22}^{\infty}}{1-2\nu}\sqrt{a^2-\xi_1^2}\\ T_{11}(\xi_1) &= \tfrac{4\nu(1-\nu)\sigma_{22}^{\infty}}{1-2\nu}\sqrt{a^2-\xi_1^2}\end{aligned}\right\} \tag{44}$$

$$\left.\begin{aligned}T_{12}(\xi_1) &= 2(1-\nu)\sigma_{12}^{\infty}\sqrt{a^2-\xi_1^2}\\ T_{21}(\xi_1) &= 2(1-\nu)\sigma_{12}^{\infty}\sqrt{a^2-\xi_1^2}\end{aligned}\right\} \tag{45}$$

It is seen from eqns (42) to (45), first, that the densities of the force doublets of a crack problem are written in the form (constant) $\times\sqrt{a^2-\xi_1^2}$. Therefore, for the two-dimensional crack problems we define the function

$$\Phi(\xi_1) = \sqrt{a^2-\xi_1^2} \tag{46}$$

as the basic density function of force doublets. Second, it is seen from eqns (42) to (45) that for the mode I crack problem the densities of the force doublets of the

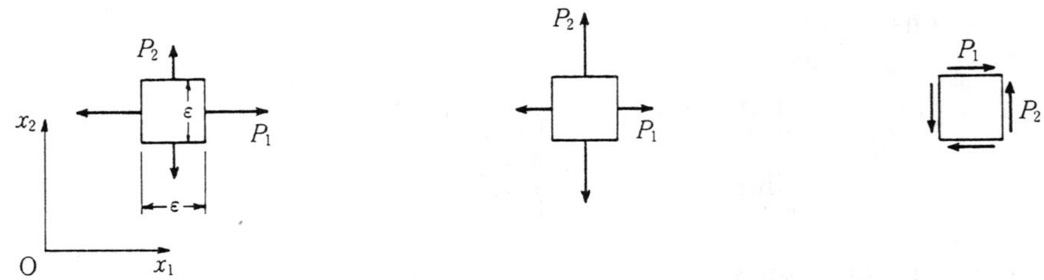

$$P_2 = \begin{cases} \nu P_1 & \text{(plane stress)} \\ \frac{\nu}{1-\nu} P_1 & \text{(plane strain)} \end{cases} \qquad P_1 = \begin{cases} \nu P_2 & \text{(plane stress)} \\ \frac{\nu}{1-\nu} P_2 & \text{(plane strain)} \end{cases} \qquad P_2 = P_1$$

(a) x_1x_1-direction $\quad$ x_2x_2-direction $\quad$ (b) x_1x_2-direction

Figure 11: Standard set of force doublets (two-dimensional case): (a) standard tensile sets of point force doublets; (b) standard shearing set of point force doublets

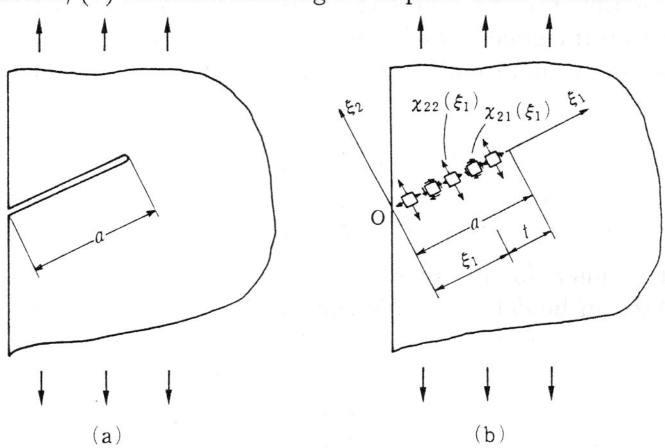

Figure 12: An edge crack in a semi-infinite plate

x_1x_1-direction (x_1: the direction parallel to crack line) always equal the ν times (plane stress) or $\nu\ /(1-\nu)$ times (plane strain) of those of the x_2x_2-direction (x_2: the direction perpendicular to crack line) and for the mode II crack problem densities of the force doublets of the x_1x_2-direction always equal those of the x_2x_1-direction. This is related with the fact that the minute elements at which the force doublets are applied do not shrink along the x_1 direction and the condition of resultant moment $= 0$ is required to be satisfied in the elements. In numerical analyses it is reasonable to represent the densities of two kinds of the force doublets (for example, those of x_2x_2-direction and of x_1x_1-direction) by one parameter (for example, by only the density of force doublets of x_2x_2-direction). Therefore, we introduce a new definition of a standard set of force doublets. 'The standard set of force doublets having unit strength' means some combination of two kinds of force doublets as shown in Fig. 11.

Consequently, based on the above discussions, in analyzing a general two-dimensional crack problem, for example, in analyzing the problem of an edge crack in a semi-infinite plate as shown in Fig. 12(a), the standard tensil and shearing sets of force doublets are distributed along the imaginary crack line as shown in Fig. 12(b). The densities of tensile sets and shearing sets are

$$\left.\begin{aligned} \chi_{22} &= W_2(\xi_1)\Phi(\xi_1) \\ \chi_{12} &= W_1(\xi_1)\Phi(\xi_1) \end{aligned}\right\} \tag{47}$$

In eqn (47), $W_2(\xi_1), W_1(\xi_1)$ are the weight functions, which are determined by numerical calculation. The change in these weight functions along an imaginary crack line is comparatively small, which improves the accuracy of the solution. In addition, it is important that when the basic density function of the force doublets such as eqn (46) is used, the stress intensity factors can be calculated simply by the values of the weight functions, $W_1(a)$ and $W_2(a)$ alone at the crack tip.

$$\left.\begin{aligned} K_I &= \begin{cases} W_2(a)\frac{(1-\nu^2)}{4}\sqrt{\pi a} & \text{(plane stress)} \\ W_2(a)\frac{(1-2\nu)}{4(1-\nu)^2}\sqrt{\pi a} & \text{(plane strain)} \end{cases} \\ K_{II} &= \begin{cases} W_1(a)\frac{(1+\nu)}{4}\sqrt{\pi a} & \text{(plane stress)} \\ W_1(a)\frac{1}{2(1-\nu)}\sqrt{\pi a} & \text{(plane strain)} \end{cases} \end{aligned}\right\} \tag{48}$$

The singularity at the crack tip is simulated by the basic density function. Actually, the simulation of singularity at the crack tip is dependent on the characteristics of the basic density function near the crack tip. Therefore, instead of eqn (46), the following function can also be used as the basic density function of the force doublets because it has the same characteristic as the function of eqn (46) at the crack tip

$$\Phi(\xi_1) = \sqrt{t} \tag{49}$$

where t is the distance from the point to the crack tip (Fig. 12(b)).

When eqn (49) is used the stress intensity factors are calculated as

$$\left.\begin{aligned} K_I &= \begin{cases} W_2(a)\frac{(1-\nu^2)}{4\sqrt{2}}\sqrt{\pi} & \text{(plane stress)} \\ W_2(a)\frac{(1-2\nu)}{4\sqrt{2}(1-\nu)^2}\sqrt{\pi} & \text{(plane strain)} \end{cases} \\ K_{II} &= \begin{cases} W_1(a)\frac{(1+\nu)}{2\sqrt{2}}\sqrt{\pi} & \text{(plane stress)} \\ W_1(a)\frac{1}{2\sqrt{2}(1-\nu)}\sqrt{\pi} & \text{(plane strain)} \end{cases} \end{aligned}\right\} \tag{50}$$

4.3 Basic density functions in three-dimensional crack problems

Figure 13 shows an elliptic crack in a uniformly stressed infinite body. When the problem is analyzed by BFM, the standard sets of force doublets are continuously distributed on the imaginary crack plane in an infinite body. The standard sets of force doublets are defined as the combination of two or three kinds of force doublets as shown in Fig. 14.

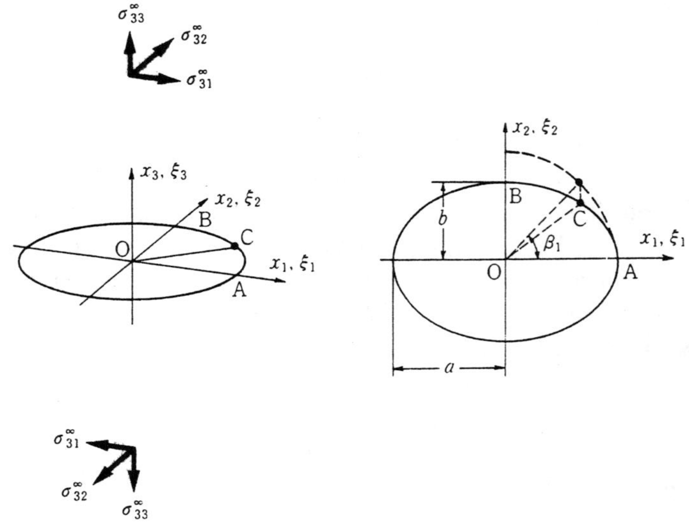

Figure 13: An elliptic crack in an infinite body subjected to a uniform tension and shear

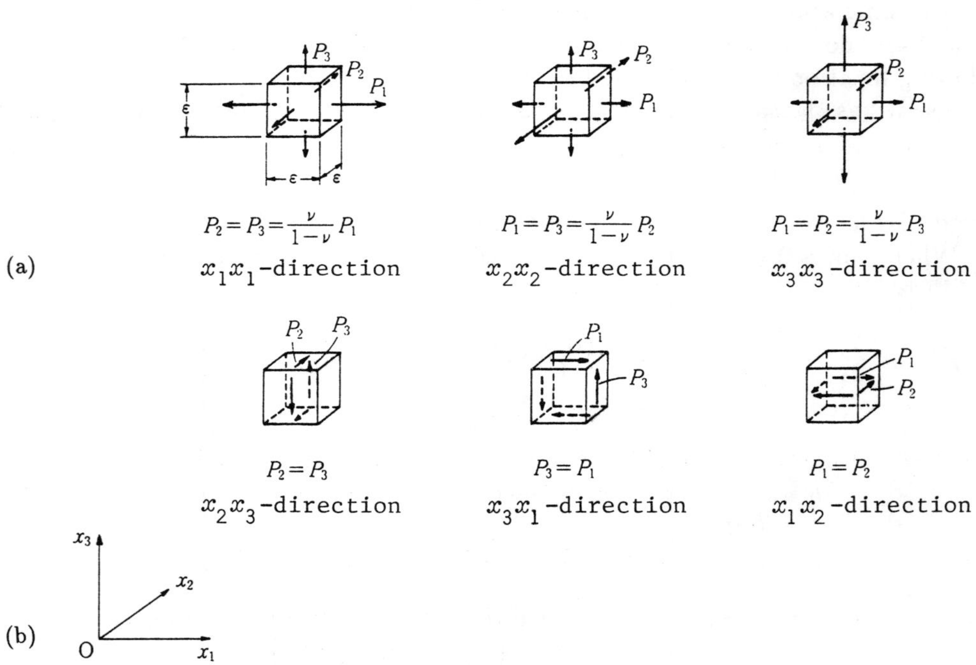

Figure 14: Standard sets of force doublets (three-dimensional case): (a) standard tensile sets of force doublets; (b) Standard shearing sets of force doublets

For the problem of Fig. 13 the densities of standard sets of force doublets can be written in closed form;
when subjected to σ_{33}^{∞}

$$\chi_{33}(\xi_1,\xi_2) = \frac{4(1-\nu)^2 b\sigma_{33}^{\infty}}{(1-2\nu)E(k)}\left(1-\frac{\xi_1^2}{a^2}-\frac{\xi_2^2}{b^2}\right)^{1/2} \tag{51}$$

when subjected to σ_{31}^{∞}

$$\chi_{13}(\xi_1,\xi_2) = \frac{2(1-\nu)bk^2\sigma_{31}^{\infty}}{(k^2-\nu)E(k)+\nu k'^2 K(t)}(1-\frac{\xi_1^2}{a^2}-\frac{\xi_2^2}{b^2})^{1/2} \tag{52}$$

when subjected to σ_{32}^{∞}

$$\chi_{23}(\xi_1,\xi_2) = \frac{2(1-\nu)bk^2\sigma_{32}^{\infty}}{(k^2+\nu k'^2)E(k)-\nu k'^2 K(t)}(1-\frac{\xi_1^2}{a^2}-\frac{\xi_2^2}{b^2})^{1/2} \tag{53}$$

where $k' = b/a =< 1$; $k_2 = 1 - b^2/a^2$

and $K(k)$ and $E(k)$ are the complete elliptic integrals of the first and second kinds, respectively.

$$\left.\begin{aligned} K(k) &= \int_0^{\frac{\pi}{2}} \frac{d\lambda}{(1-k^2\sin^2\lambda)^{1/2}} \\ E(k) &= \int_0^{\frac{\pi}{2}} (1-k^2\sin^2\lambda)^{1/2} d\lambda \end{aligned}\right\}$$

It is seen from eqns (51) and (53) that in all the cases the densities of force doublets are expressed in the following form.

$$\text{density} = \text{constant} \times (1-\frac{\xi_1^2}{a^2}-\frac{\xi_2^2}{b^2})^{1/2} \tag{54}$$

Accordingly, if the following function

$$\Phi(\xi_1,\xi_2) = (1-\frac{\xi_1^2}{a^2}-\frac{\xi_2^2}{b^2})^{1/2} \tag{55}$$

is taken as the basic density function of the force doublets, then for the general three-dimensional crack problems, the densities of the standard sets of force doublets can be expressed in the form of (weight functions) × (basic density function) as follows.

$$\left.\begin{aligned} \chi_{33}(\xi_1,\xi_2) &= W_3(\xi_1,\xi_2)\Phi(\xi_1,\xi_2) \\ \chi_{13}(\xi_1,\xi_2) &= W_1(\xi_1,\xi_2)\Phi(\xi_1,\xi_2) \\ \chi_{23}(\xi_1,\xi_2) &= W_2(\xi_1,\xi_2)\Phi(\xi_1,\xi_2) \end{aligned}\right\} \tag{56}$$

where $W_3(\xi_1,\xi_2), W_1(\xi_1,\xi_2), W_2(\xi_1,\xi_2)$ are the weight function.

The stress intensity factors at the crack front are calculated from the values of

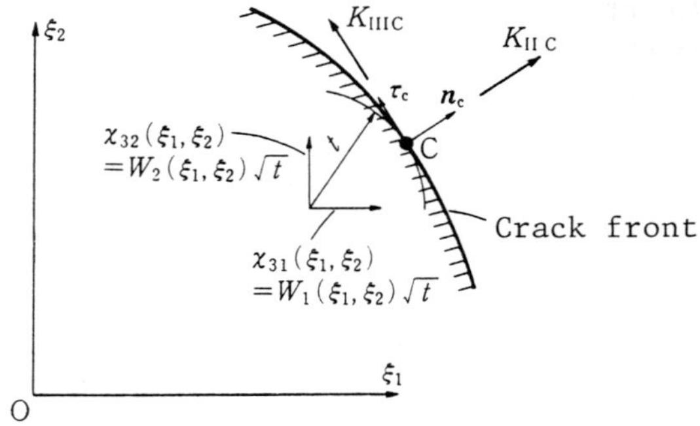

Figure 15: Crack front of arbitrary shape

weight functions at the point alone.

$$\left.\begin{aligned} K_{IC} &= \tfrac{1-2\nu}{4(1-\nu)^2}\sqrt{\pi/ab}(a^2\sin^2\beta_1 + b^2\cos^2\beta_1)^{1/4}W_3(C) \\ K_{IIC} &= \tfrac{1}{2(1-\nu)(a^2\sin^2\beta_1+b^2\cos^2\beta_1)^{1/4}}\left[\sqrt{\pi a/b}\sin\beta_1 W_2(C) + \sqrt{\pi b/a}\cos\beta_1 W_1(C)\right] \\ K_{IIIC} &= \tfrac{1}{2(a^2\sin^2\beta_1+b^2\cos^2\beta_1)^{1/4}}\left[\sqrt{\pi b/a}\cos\beta_1 W_2(C) - \sqrt{\pi a/b}\sin\beta_1 W_1(C)\right] \end{aligned}\right\} \quad (57)$$

The basic density function can also be defined as

$$\Phi(\xi_1,\xi_2) = \sqrt{t} \quad (58)$$

where t is the minimum distance to the crack front from a point as shown in Fig. 15.

The basic density function of eqn (58) can be used for the crack of arbitrary shape. The stress intensity factors are given by the following equations.

$$\left.\begin{aligned} K_{IC} &= \tfrac{(1-2\nu)}{4(1-\nu)^2}\sqrt{\pi/2}W_3(C) \\ K_{IIC} &= \tfrac{1}{(1-\nu)}\sqrt{\pi/2}\,[\cos(\mathbf{n}_C,\xi_1)W_1(C)] \\ &\quad + [\cos(\tau_C,\xi_1)W_1(C) + \cos(\tau_C,\xi_2)W_2(C)] \end{aligned}\right\} \quad (59)$$

5 Other techniques of high accuracy and efficiency

Sine the basic conception of BFM is simple and intuitive, the various techniques for improving the solution accuracy are easy to understand. Besides the idea of the basic density functions described in the previous section there are some other techniques, which will be summarized in the present section.

5.1 Applying appropriate forces to auxiliary region

The reason why the accurate solution can be obtained by using the idea of basic density functions is because in this case the variation in the unknown function (weight function) along an imaginary boundary become comparatively small. As mentioned in Section 2, the loading conditions in the auxiliary region can be arbitrarily specified,

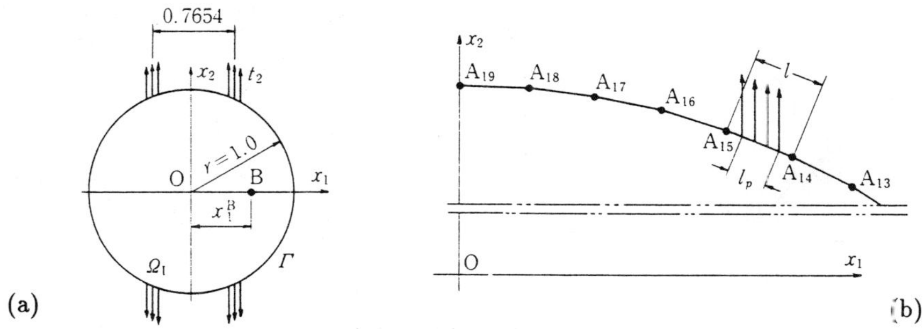

Figure 16: A circular plate subjected to tensile forces at four places: (a) analyzed problem; (b) subdivision of boundary

Table 2: Stresses σ_{22} at point B when $q = 0.001$

x_1^B/r	0.0	0.2	0.4	0.6	0.8
Method B	1.413	1.384	1.245	0.945	0.599
Method A	1.612	1.582	1.400	0.992	0.491
Accurate values	1.420	1.394	1.256	0.953	0.560

and according to the loading state in the auxiliary region the distribution of the point forces or the discrepancies along the imaginary boundary may be changed. Therefore, by specifying appropriately the loading conditions in an auxiliary region the changes in the density function may become comparatively small, and thus the solution accuracy may be improved.

Figure 16 shows a circular plate subjected to four distributed forces at four places on the circumference. We define the ratio of the part, where the tractions are applied, to the length of an element, as $q = l_p/l$. Table 2 shows the numerical results of the stress σ_{22} at the point B when $q = 0.001$.

The numerical results are obtained by two kinds of methods: method A and method B. By method A, eqn (13) (see Section 2) is used. That is, the unknowns are the densities of discrepancies, which correspond to the gaps between the region in question and the auxiliary region. The auxiliary region is not subjected to any forces. Therefore, as shown in Fig. 17(a), the change in the gap is severe, for the region in question is subjected to discontinuous tractions(q =0.001).

By method B, the following equation is used.

$$\sigma_{ij}(P) = \int_\Gamma \sigma^*_{ij}(P, Q^n_\eta)\psi_\eta(Q)\mathrm{d}\Gamma(Q) + 2\int_\Gamma \sigma_{ij}(P, Q_\eta)t^{n_I}_\eta(Q)\mathrm{d}\Gamma \quad (P \in \Omega_I) \qquad (60)$$

Equation (60) is obtained from the general equation (eqn (5) in Section 2) by putting $t^{n_E}_\eta(Q) = t^{n_I}_\eta(Q)$; $Q \in \Gamma$. That is, in method B the unknowns are the densities of discrepancies as in the case of method A. However, in this case the auxiliary region is subjected to the same surface force as in the region in question. So, the

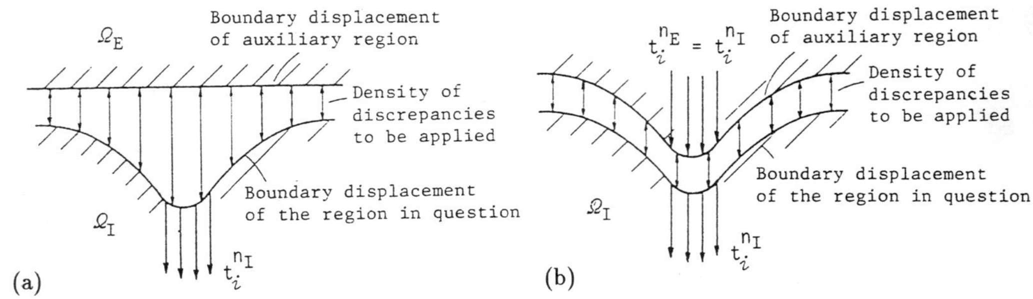

Figure 17: Gaps between the region in question and the auxiliary region

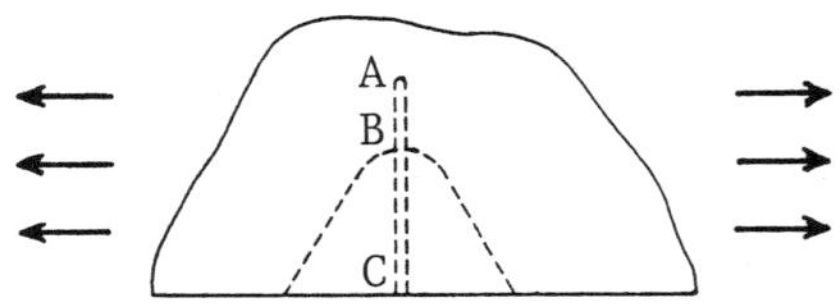

Figure 18: Semi-infinite plate having a notch with a crack emanating from its apex

change in the gap becomes very small as shown in Fig. 17(b), and thus the solution accuracy is high as shown in Table 2.

In the above example we improved the unknown function involved in the boundary condition equations by applying an appropriate traction to the surface of the auxiliary region. In some cases, a similar effect is attained by applying the appropriate forces in an internal portion of the auxiliary region. As an example, Fig. 18 shows the problem of a semi-infinite plate having a notch with a crack emanating from its apex. In solving this problem, the force doublets are applied along the part BC (i.e. in the internal portion of the auxiliary region) in addition to the portion AB which should become a crack, because it makes the shear stress small at B and consequently the boundary conditions can be satisfied easily.

5.2 Boundary conditions in terms of resultant forces

In numerical analysis, the unknowns, the densities of distributed point forces or discrepancies, are usually determined from the boundary conditions at a finite number of points; e.g. at the mid-points of each element. However, it is shown that in the stress boundary-value problems, if the stress conditions at the mid-points of each element are replaced by the resultant force conditions within each element, the accuracy of the numerical results would be greatly improved according to Saint Venant's principle.

Figure 19 shows the case of tension of an infinite plate with two cracks. Table 3 shows the numerical results obtained by using the stress boundary conditions and the results by the resultant force boundary conditions.[21] M is the number of subdivisions of a crack. It is seen from Table 3 that the resultant force conditions give the results of excellent accuracy in the whole range of the parameter including $a/b = 1$ (the case of extremely close cracks).

In two-dimensional problems, the resultant forces are calculated directly from the

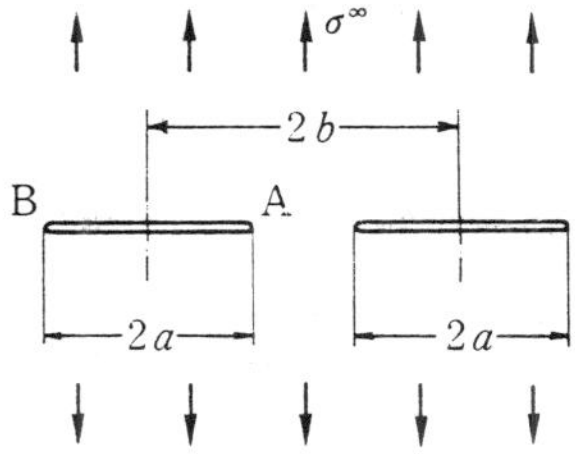

Figure 19: Tension of an infinite plate with two equal cracks

Table 3: F_I of cracks in Fig. 19 $F_I = \frac{K_I}{\sigma_\infty\sqrt{\pi a}}$

	$a/b = 0.8$				$a/b = 1.0$	
	F_{IA}		F_{IB}		F_{IB}	
M	Resultant force condition	Stress condition	Resultant force condition	Stress condition	Resultant force condition	Stress condition
5	1.22417	1.1983	1.08078	1.0854	1.40773	1.2182
10	1.22768	1.2150	1.08101	1.0831	1.41288	1.2368
15	1.22836	1.2204	1.08104	1.0823	1.41366	1.2465
20	1.22861	1.2229	1.08106	1.0820	1.41391	1.2723
25	1.22873	1.2244	1.08106	1.0817	1.41402	1.3110
∞	1.22894		1.08107		$\sqrt{2} = 1.41421$	

complex potentials $\phi(z)$ and $\psi(z)$. That is, in order to obtain the values of resultant forces the special integral is not necessary. So this technique was applied to the various two-dimensional problems of notches or cracks.

In three-dimensional problems, however, this techniques has some difficulty with regard to calculating time because the double integrals are necessary in order to obtain the resultant force (this needs integrating once) due to the continuously distributed point forces or discrepancies (this needs integrating once more).

In order to avoid the above difficulty a new technique was developed recently.[22] According to the new technique, the point forces to be continuously distributed on each element of the imaginary boundary are approximated by a single point force acting at the outside of the imaginary boundary (as shown in Fig. 20) and the resultant force boundary conditions are used. The resultant force of each element due to a point force in this case can be obtained by a single integral. Therefore, the solution with comparatively high accuracy can be obtained within a short calculation time. Details about this technique are found in Ref. 22.

5.3 Using special fundamental solutions

So far, in the equations of BFM we have used the fundamental solutions (the elastic fields due to a point force or a concentrated discrepancy) for an infinite plate or for an

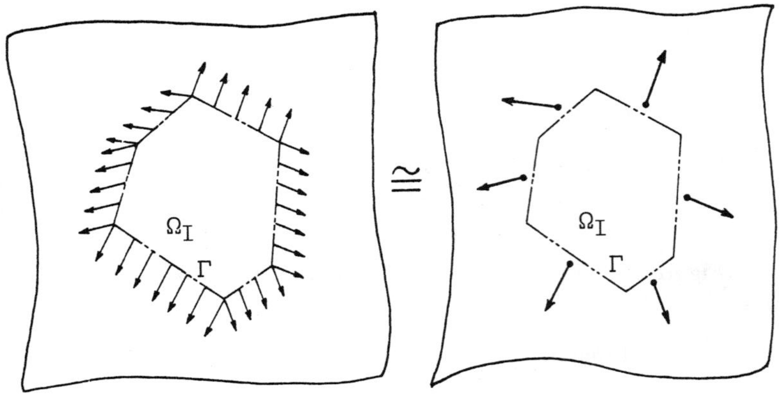

Figure 20: Distributed forces are approximated by discrete forces

infinite body. However, it is obvious that there is no trouble when the fundamental solutions other than those for an infinite plate or for an infinite body are used, because BFM is based on the principle of superposition.

Although all problems can be solved in principle by using the fundamental solutions for an infinite plate (or body) based on the present method, the use of other special fundamental solutions sometimes makes the method much more effective. Generally, the special fundamental solutions are used for the two following objectives.

(1) The boundary conditions in a part of the boundary are to be satisfied *a priori* without distributing point forces or discrepancies over the part.

For example, for a semi-infinite plate (or body), or for an infinite plate containing a traction-free crack or hole, the fundamental solutions can be written in a closed form. So, in solving the problems of a semi-infinite plate (or body) or in solving the problems of a crack or an elliptic hole, if the above special fundamental solutions are used, then on the edges of the correspondent part of the domain the traction-free boundary conditions are accurately satisfied automatically, and thus the accuracy of numerical solutions is improved.

(2) The dimensions of unknowns can be decreased if the given problem is symmetrical.

For example, in the problems of an axisymmetric body under torsion or tension, not only the shape of the body but also the stress distribution is symmetric about the axis. Therefore, in these problems, the stress field due to the forces acting on a ring in an infinite body, as shown in Fig. 21, can be used as the fundamental functions. So, both the unknown densities of the point forces and the boundary conditions are independent of the parameter θ. It means that the three-dimensional problems are reduced to the two-dimensional problems as shown in Fig. 22.

In the problem of an axisymmetric body under bending, the analysis is more difficult than torsion and tension problems. However, if the appropriate fundamental solutions (see Fig. 32) are used, the calculation procedure becomes almost similar to that in the cases of torsion and tension.

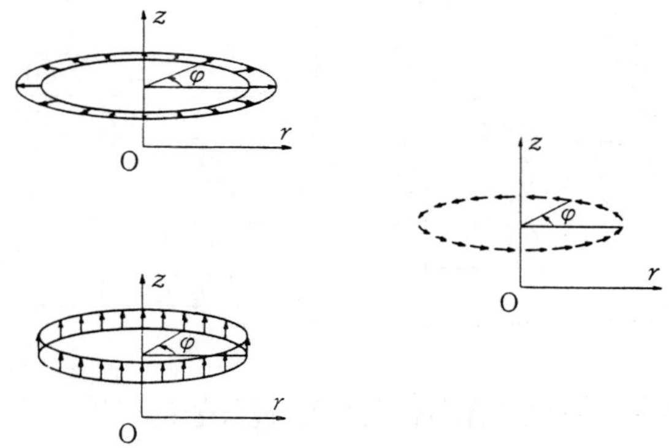

Figure 21: Fundamental solutions for axisymmetric problems

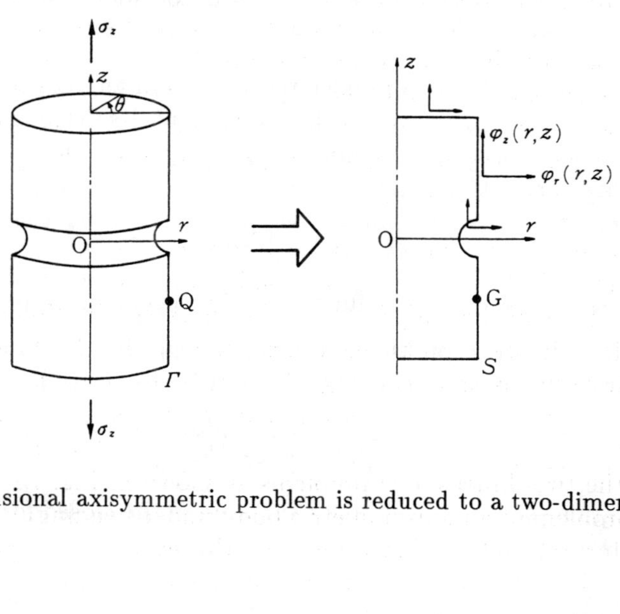

Figure 22: Three-dimensional axisymmetric problem is reduced to a two-dimensional problem

Table 4: Numerical results of Fig. 24

Number of division N	Number of unknowns	σ_A/σ_0	Error (%)
2	12	6.2604	2.008
4	20	6.3716	0.268
8	36	6.3869	0.028
16	68	6.3885	0.003
20	84	6.3887	0.000
∞[24]		6.3887	-

6 Some examples of numerical results

BFM has furnished a lot of useful and highly accurate solutions in practice since its development in 1967. In the present section some examples of numerical results will be shown.

Example 1. Versatile program of analysis of two-dimensional elastic problem

BFM was applied to various two-dimensional notch or crack problems in the early stages. Recently, a versatile program for two-dimensional problems was presented. The program gives the solutions with excellent accuracy and a short calculation time, because the following points were taken into consideration for the versatility and analytical accuracy.

1. The elastic fields due to a point force applied in an infinite plate with a traction-free circular hole are used as the fundamental solutions. By using these fundamental solutions, many kinds of stress concentration problems can be solved with a high accuracy, because in many important stress concentration problems the notch root is usually able to be approximated by a circular arc. See Section 5.3 .
2. For cracks the basic density functions such as eqn (46) are used. See Section 4.2.
3. When the tractions on the boundary are given as a boundary condition, the boundary condition is expressed in terms of resultant forces. See Section 5.2.
4. Near the corners, the point forces are also applied on the extended portion of the boundary, as shown in Fig. 23. See Section 5.1.
5. All integrals used here are those expressed in the closed form. That is, numerical integrals are not used at all.

All the numerical results were obtained by using a personal computer. In Tables 4-7, the numerical results obtained from this program and the previous solutions are compared, from which the high accuracy of the method is recognized.

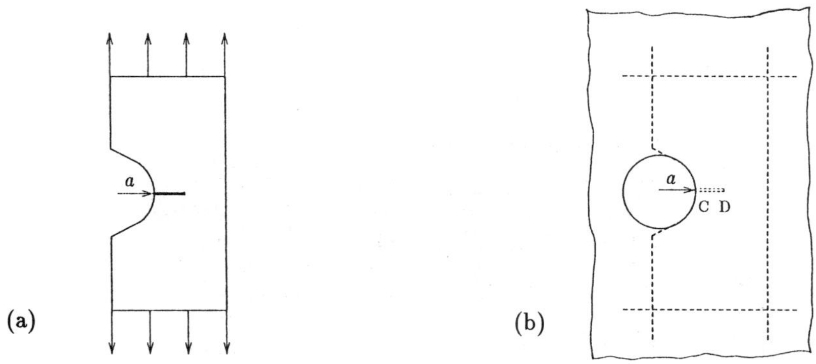

Figure 23: Imaginary boundary on which point forces are applied: (a) problem; (b) imaginary boundary

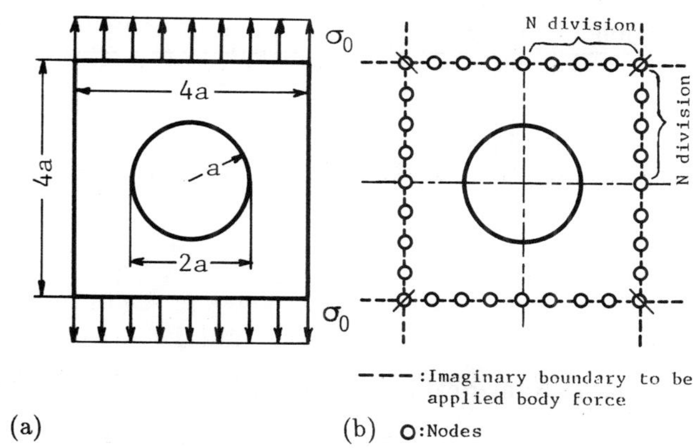

Figure 24: A rectangular plate with a hole: (a) given problem; (b) method of division

Table 5: Numerical results of Fig. 25

		$\theta = 30°$		$\theta = 45°$	
N	M	F_{IB}	F_{IIB}	F_{IB}	F_{IIB}
4	2	0.8288	-0.3955	0.6382	-0.5162
	4	0.8300	-0.3956	0.6394	-0.5160
	6	0.8301	-0.3956	0.6395	-0.5159
8	2	0.8223	-0.3902	0.6328	-0.5076
	4	0.8250	-0.3905	0.6345	-0.5078
	6	0.8253	-0.3905	0.6347	-0.5078
12	2	0.8221	-0.3893	0.6317	-0.5061
	4	0.8242	-0.3896	0.6335	-0.5064
	6	0.8245	-0.3896	0.6338	-0.5064
∞[25]		0.8245	-0.3895	0.6339	-0.5053

Table 6: Numerical results of Fig. 26

b/a	Present analysis	Isida[26] Noguchi	Isida[21] (BFM)	Kitagawa[25] Yuuki
0.005	0.675	0.669	-	-
0.01	0.656	0.652	-	-
0.05	0.594	0.593	0.593	0.59
0.1	0.560	0.560	0.560	0.56
0.2	0.528	0.528	0.528	0.54
0.3	0.512	0.512	0.512	-
0.4	0.504	0.504	0.504	0.51
0.5	0.500	0.500	0.500	-

Table 7: Numerical results of Fig. 27

$F_I = K_I/\sigma_0\sqrt{\pi C}$						
c/a	Present analysis			Isida *et al.*[9]		
θ	0.05	0.10	0.20	0.05	0.10	0.20
0°	3.037	2.772	2.374	3.037	2.772	2.374
15°	2.955	2.702	2.314	2.955	2.702	2.314
30°	2.720	2.502	2.140	2.720	2.501	2.140
45°	2.368	2.202	1.870	2.368	2.203	1.870

$F_{II} = K_{II}/\sigma_0\sqrt{\pi C}$						
c/a	Present analysis			Isida *et al.*[9]		
θ	0.05	0.10	0.20	0.05	0.10	0.20
0°	0.000	0.000	0.000	0.000	0.000	0.000
15°	0.451	0.396	0.327	0.452	0.396	0.327
30°	0.844	0.740	0.611	0.843	0.739	0.610
45°	1.126	0.990	0.815	1.126	0.990	0.815

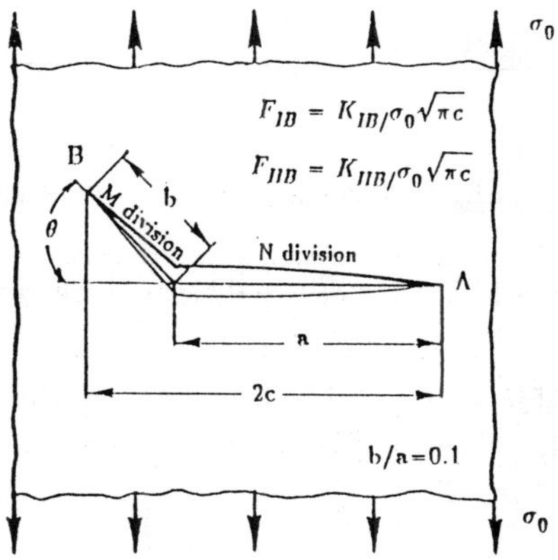

Figure 25: Tension of an infinite plate with a bent crack

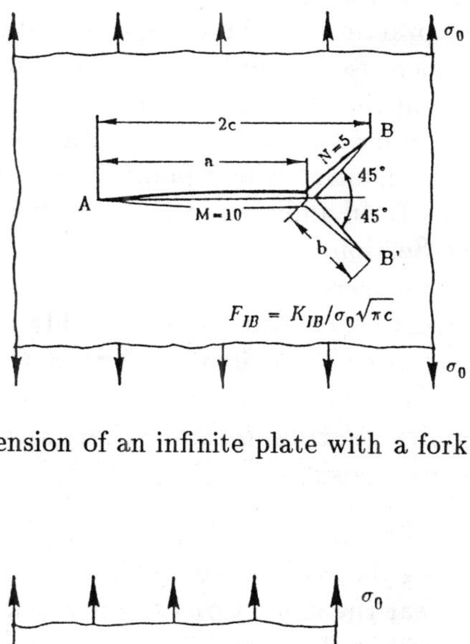

Figure 26: Tension of an infinite plate with a fork type 45° crack

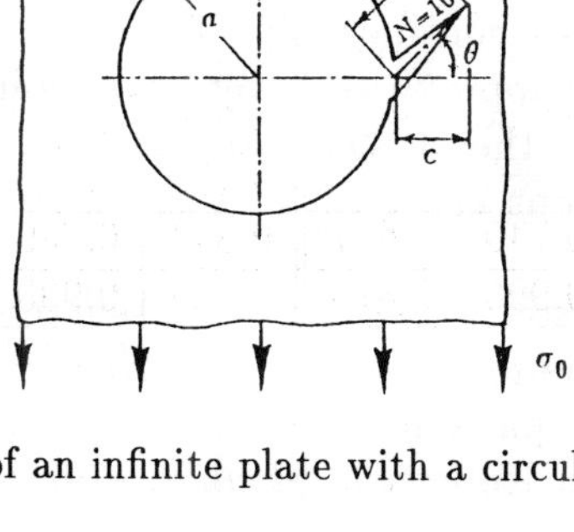

Figure 27: Tension of an infinite plate with a circular hole having an edge crack

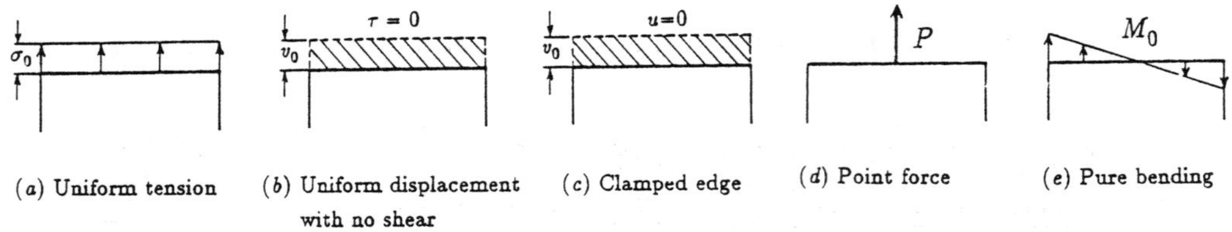

Figure 28: Five loading conditions were treated

In Tables 8 to 13, the effects of specimen length on the stress concentration (intensity) factors are shown for five loading conditions shown in Fig. 28 and for six types of specimen shown in Fig. 29.

Example 2. Stress concentration of a cylindrical bar with a V-shaped circumferential groove under torsion, tension or bending.[32]

The stress concentration problem of a cylindrical bar with a circumferential groove (Fig. 30) is mainly used in practice for the design of shafts. This problem is also important for the test specimen used in order to investigate the fatigue strength of metals. In solving this problem the stress fields due to the ring forces in an infinite body, as shown in Fig. 32, are used as the fundamental solutions (see Section 5.3). Figure 31 shows imaginary boundaries where point forces are applied. Point forces are applied along the part BA'B' in addition to the part BAB' which should become a circumferential groove (see Section 5.1).

Figures 33-35 show the numerical results. As a result of the systematic calculations, it is found that the stress concentration factors obtained by Neuber's trigonometric rule have non-conservative errors of about 10% for a wide range of notch depths.

Example 3. Stress intensity factors for an internal semi-elliptical surface crack in cylindrical pressure vessels.[22] (Fig. 36)

In the present analysis, the different techniques are used for a crack surface and for cylindrical surfaces, considering that the values to be obtained are the stress intensity factors, i.e. the information near the crack tip. For the crack surface, the tensile type doublets are continuously distributed on the imaginary crack surface. The density of doublets is expressed as (weight) × (basic density) in order to obtain the solutions with higher accuracy under a small number of elements (see Section 4.3). For cylindrical boundaries, the point forces to be continuously distributed on each element of the imaginary cylindrical surfaces are approximated by single point forces acting at points near the outside of the center of the element, and as the boundary conditions the resultant force boundary conditions are used for the elements of cylindrical surfaces (see Section 5.2).

Figure 37 gives the comparison between the present results and those of Raju & Newman.[33] It is found that the results of Raju & Newman have non-conservative errors. Figure 38 gives the comparison between the present results and those of Kobayashi *et al.*[35] The results of Kobayashi *et al.* are in good agreement with the

Table 8: S.C.F. of SCN, ((a) ~ (e); see Fig. 28)

(a) ρ/W=0.1

L/W	(a)	(b)	(c)	(d)	(e)
0.6	2.259	2.015	1.914	1.152	2.321
0.8	2.228	2.063	2.015	1.865	2.286
1.0	2.224	2.092	2.073	2.148	2.282
1.2	2.224	2.113	2.107	2.233	2.281
1.4	2.224	2.128	2.127	2.243	2.281
1.6	2.223	2.139	2.139	2.229	2.281
1.8	2.223	2.148	2.148	2.228	2.281
2.0	2.223	2.156	2.156	2.226	2.281
3.0	2.223	2.178	2.178	2.224	2.281
∞	2.223[27]				2.281[27]

(b) ρ/W=0.5

L/W	(a)	(b)	(c)	(d)	(e)
0.6	1.238	0.418	0.419	1.259	1.233
0.8	1.242	0.463	0.464	1.253	1.237
1.0	1.244	0.502	0.503	1.250	1.241
1.2	1.245	0.538	0.538	1.248	1.243
1.4	1.245	0.571	0.571	1.246	1.243
1.6	1.245	0.601	0.601	1.245	1.243
1.8	1.245	0.628	0.628	1.245	1.243
2.0	1.245	0.653	0.653	1.245	1.243
3.0	1.245	0.752	0.752	1.245	1.243
∞	1.244[27]				1.243[27]

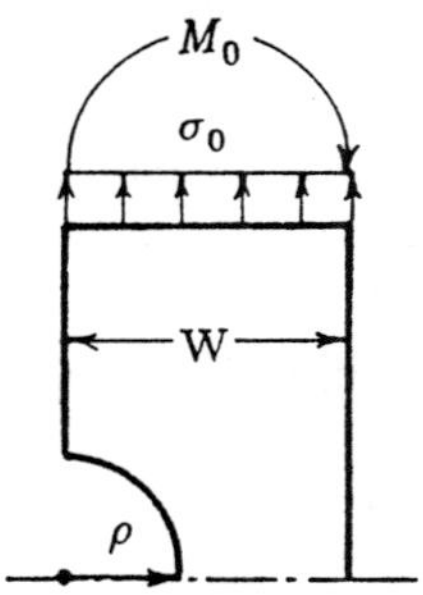

$$K_t = \sigma_{max}/\sigma_n$$

$$\sigma_n = \frac{W}{W-\rho}\left(1+\frac{3\rho}{W-\rho}\right)\sigma_0 \quad \text{for (a)}\sim\text{(d)}$$

$$\sigma_n = \frac{6M_0}{(W-\rho)^2} \quad \text{for (e)}$$

σ_0; average stress

Table 9: S.C.F. of SVN, ((a) ~ (e); see Fig. 28)

(a) ρ/W=0.1 , t/W=0.2

L/W	(a)	(b)	(c)	(d)	(e)
0.6	2.357	1.699	1.651	1.631	2.450
0.8	2.292	1.810	1.786	2.066	2.370
1.0	2.282	1.885	1.875	2.233	2.358
1.2	2.281	1.939	1.935	2.284	2.358
1.4	2.281	1.981	1.979	2.293	2.358
1.6	2.281	2.013	2.013	2.287	2.357
1.8	2.281	2.039	2.039	2.285	2.357
2.0	2.281	2.061	2.061	2.284	2.357
3.0	2.280	2.129	2.129	2.282	2.357
∞	2.280[27]				2.356[27]

(b) ρ/W=0.1, t/W=0.4

L/W	(a)	(b)	(c)	(d)	(e)
0.6	2.141	0.991	0.998	2.041	2.191
0.8	2.101	1.103	1.108	2.089	2.138
1.0	2.096	1.197	1.199	2.102	2.136
1.2	2.095	1.275	1.275	2.103	2.136
1.4	2.096	1.341	1.340	2.100	2.136
1.6	2.096	1.397	1.395	2.097	2.136
1.8	2.096	1.445	1.443	2.096	2.136
2.0	2.096	1.488	1.488	2.096	2.136
3.0	2.096	1.676	1.676	2.096	2.136
∞	2.096[27]				2.136[27]

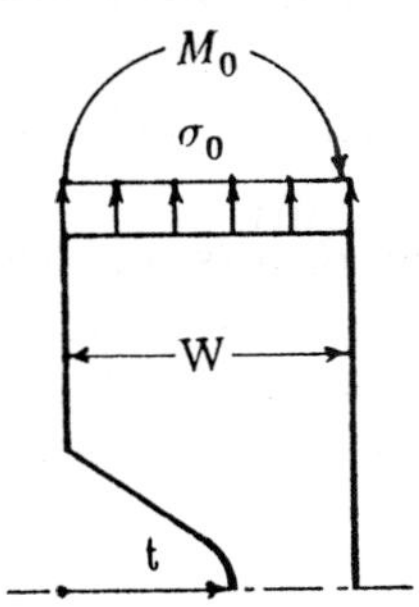

$$K_t = \sigma_{max}/\sigma_n$$

$$\sigma_n = \frac{W}{W-t}\left(1+\frac{3t}{W-t}\right)\sigma_0 \quad \text{for (a)}\sim\text{(d)}$$

$$\sigma_n = \frac{6M_0}{(W-t)^2} \quad \text{for (e)}$$

σ_0; average stress

Table 10: S.I.F. of SEN, ((a) ~ (e); see Fig. 28)

(a) c/W=0.1

L/W	(a)	(b)	(c)	(d)	(e)
0.6	1.205	1.097	1.036	0.549	1.062
0.8	1.191	1.119	1.090	0.971	1.049
1.0	1.189	1.132	1.120	1.142	1.047
1.2	1.189	1.141	1.137	1.194	1.047
1.4	1.189	1.147	1.147	1.209	1.047
1.6	1.189	1.152	1.152	1.191	1.047
1.8	1.189	1.156	1.156	1.190	1.047
2.0	1.189	1.160	1.160	1.189	1.047
3.0	1.189	1.169	1.169	1.189	1.047
∞	1.189[17]				1.047[28]

(b) c/W=0.5

L/W	(a)	(b)	(c)	(d)	(e)
0.6	2.895	1.039	1.050	2.742	1.475
0.8	2.834	1.178	1.182	2.786	1.504
1.0	2.825	1.298	1.298	2.814	1.498
1.2	2.824	1.401	1.400	2.825	1.497
1.4	2.824	1.492	1.490	2.825	1.497
1.6	2.824	1.571	1.569	2.825	1.497
1.8	2.824	1.642	1.640	2.825	1.497
2.0	2.824	1.705	1.703	2.825	1.498
3.0	2.824	1.940	1.939	2.825	1.498
∞	2.824[17]				1.495[29]

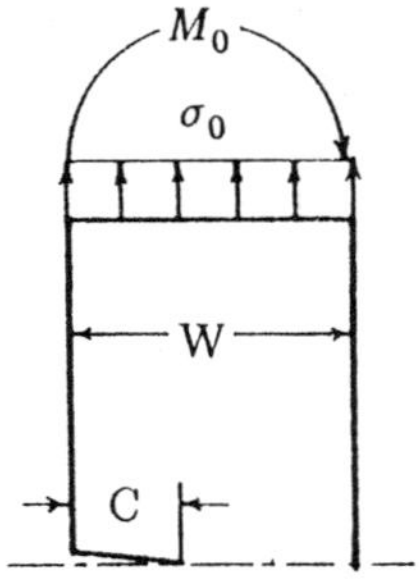

$$F_I = K_I/\sigma_n\sqrt{\pi C}$$

$\sigma_n = \sigma_0$ for (a)~(d)

$\sigma_n = \dfrac{6M_0}{W^2}$ for (e)

σ_0; average stress

Table 11: S.C.F. of DCN, ((a) ~ (e); see Fig. 28)

(a) $2\rho/W$=0.2

L/W	(a)	(b)	(c)	(d)	(e)
0.6	2.499	2.433	2.297	1.293	1.988
0.8	2.438	2.432	2.372	2.049	1.986
1.0	2.430	2.430	2.407	2.348	1.986
1.2	2.429	2.429	2.422	2.439	1.985
1.4	2.429	2.429	2.428	2.453	1.985
1.6	2.429	2.429	2.429	2.444	1.985
1.8	2.429	2.429	2.429	2.437	1.985
2.0	2.429	2.429	2.429	2.432	1.985
3.0	2.429	2.429	2.429	2.430	1.985
∞	2.429[30]				1.985[30]

(b) $2\rho/W$=0.6

L/W	(a)	(b)	(c)	(d)	(e)
0.6	1.415	1.432	1.414	1.450	1.235
0.8	1.417	1.422	1.421	1.457	1.236
1.0	1.420	1.420	1.421	1.444	1.236
1.2	1.421	1.420	1.420	1.433	1.236
1.4	1.421	1.420	1.420	1.426	1.235
1.6	1.421	1.420	1.420	1.422	1.235
1.8	1.421	1.420	1.420	1.421	1.235
2.0	1.421	1.420	1.420	1.420	1.235
3.0	1.421	1.420	1.420	1.420	1.235
∞	1.420[30]				1.235[30]

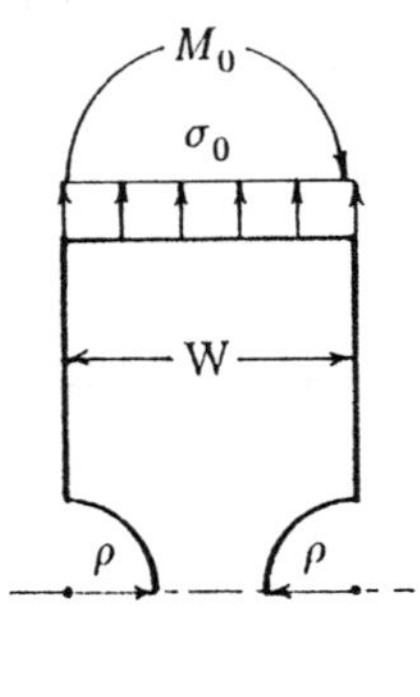

$$K_t = \sigma_{max}/\sigma_n$$

$\sigma_n = \dfrac{W}{W-2\rho}\sigma_0$ for (a)~(d)

$\sigma_n = \dfrac{6M_0}{(W-2\rho)^2}$ for (e)

σ_0; average stress

Table 12: S.C.F. of DVN, ((a) ~ (e); see Fig. 28)

(a) $2\rho/W$=0.2,$2t/W$=0.4

L/W	(a)	(b)	(c)	(d)	(e)
0.6	2.534	2.429	2.337	1.861	1.853
0.8	2.439	2.428	2.393	2.243	1.852
1.0	2.427	2.426	2.414	2.387	1.851
1.2	2.426	2.424	2.421	2.429	1.850
1.4	2.426	2.424	2.424	2.237	1.850
1.6	2.425	2.424	2.424	2.433	1.850
1.8	2.424	2.424	2.424	2.428	1.850
2.0	2.424	2.424	2.424	2.427	1.850
3.0	2.424	2.424	2.424	2.425	1.850
∞	2.424[30]				1.850[30]

(b) $2\rho/W$=0.2,$2t/W$=0.6

L/W	(a)	(b)	(c)	(d)	(e)
0.6	2.129	2.103	2.072	2.015	1.517
0.8	2.098	2.098	2.090	2.091	1.517
1.0	2.096	2.096	2.095	2.106	1.517
1.2	2.096	2.096	2.096	2.104	1.516
1.4	2.096	2.096	2.096	2.101	1.516
1.6	2.096	2.096	2.096	2.098	1.516
1.8	2.096	2.096	2.096	2.097	1.516
2.0	2.096	2.096	2.096	2.096	1.516
3.0	2.096	2.096	2.096	2.096	1.516
∞	2.096[30]				1.516[30]

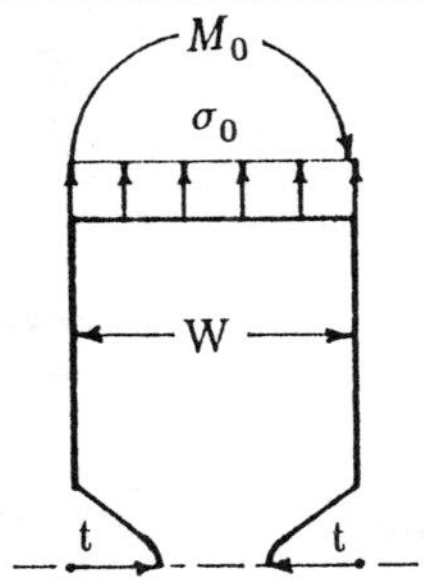

$$K_t = \sigma_{max}/\sigma_n$$

$$\sigma_n = \frac{W}{W-2t}\sigma_0 \quad \text{for (a)}\sim\text{(d)}$$

$$\sigma_n = \frac{6M_0}{(W-2t)^2} \quad \text{for (e)}$$

σ_0; average stress

Table 13: S.I.F. of DEC, ((a) ~ (e); see Fig. 28)

(a) $2c/W$=0.2

L/W	(a)	(b)	(c)	(d)	(e)
0.6	1.140	1.114	1.045	0.522	1.131
0.8	1.116	1.113	1.083	0.909	1.130
1.0	1.112	1.112	1.100	1.067	1.130
1.2	1.112	1.112	1.108	1.115	1.130
1.4	1.112	1.112	1.111	1.122	1.130
1.6	1.112	1.112	1.112	1.117	1.130
1.8	1.112	1.112	1.112	1.115	1.130
2.0	1.112	1.112	1.112	1.112	1.130
3.0	1.112	1.112	1.112	1.112	1.130
∞	1.112[3]				1.102[31]

(b) $2c/W$=0.6

L/W	(a)	(b)	(c)	(d)	(e)
0.6	1.299	1.239	1.200	1.029	2.154
0.8	1.246	1.238	1.223	1.154	2.152
1.0	1.237	1.237	1.231	1.210	2.152
1.2	1.236	1.236	1.234	1.232	2.152
1.4	1.236	1.236	1.236	1.238	2.152
1.6	1.236	1.236	1.236	1.238	2.151
1.8	1.236	1.236	1.236	1.237	2.151
2.0	1.236	1.236	1.236	1.236	2.151
3.0	1.236	1.236	1.236	1.236	2.151
∞	1.236[3]				2.126[31]

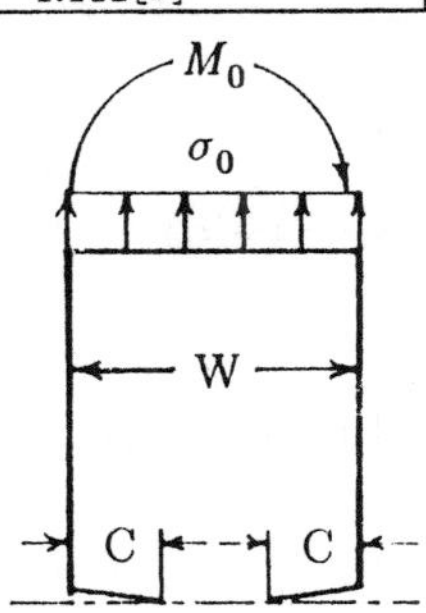

$$F_I = K_I/\sigma_n\sqrt{\pi C}$$

$$\sigma_n = \sigma_0 \quad \text{for (a)}\sim\text{(d)}$$

$$\sigma_n = \frac{6M_0}{W^2} \quad \text{for (e)}$$

σ_0; average stress

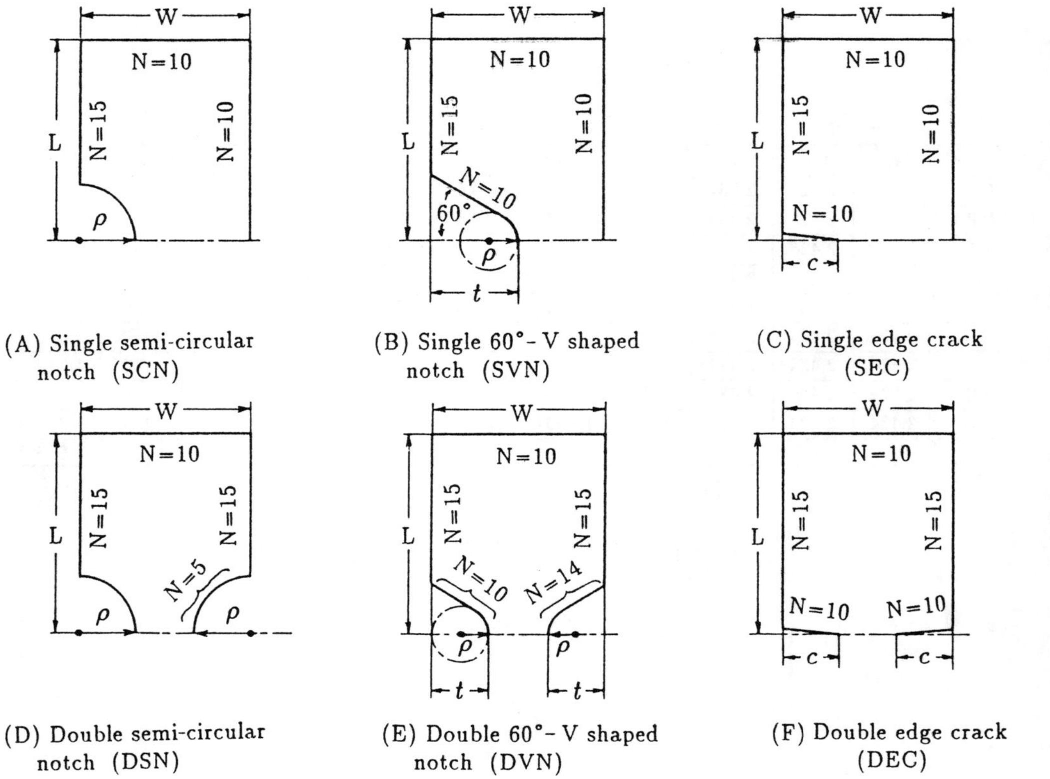

Figure 29: Six types of specimens were treated

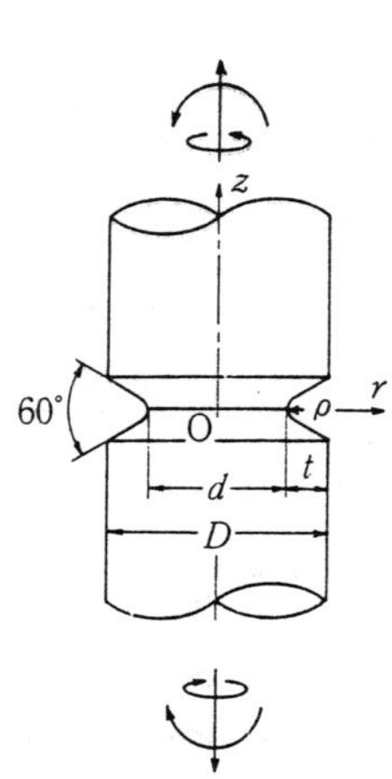

Figure 30: A cylindrical bar with a V-shaped circumferential groove

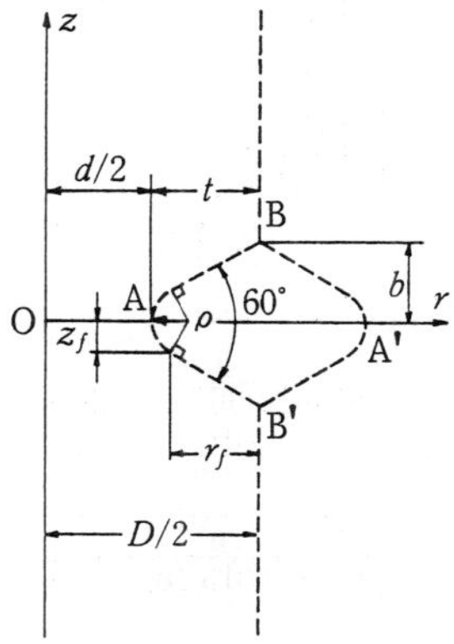

Figure 31: Imaginary boundaries where body forces are distributed

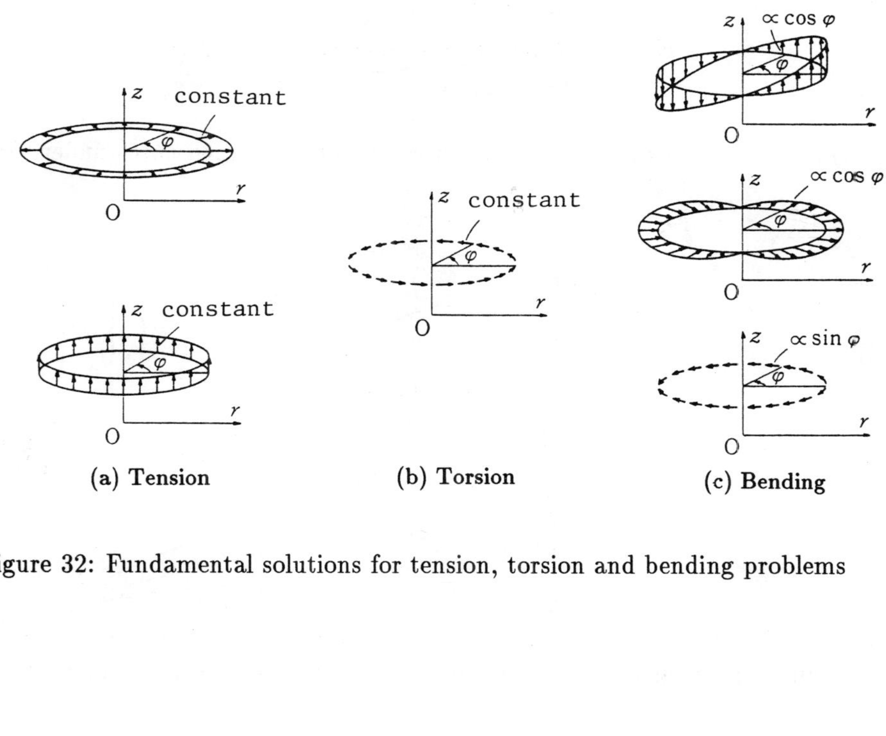

Figure 32: Fundamental solutions for tension, torsion and bending problems

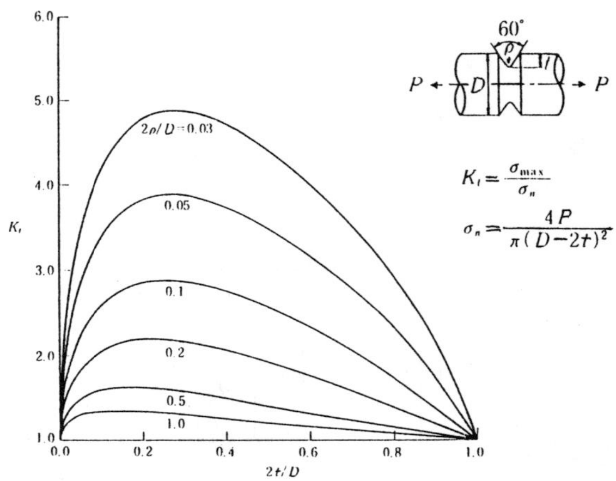

Figure 33: Stress concentration factors of a 60° V-shaped notch under tension

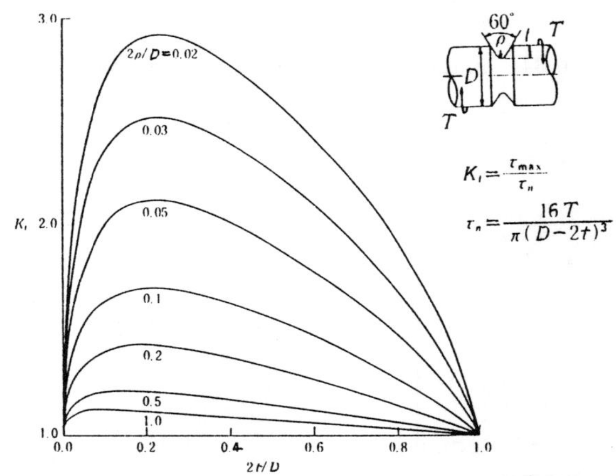

Figure 34: Stress concentration factors of a 60° V-shaped notch under torsion

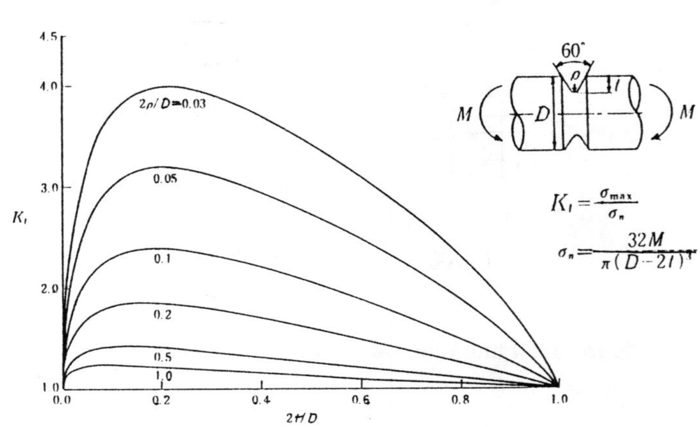

Figure 35: Stress concentration factors of a 60° V-shaped notch under bending

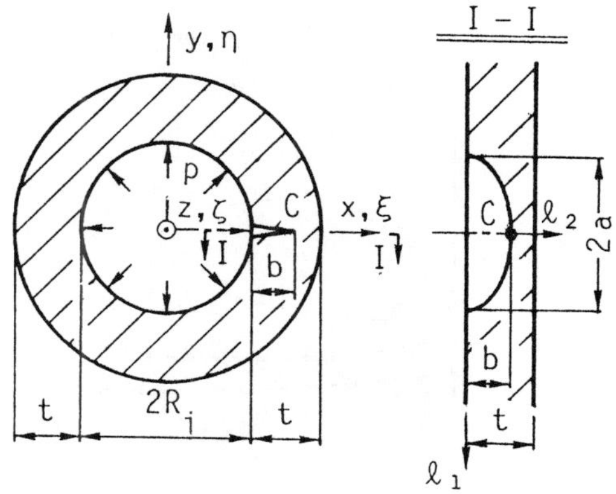

Figure 36: An internal semi-elliptical surface crack in a cylindrical pressure vessel

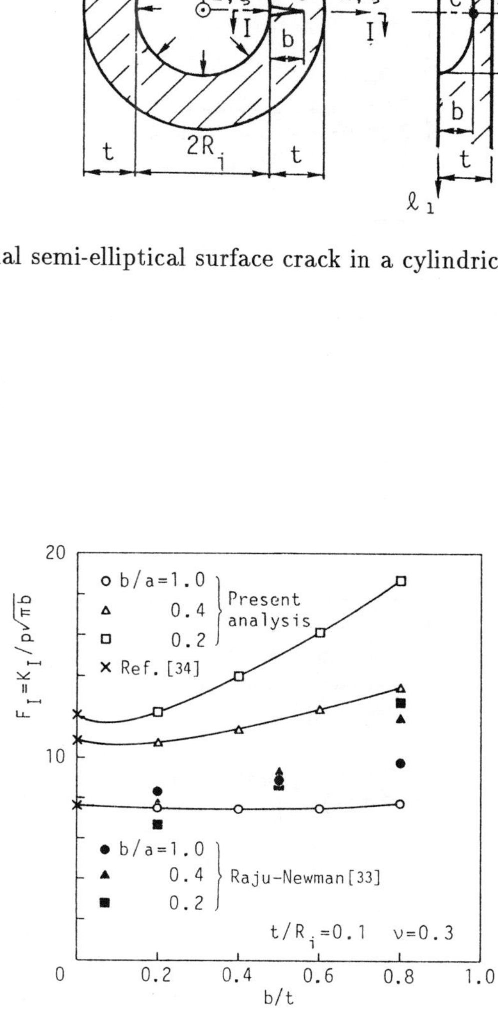

Figure 37: Comparison with results of Raju & Newman

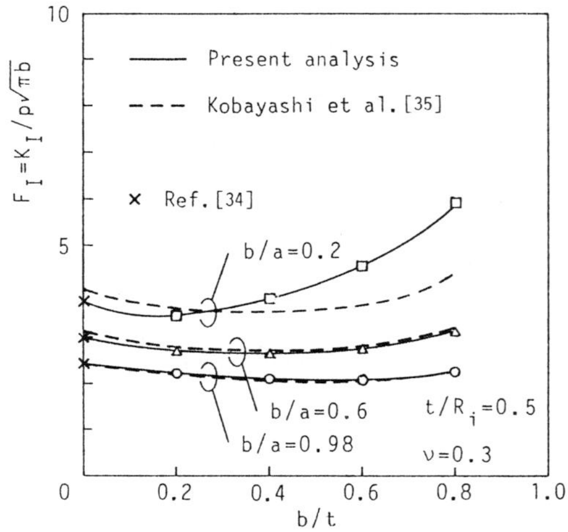

Figure 38: Comparison with results of Kobayashi *et al.*[35]

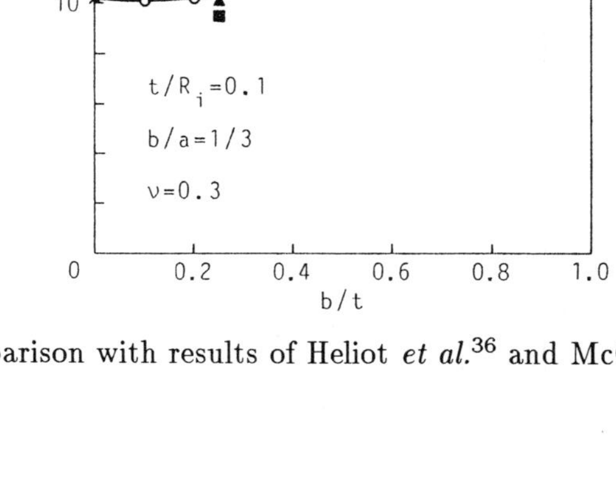

Figure 39: Comparison with results of Heliot *et al.*[36] and McGown & Raymund[37]

Table 14: F_I for the case when the internal pressure is applied to the crack surface

t/R_i	b/a \ b/t	0	0.2	0.4	0.6	0.8
0.1	1.0	7.59*	7.45	7.42	7.49	7.71
	0.5	10.20*	10.06	10.45	11.11	11.90
	0.25	11.79*	11.86	13.24	15.09	17.01
0.2	1.0	4.31*	4.15	4.07	4.07	4.24
	0.5	5.79*	5.59	5.70	6.06	6.71
	0.25	6.70*	6.57	7.23	8.31	9.83
0.3	1.0	3.23*	3.05	2.95	2.94	3.09
	0.5	4.34*	4.09	4.13	4.39	4.99
	0.25	5.01*	4.81	5.24	6.07	7.43
0.4	0.1	2.69*	2.50	2.39	2.37	2.52
	0.5	3.61*	3.34	3.34	3.56	4.13
	0.25	4.18*	3.93	4.24	4.93	6.17
0.5	1.0	2.37*	2.17	2.05	2.03	2.18
	0.5	3.19*	2.89	2.86	3.06	3.60
	0.25	3.68*	3.39	3.63	4.24	5.36

*:Ref.[34]

present results for the case when $b/a > 0.6$. Figure 39 also gives the comparison among the present results and those of Heliot, Labbens, and Pellissier-Tanon,[36] and McGown & Raymund.[37] The results of Heliot *et al.* by the boundary integral method are more accurate than McGown & Raymund's by the finite element method.

The results of the systematic calculations ($\nu = 0.3$) are shown in Tables 14 and 15, and Figs 40-42. The values for the case when $b/t = 0$ were taken from the results of a semi-elliptical surface crack in a semi-infinite body.

Example 4. Tension and bending of plates of finite thickness with a semi-elliptical surface crack (Figs 43 and 44)[34].

The plate having a semi-elliptical surface crack is assumed to be infinite in the x and z direction (Fig. 43) and to be subjected to the linearly varying stress at $|z| \to \infty$, which is given by

$$\sigma_z^\infty = \sigma_T + (1 - \frac{2y}{h})\sigma_B \tag{61}$$

where σ_T and σ_B are the uniform and bending components of the applied stress as shown in Fig. 44.

In this analysis, the boundary conditions of the front surface are completely satisfied by using the stress fields due to a point force acting in a semi-infinite body as fundamental solutions (see Section 5.3), and the boundary conditions of the back surface are processed in terms of the resultant forces instead of the stress, which greatly improved the accuracy of the numerical results (see Section 5.2). Numerical results

Table 15: F_I^* for the case when the internal pressure is not applied to the crack surface

t/R_i	b/a \ b/t	0	0.2	0.4	0.6	0.8
	1.0	6.94*	6.79	6.76	6.82	7.01
0.1	0.5	9.31*	9.17	9.52	10.11	10.82
	0.25	10.77*	10.82	12.06	13.74	15.47
	1.0	3.65*	3.50	3.42	3.40	3.52
0.2	0.5	4.91*	4.71	4.79	5.07	5.59
	0.25	5.67*	5.55	6.08	6.96	8.21
	1.0	2.57*	2.40	2.30	2.27	2.37
0.3	0.5	3.45*	3.23	3.22	3.41	3.84
	0.25	3.99*	3.79	4.09	4.71	5.73
	0.1	2.03*	1.85	1.74	1.70	1.79
0.4	0.5	2.73*	2.48	2.44	2.57	2.95
	0.25	3.15*	2.92	3.11	3.58	4.43
	1.0	1.71*	1.52	1.41	1.36	1.44
0.5	0.5	2.30*	2.04	1.97	2.07	2.40
	0.25	2.66*	2.40	2.52	2.89	3.61

*:Ref.[34]

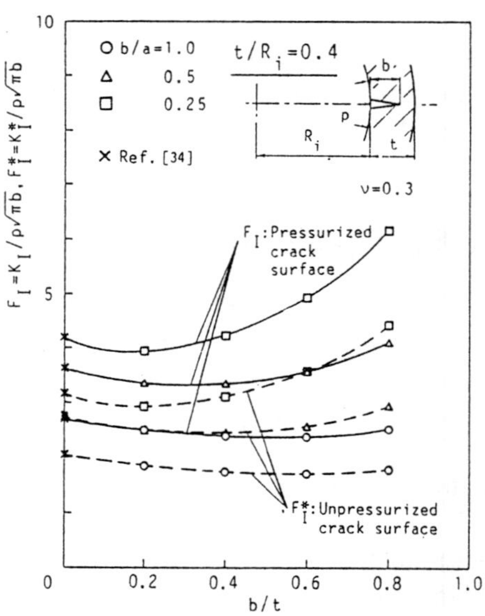

Figure 40: F_I and F_I^* versus b/t for various $b/a (t/R = 0.1)$

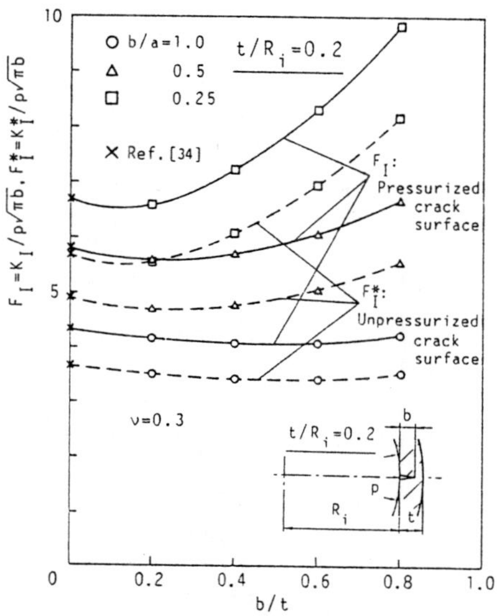

Figure 41: F_I and F_I^* versus b/t for various $b/a(t/R = 0.2)$

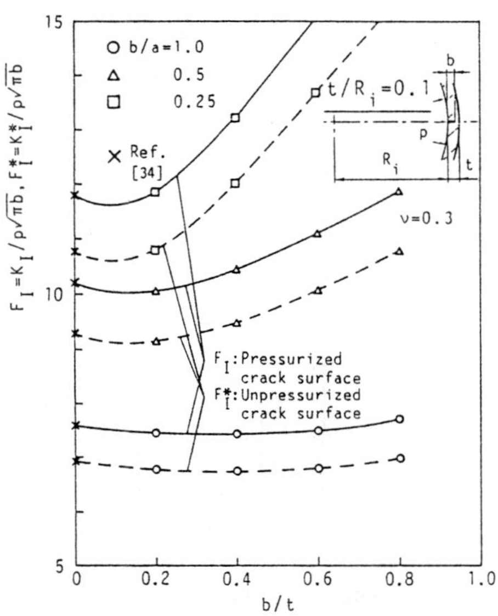

Figure 42: F_I and F_I^* versus b/t for various $b/a(t/R = 0.4)$

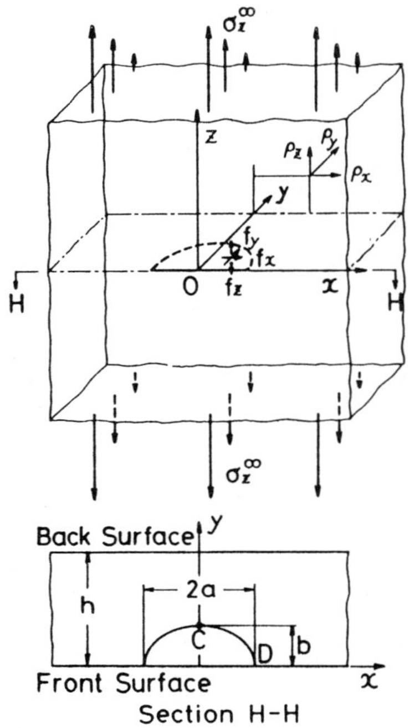

Figure 43: Finite thickness plate with a semi-elliptical crack

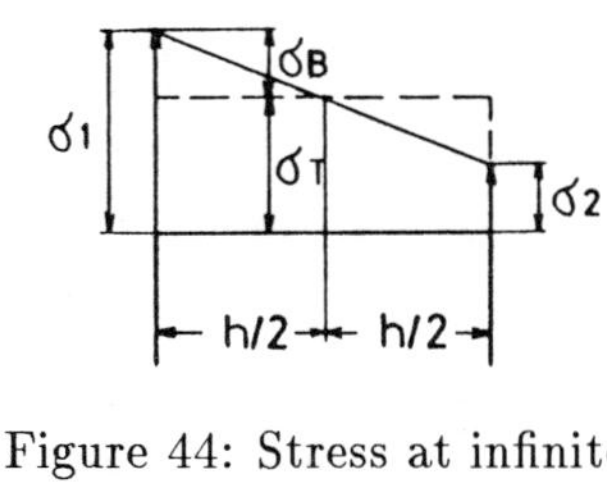

Figure 44: Stress at infinite

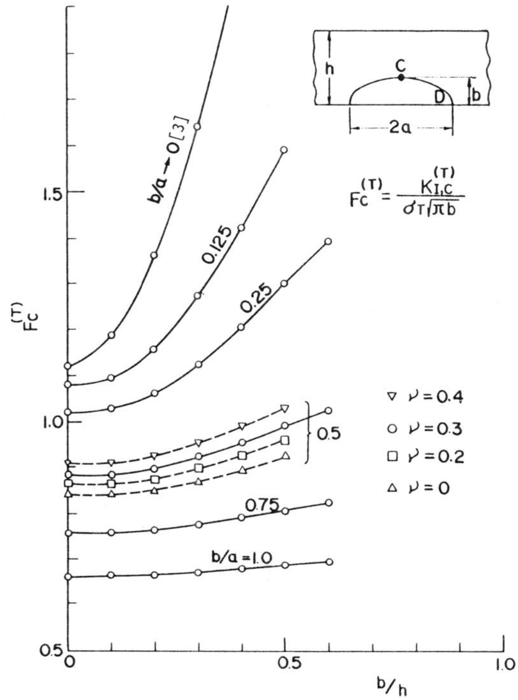

Figure 45: $F_C^{(T)}$ versus b/h for various b/a

are shown in Figs 45 and 46.

7 Application to elastic-plastic problems

In this section it is shown that the BFM can be extended so as to be applicable to the elastic-plastic problems. The basic principle of analysis is the same as in the case of elastic problems. That is, the solution of a problem is obtained by superposing the fundamental elastic fields, and for this purpose the problem is replaced by an infinite body without any plastic zone. It is seen from Section 2 that by inserting the region in question into an auxiliary region the problem can be transformed into the problem of an infinite body. Therefore, the key point for analyzing a plastic problem is how to replace a plastic zone by an elastic element.

As an example let us consider a problem with plastic zone as shown in Fig. 47(a). In order to replace the plastic zone by the elastic elements the following is considered.

Let us, first, cut out a small element ω from the plastic zone as shown in Fig. 47(b). To compensate the stress state we must apply a traction $-\sigma_{ij}(\omega)\times n_j(Q), (Q) \in \Gamma(\omega)(\Gamma(\omega)$: boundary of element $\omega, n_j(Q)$: outward normal vector of surface $\Gamma(\omega)$ of element ω) onto the inside surface of the cut cavity so that the elastic-plastic field is not disturbed by cutting. Before cutting out the element the total strain in the element ω is

$$\varepsilon_{ij}(\omega) = \varepsilon_{ij}^{e}(\omega) + \varepsilon_{ij}^{p}(\omega) \tag{62}$$

where $\varepsilon_{ij}^{e}(\omega)$ and $\varepsilon_{ij}^{p}(\omega)$ are the elastic strain and the plastic strain in the element,

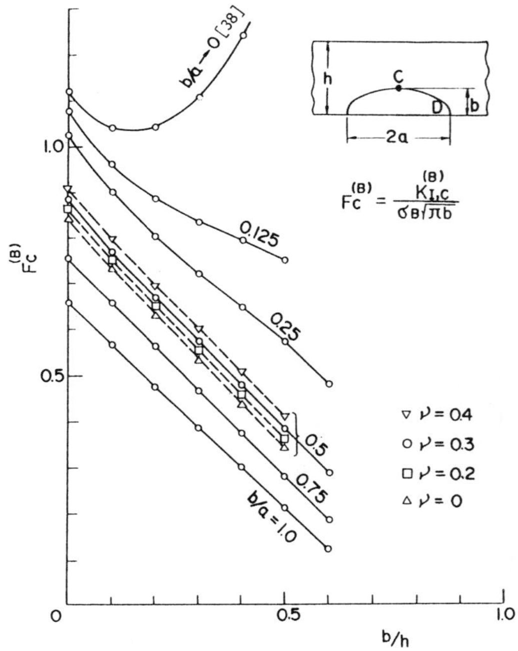

Figure 46: $F_C^{(B)}$ versus b/h for various b/a

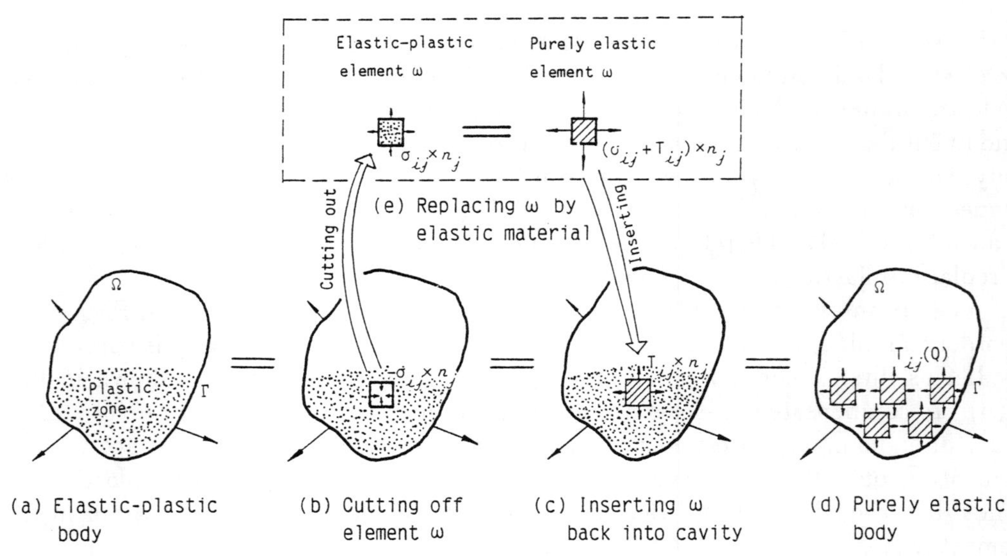

Figure 47: Elastic-plastic fields can be replaced by the elastic fields due to force doublets

respectively.

$$\varepsilon_{ij}^{\mathrm{e}}(\omega) = C_{ijkl} \cdot \sigma_{kl}(\omega) \tag{63}$$

$[C_{ijkl}(\omega)$: elastic compliance tensor$]$

Secondly, the material of the element ω is replaced by an imaginary material, whose elastic constants G, ν are the same as those before cutting out but whose yield stress is assumed to be infinite. At the same time a traction $\{\sigma_{ij}(\omega) + T_{ij}(\omega)\} \times n_j(Q)$ is applied to the surface of the element ω as shown in Fig. 47(e). The $T_{ij}(\omega)$ is assumed to be

$$T_{ij}(\omega) = D_{ijkl} \cdot \varepsilon_{kl}^{\mathrm{P}}(\omega) \tag{64}$$

where D_{ijkl} is the elastic stiffness tensor.

So, the total strain $\varepsilon'_{ij}(\omega)$ in the element ω is the same as before cutting out

$$\begin{aligned} \varepsilon'_{ij}(\omega) &= C_{ijkl}\{\sigma_{kl}(\omega) + T_{kl}(\omega)\} \\ &= \varepsilon_{ij}^{\mathrm{e}}(\omega) + \varepsilon_{ij}^{\mathrm{P}}(\omega) \\ &= \varepsilon_{ij}(\omega) \end{aligned} \tag{65}$$

Therefore this element ω replaced by the imaginary elastic material can be inserted back into the cavity without any gap. By this way the plastic element ω is replaced by an elastic element and the total strain field has not been disturbed. After insertion, however, there remain the forces $T_{ij}(\omega) \times n_j(Q)$ acting at the surface of ω as shown in Fig. 47(c). It is seen that these forces are equivalent to the force doublets whose density is $T_{ij}(\omega)$.

By replacing the plastic elements by the elastic elements one after another in this way, the problem with plastic zone can be finally reduced to the problem of a perfect elastic body without any plastic zone as shown in Fig. 47(d). In Fig. 47(d) the force doublets with density

$$T_{ij}(Q) = D_{ijkl} \cdot \varepsilon_{kl}^{\mathrm{P}}(Q) \qquad (Q \in \Omega) \tag{66}$$

are distributed over the region corresponding to the plastic zone in order to compensate the plastic strain lost when exchanging the real material by an imaginary elastic material. It is seen that the displacement field of Fig. 47(d) is the same as those of Fig. 47(a), and the stress field of Fig. 47(d) is equal to the sum of the stress field of Fig. 47(a) and $T_{ij}(Q)$. For the displacement and stress fields of Fig. 47(d) we have eqns (4) and (5) or eqns (8) and (9), etc., deduced from eqns (4) and (5). However, for simplicity we use eqns (8) and (9) here. That is, only the point forces applied along an imaginary boundary. So, the displacement field and the stress field in Fig. 47(d) are given by summing simply all the effects due to point forces and force doublets, whose densities are $\phi_\eta(Q)$ and $T_{\xi,\eta}(Q)$, respectively. Thus, the displacement field $u_i(P)$ and the stress field $\sigma_{ij}(P)$ in Fig. 47(a) can be expressed as follows

$$u_i(P) = \int_\Gamma u_i(P, Q_\eta)\phi_\eta(Q)\mathrm{d}\Gamma(Q) + \int_\Omega u_i(P, Q_{\xi,\eta})T_{\xi\eta}(Q)\mathrm{d}\Omega(Q) \quad (P \in \Omega) \tag{67}$$

$$\sigma_{ij}(P) = \int_\Gamma \sigma_{ij}(P, Q_\eta)\phi_\eta(Q)\mathrm{d}\Gamma(Q) + \int_\Omega \sigma_{ij}(P, Q_{\xi,\eta})T_{\xi\eta}(Q)\mathrm{d}\Omega(Q) \quad (P \in \Omega) \tag{68}$$

where $u_i(P, Q_{\xi,\eta})$ and $\sigma_{ij}(P, Q_{\xi,\eta})$ are the i-component of displacement and the ij-component of stress at a point P due to the $\xi\eta$-directional unit force doublets acting at a point Q, respectively.

It has become clear from the above discussion that if the density of the point forces distributed along the imaginary boundary and the density of the force doublets distributed in the imaginary plastic zone are determined, the displacements and stresses at any point P can be calculated by eqns (67) and (68). That is, in the present method the analysis of an elastic-plastic problem is reduced to calculating the densities $\phi_\eta(Q)$ and $T_{\xi\eta}(Q)$.

The densities $\phi_\eta(Q)$ and $T_\eta(Q)$ are determined so that the given boundary conditions and the given elastic-plastic constitutive equation can be satisfied.

When calculated point P is moved to the boundary, eqns (69) and (70)

$$u_i(P) = \int_\Gamma u_i(P, Q_\eta)\phi_\eta(Q)\mathrm{d}\Gamma(Q) + \int_\Omega u_i(P, Q_{\xi,\eta})T_{\xi\eta}(Q)\mathrm{d}\Omega(Q) \quad (P \in \Gamma) \quad (69)$$

$$\sigma_{ij}(P) = \int_\Gamma \sigma_{ij}(P, Q_\eta)\phi_\eta(Q)\mathrm{d}\Gamma(Q) + \int_\Omega \sigma_{ij}(P, Q_{\xi,\eta})T_{\xi\eta}(Q)\mathrm{d}\Omega(Q) \quad (P \in \Gamma) \quad (70)$$

can be used as the expressions of the boundary conditions.

For the elastic problem the unknowns to be determined are the densities of the point forces $\phi_\eta(Q)$ alone, and they can be determined from the boundary condition equation alone. For the elastic-plastic problem, however, the unknowns to be determined are the densities of the point forces $\phi_\eta(Q)$ and the densities of the force doublets $T_{\xi\eta}(Q)$. Then, besides the boundary condition eqn (69) or (70) the elastic-plastic constitutive equation has to be taken into account.

It has been well-known that the plastic deformations should be treated by the incremental analysis. The total strain-increment is expressed as the sum of the elastic strain-increment and the plastic strain-increment.

$$\mathrm{d}\varepsilon_{ij} = \mathrm{d}\varepsilon_{ij}^{\mathrm{e}} + \mathrm{d}\varepsilon_{ij}^{\mathrm{P}} \quad (71)$$

The elastic strain-increment $\mathrm{d}\varepsilon_{ij}^{\mathrm{e}}$ is related to the stress-increment $\mathrm{d}\sigma_{ij}^{\mathrm{P}}$ by Hooke's law

$$\mathrm{d}\varepsilon_{ij}^{\mathrm{e}} = C_{ijkl} \cdot \mathrm{d}\sigma_{kl} \quad (72)$$

According to the Prandtl-Reuss equation, the plastic strain-increment $\mathrm{d}\varepsilon_{ij}^{\mathrm{P}}$, which is represented by the increment of the force doublets here as

$$\mathrm{d}\varepsilon_{ij}^{\mathrm{P}} = C_{ijkl} \cdot \mathrm{d}T_{kl} \quad (73)$$

also is related to $\mathrm{d}\sigma_{ij}$

$$\mathrm{d}\varepsilon_{ij}^{\mathrm{P}} = S_{ijkl} \cdot \mathrm{d}\sigma_{kl} \quad (74)$$

where S_{ijkl} is the plastic compliance tensor and can be derived as follows.

It is generally postulated, based on experiments, that yielding can occur only if the stresses σ_{ij} satisfy the general yield criterion

$$f(\sigma_{ij}) = C \quad (75)$$

[C: hardening parameter]

and that the plastic strain-increment is expressed as

$$\mathrm{d}\varepsilon_{ij}^{\mathrm{p}} = h\frac{\partial f}{\partial \sigma_{ij}}\mathrm{d}f \tag{76}$$

where h is a proportionally constant, as yet undetermined.

When von Mises yield criterion is assumed, eqn (75) is written as

$$\begin{aligned} f(\sigma_{ij}) &= \sqrt{1/2}\left\{(\sigma_{11}-\sigma_{22})^2+(\sigma_{22}-\sigma_{33})^2+(\sigma_{33}-\sigma_{11})^2 \right. \\ &\quad \left. + 6(\sigma_{12}^2+\sigma_{23}^2+\sigma_{31}^2)\right\}^{1/2} \\ &= \overline{\sigma} \\ &= \sigma_s \end{aligned} \tag{77}$$

$$\overline{\sigma} = H(\int \mathrm{d}\overline{\varepsilon}^{\mathrm{p}}) \tag{78}$$

[$\overline{\sigma}$: equivalent stress, $\mathrm{d}\overline{\varepsilon}^{\mathrm{p}}$: equivalent plastic strain-increment]

Noting the equality

$$\sigma_{mn}\frac{\partial f}{\partial \sigma_{mn}} = \overline{\sigma} \tag{79}$$

and using eqn (76), we obtain

$$h\overline{\sigma}\mathrm{d}f = h\sigma_{mn}\frac{\partial f}{\partial \sigma_{mn}}\mathrm{d}f = \sigma_{mn}\mathrm{d}\varepsilon_{mn}^{\mathrm{p}} = \overline{\sigma}\mathrm{d}\overline{\varepsilon}^{\mathrm{p}} \tag{80}$$

By using eqns (78) and (77), $\mathrm{d}\overline{\varepsilon}^{\mathrm{p}}$ can be written as

$$\mathrm{d}\overline{\varepsilon}^{\mathrm{p}} = \frac{\mathrm{d}\overline{\sigma}}{H'} = \frac{\partial f}{H'\partial \sigma_{kl}}\mathrm{d}\sigma_{kl} \tag{81}$$

where $H' = \mathrm{d}\overline{\sigma}/\mathrm{d}\overline{\varepsilon}^{\mathrm{p}}$ corresponds to the slope of the equivalent stress $\overline{\sigma}$ – the plastic strain ($\int \mathrm{d}\overline{\varepsilon}^{\mathrm{p}}$) curve.

Now substituting eqns (80) and (81) into eqn (76), the equation expressing the relation between $\mathrm{d}\varepsilon^{\mathrm{p}}$ and $\mathrm{d}\sigma_{ij}$ can be obtained.

$$\mathrm{d}\varepsilon_{ij}^{\mathrm{p}} = \frac{1}{H'}\cdot\frac{\partial f}{\partial \sigma_{ij}}\cdot\frac{\partial f}{\partial \sigma_{kl}}\cdot \mathrm{d}\sigma_{kl} \tag{82}$$

That is,

$$S_{ijkl} = \frac{1}{H'}\cdot\frac{\partial f}{\partial \sigma_{ij}}\cdot\frac{\partial f}{\partial \sigma_{kl}} \tag{83}$$

When the material is assumed to obey Tresca's yield criterion, eqn (75) is written as

$$f(\sigma_{ij}) = (\sigma_I - \sigma_{III})/2 = \tau_S \tag{84}$$

$(\sigma_I, \sigma_{II}, \sigma_{III}$: principal stress, $\sigma_I \geq \sigma_{II} \geq \sigma_{III})$

$$\tau_S = F\left(\int \mathrm{d}\lambda^{\mathrm{p}}\right) \tag{85}$$

Proceeding as before, S_{ijkl} for the material obeying Tresca's yield criterion can be derived as

$$S_{ijkl} = \frac{1}{F'} \cdot \frac{\partial f}{\partial \sigma_{ij}} \cdot \frac{\partial f}{\partial \sigma_{kl}} \tag{86}$$

Consequently, from the above discussions it turned out that the densities $\phi_\eta(Q)$ and $T_{\xi\eta}(Q)$, the unknowns to be determined in the present elastic-plastic analysis, are determined from the following four conditions:

1. Boundary condition eqn (69) (in the case of displacement boundary-value problems) or eqn (70) (in the case of stress boundary-value problems)
2. Equation of stress in the plastic zone (68)
3. Relation between force doublets and plastic strains (73)
4. Elastic-plastic constitutive equation (74)

When the von Mises yield criterion is assumed eqn (83) is used for S_{ijkl}, and when Tresca's yield criterion is assumed eqn (86) is used for S_{ijkl}.

The densities $\phi_\eta(Q)$ and $T_{\xi\eta}(Q)$ usually have to be determined by numerical calculation. The numerical calculation here is referred to the following two procedures.

(1) Because the plastic compliance tensor S_{ijkl} is dependent on the stress state, the analysis is non-linear. Therefore, in the incremental analysis, some material is treated as quasi-elastic with a constant S_{ijkl} value which may correspond to either the initial stress values of the increment or the average stresses in the increment. This process is essentially the same as those used in the finite element method for solving elastic-plastic problems.[39,40]

(2) In order to solve the integral equation numerically, the imaginary boundary and the imaginary plastic zone are subdivided and the densities $\phi_\eta(Q)$ and $T_{\xi\eta}(Q)$ are discretized into a series of unknowns by assuming $\phi_\eta(Q)$ and $T_{\xi\eta}(Q)$ to be constant, linear or quadratic, etc., in each subdivided element. So the integral equation is approximated by simultaneous equations. The coefficients matrix of the simultaneous equations is found by integrating $u_i(P, Q_\eta)$ and $\sigma_{ij}(P, Q_\eta)$ over the boundary elements and integrating $u_i(P, Q_\eta)$ and $\sigma_{ij}(P, Q_\eta)$ over the zone elements. It is not necessary to demonstrate the details here, because these procedures are essentially similar to those in the case of pure elastic problems. It should be noted that, however, when the point P is involved in the integral region the integral $\int_\Omega \sigma_{ij}(P, Q_{\xi,\eta}) T_{\xi\eta}(Q) \mathrm{d}\Omega(Q)$ is strongly singular.

Based on the intuitive consideration for the plastic strain shown above, the problem of finding the principal value of the singular integral $\int_\Omega \sigma_{ij}(P, Q_{\xi,\eta}) T_{\xi\eta}(Q) \mathrm{d}\Omega(Q)$ can be overcome in a simple and elegant fashion.

The integral $\int_\Omega \sigma_{ij}(P, Q_{\xi,\eta}) T_{\xi\eta}(Q) \mathrm{d}\Omega(Q)$ is originally based on the consideration that the plastic zone can be represented by piling up the minute cubes which are subjected to a traction correspondent for force doublets. As seen in Fig. 48, the plastic zone in Fig. 48(a) is replaced by Fig. 48(b) where the elastic minute cubes subjected to proper tractions $T_{\xi\eta}(Q)$ are piled up. If the plastic deformation in the

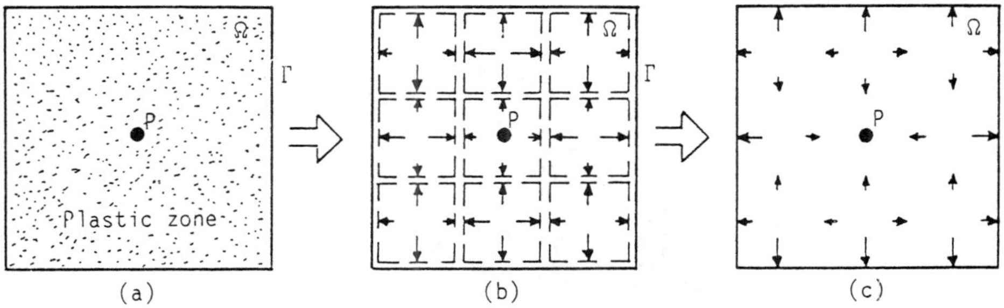

Figure 48: Treatment of singular integral

region Ω are uniform, the force doublets applied at the minute cubes becomes equal. In this case, all the forces applied to the touching faces of adjacent cubes cancel each other during the piling up process, and only the forces which are applied to the outside surface Γ of the plastic zone remain. Its density along Γ is:

$$a_\xi(Q) = T_{\xi\eta}(Q)\, n_\eta(Q) \tag{87}$$

$[n_\eta(Q)$: component of outward normal vector of $\Gamma]$

If the plastic deformations are not uniform, then, besides $a_\xi(Q)$ there are forces acting at the touching faces of the adjacent cubes as the remainder of tractions, as shown in Fig. 48(c). These forces can be regarded as the body forces distributed in the region Ω and their density $b_\xi(Q)$ is:

$$b_\xi(Q) = -\frac{\partial T_{\xi\eta}(Q)}{\partial Q_\eta} \tag{88}$$

Consequently, we can evaluate the integral $\int_\Omega \sigma_{ij}(P, Q_{\xi,\eta})T_{\xi\eta}(Q)\mathrm{d}\Omega(Q)$ as follows:

$$\begin{aligned} &\int_\Omega \sigma_{ij}(P, Q_{\xi,\eta})T_{\xi\eta}(Q)\mathrm{d}\Omega(Q) \\ &= \int_\Gamma \sigma_{ij}(P, Q_\xi)a_\xi(Q)\mathrm{d}\Gamma(Q) + \int_\Omega \sigma_{ij}(P, Q_\xi)b_\xi(Q)\mathrm{d}\Omega(Q) \end{aligned} \tag{89}$$

As an example, the elastic-plastic problem of an infinite plate with a crack is analyzed.[13]

Figure 49 shows the extent of the Mises plastic zone under a plane stress state. In the figure the yield zones based on elastic stress fields are also shown. The shapes and the dimensions of plastic zone based on the elastic-plastic analyses are similar to those of yield zone based on the pure elastic fields, but the former tilts slightly toward the x-axis as compared with the latter.

If Tresca's criterion is used, the plastic zone shape turns out to be different greatly from those based on the pure elastic fields. Figure 50 shows the extent of Tresca's plastic zone under the plane stress state. The plastic zone near a crack tip is found to be in the shape '←', i.e. one part of the zone satisfying the yield condition

$$|\sigma_1 - \sigma_2| = \sigma_s \quad , \quad \left\{ \begin{array}{l} \sigma_1 \\ \sigma_2 \end{array} \right. = \frac{\sigma_x + \sigma_y}{2} \pm \sqrt{\{(\sigma_x - \sigma_y)/2\}^2 + \tau_{xy}^2} \tag{90}$$

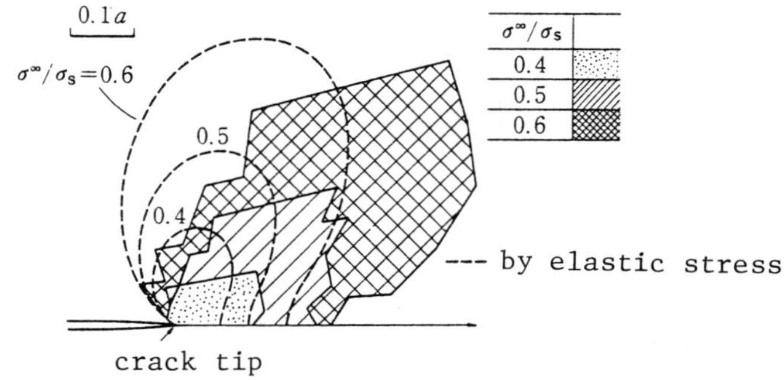

Figure 49: Extent of the Mises plastic zone near a crack tip (plane stress, $H'/E = 0.01$)

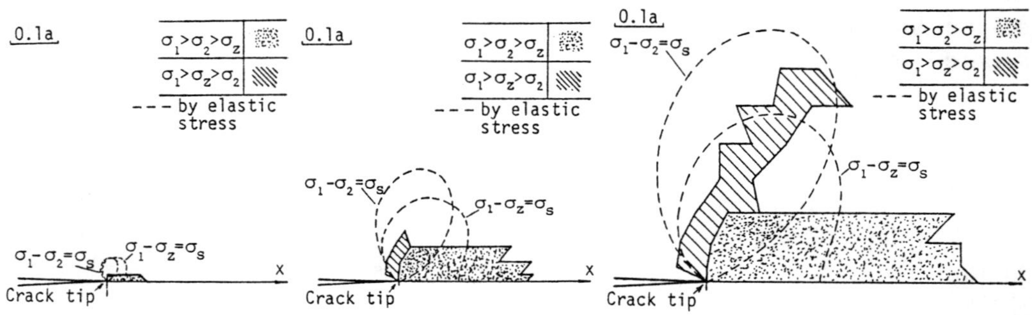

Figure 50: Extent of Tresca's plastic zone near a crack tip (plane stress, $H'/E = 0.01$)

extends slenderly along the x-axis including the crack, and another part satisfying the yield condition $|\sigma_1 - \sigma_2| = \sigma_s$ extends at about 60° from the x-axis. Moreover, the extents of Tresca's plastic zone under the same loading ($\sigma_\infty/\sigma_s = 0.45$) for the different hardening are compared in Fig. 51. In strain hardening materials, even when the plastic zone is extended ahead of a crack along the x=axis, the stresses on both sides of the plastic zone can still be raised by increasing loading because of strain hardening and can be yielded. When the rate of the strain hardening is very low, the plastic zone becomes a strip as in the case of the Dugdale model applied to a non-hardening material.

8 Application to elastodynamic problems

Based on the principle of superposition, BFM can be extended so as to be applicable even to elastodynamic problems.

Let us consider an elastodynamic problem of region Ω_I which is subjected to a given body force $f_i(Q,t); Q \in \Omega_I, t > 0$ and a given surface force $t_i^{nI}(Q,t); Q \in \Gamma\ , t > 0$, in which t is the time. At a time $t = 0$, the initial displacement and the initial

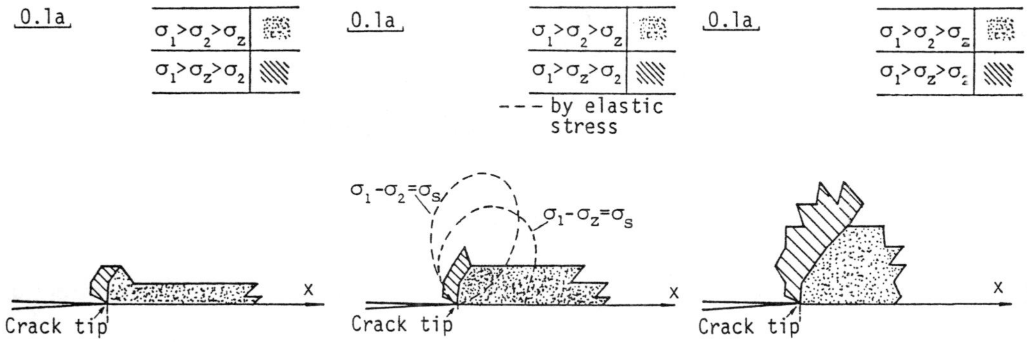

Figure 51: Comparison of Tresca's plastic enclaves under the same loading ($\sigma_\infty/\sigma_s = 0.45$) for different hardening (plane stress)

velocity are assumed to be given as follows.

$$\left.\begin{aligned} u_i(Q,0) &= u_{0i} \\ u_{i,t}(Q,0) &= V_{0i} \end{aligned}\right\} \quad (Q \in \Omega_I) \tag{91}$$

In order to construct an integral equation expressing the required elastodynamic field, we insert the region Ω_I in question into an auxiliary region Ω_E as in the case of the elastostatic problem. However, it is seen that the gaps between the regions Ω_I and Ω_E change with the time in this case. Therefore the density of discrepancies to be embedded along the imaginary boundary has to be a function of time t.

$$\psi_\eta(Q,t) = u_\eta^E(Q,t) - u_\eta^I(Q,t) \quad (Q \in \Gamma, t \geq 0) \tag{92}$$

In eqn (92), $u_i^E(Q,t)$ and $u_i^I(Q,t); Q \in \Gamma\ , t \geq 0$, are the surface displacements of the regions Ω_E and Ω_I, respectively. Similarly, the density of the point forces to be distributed along the imaginary boundary also becomes a function of time t.

$$\phi_\eta(Q,t) = t_\eta^{n_I}(Q,t) + t_\eta^{n_E}(Q,t) \quad (Q \in \Gamma, t \geq 0) \tag{93}$$

That is, the replacement of the problems of a finite region with the problem of an infinite body is still possible, as in the case of the elastostatic problem, although the point forces and the discrepancies to be embedded change with the time. Therefore the elastodynamic field in the region Ω_I at a time t can be obtained by superposing the fundamental elastodynamic fields in an infinite body due to all the embedded point forces and discrepancies applied from $t = 0$ up to $t = t$. Consequently, $u_i(P,t)$ (the i-direction displacement at a point $P; P \in \Omega_I$ and a time $t; t > 0$) and $\sigma_{ij}(P,t)$ (the stress at a point $P; P \in \Omega_I$ and time $t; t > 0$) can be expressed as follows.

$$\begin{aligned} u_i(P,t) = &\int_0^t \mathrm{d}\tau \int_\Gamma u_i(P,Q_\eta,t-\tau)\phi_\eta(Q,\tau)\mathrm{d}\Gamma(Q) \\ &+ \int_0^t \mathrm{d}\tau \int_\Gamma u_i^*(P,Q_\eta^n,t-\tau)\psi_\eta(Q,\tau)\mathrm{d}\Gamma(Q) \\ &+ \int_0^t \mathrm{d}\tau \int_{\Omega_I} u_i(P,Q_\eta,t-\tau)F_\eta(Q,\tau)\mathrm{d}\Omega(Q) \quad (P \in \Omega_I, t > 0) \end{aligned} \tag{94}$$

$$\sigma_{ij}(P,t) = \int_0^t d\tau \int_\Gamma \sigma_{ij}(P,Q_\eta,t-\tau)\phi_\eta(Q,\tau)d\Gamma(Q)$$
$$+\int_0^t d\tau \int_\Gamma \sigma^*_{ij}(P,Q^n_\eta,t-\tau)\psi_\eta(Q,\tau)d\Gamma(Q)$$
$$+\int_0^t d\tau \int_{\Omega_I} \sigma_{ij}(P,Q_\eta,t-\tau)F_\eta(Q,\tau)d\Omega(Q) \qquad (P\in\Omega_I, t>0) \tag{95}$$

[$u_i(P,Q_\eta,t-\tau)$: i-direction displacement at point P and time t due to η-direction unit point force acting at a point Q and time τ ,

$\sigma_{ij}(P,Q_\eta,t-\tau)$: stress σ_{ij} at a point P and time t due to η-direction unit point force acting at a point Q and time τ,

$u^*_i(P,Q^n_\eta,t-\tau)$: i-direction displacement at a point P and time t due to η-direction unit concentrated discrepancy of faces, whose normal vector is of n-direction, acting at a point Q and time τ,

$\sigma^*_{ij}(P,Q^n_\eta,t-\tau)$: stress σ_{ij} at a point P and time t due to η-direction unit concentrated discrepancy of faces, whose normal vector is of n-direction, acting at a point Q and time τ]

In eqns (94) and (95), $F_\eta(Q,t)$ is the density of the body force to be embedded in the region Ω_I. When $t>0, F_\eta(Q,t)$ is equal to the given body force $f_\eta(Q,t)$. When $t=0$, however, $F_\eta(Q,t)$ is dependent on the effect of the initial displacement and the initial velocity. A detailed discussion concerning this point follows.

Providing an initial velocity $V_{0\eta}(Q)$ to an elastic body can be regarded as providing a velocity increment $= u_{\eta,t}(Q,\varepsilon)-0 = V_{0\eta}(Q)$ to the elastic body during an interval of time $t=0\sim\varepsilon(\varepsilon\to 0)$. Based on the Newton's law of motion (force $\times$ interval of time $=$ mass $\times$ increment of velocity), it is shown that the external force $F^V_\eta(Q,t)\Delta\Omega\ (Q)$ for making an increment of velocity $V_{0\eta}(Q)$ in a small element $\Delta\Omega(Q)$ has to satisfy the following equation.

$$\int_0^\varepsilon F^V_\eta(Q,t)\Delta\Omega(Q)dt = \rho\Delta\Omega(Q)\times V_{0\eta}(Q) \tag{96}$$

ρ : mass density of elastic body)

Therefore, by considering the limit of $\varepsilon\to 0$, it is conceivable that the provision of the initial velocity $V_{0\eta}(Q)$ corresponds to the embedding of a body force with density $F^V_\eta(Q,t)$

$$F^V_\eta(Q,t) = \rho V_{0\eta}(Q)\delta(t-0^+) \tag{97}$$

($\delta(\cdots)$: Dirac delta function)

as shown in Fig. 52.

Moreover, giving an initial displacement $u_{0\eta}(Q)$ can be regarded as giving the increments of velocity $u_{0\eta}(Q)/\varepsilon$ and $-u_{0\eta}(Q)/\varepsilon$ at $t=0^+$ and $t=0^+ +\varepsilon(\varepsilon\to 0)$, respectively. Therefore, based on the fact illustrated in Fig. 52 giving an initial displacement $u_{0\eta}(Q)$ corresponds to applying a pair of body forces; the body forces

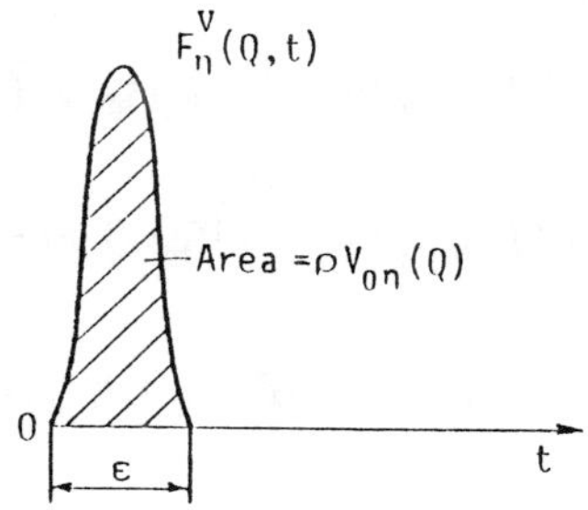

Figure 52: Body force $F_\eta(Q,t)$ equivalent to initial velocity

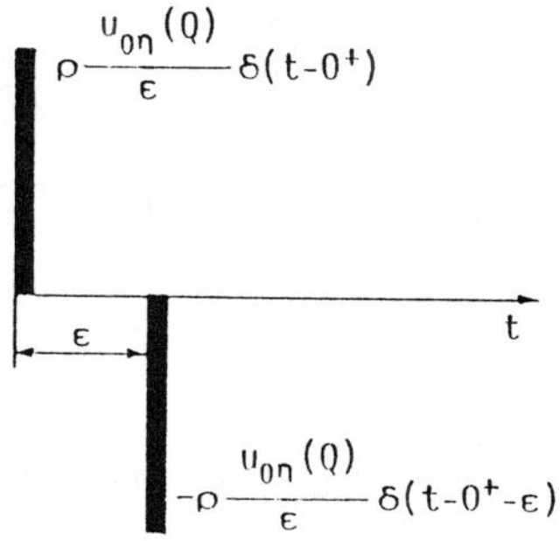

Figure 53: Body force $F_\eta(Q,t)$ equivalent to initial displacement

$\rho\{u_{0\eta}(Q)/\varepsilon\}\delta(t-\tau)|_{\tau=0^+}$ at $t=0^+$ and the body force $-\rho\{u_{0\eta}(Q)/\varepsilon\}\delta[t-(\tau+\varepsilon)]|_{\tau=0^+}$ at $t=0^+ + \varepsilon$ as shown in Fig. 53. That is, the body forces $F_\eta^u(Q,t)$ which are equivalent to the initial displacement are expressed as follows.

$$\begin{aligned} F_\eta^u(Q,t) &= \lim_{\varepsilon\to 0}\left\{\rho\frac{u_{0\eta}(Q)}{\varepsilon}\delta(t-\tau)|_{\tau=0^+} - \rho\frac{u_{0\eta}(Q)}{\varepsilon}\right. \\ &\times\ \left.\delta[t-(\tau+\varepsilon)]|_{\tau=o^+}\right\} \\ &= \lim_{\varepsilon\to 0}\left\{-\rho\frac{u_{0\eta}(Q)}{\varepsilon}\varepsilon\delta_{,\tau}(t-\tau)|_{\tau=0^+}\right\} \\ &= \rho u_{0\eta}(Q)\delta_{,t}(t-0^+) \end{aligned} \tag{98}$$

By taking eqns (97) and (98) into account, the displacement term depending upon $F_\eta(Q,t)$ in eqn (94) can be expressed by $f_\eta(Q,t)$ and the initial conditions as follows (denoted by $R_i(P,t)$).

$$\begin{aligned} R_i(P,t) &= \int_0^t d\tau\int_{\Omega_I} u_i(P,Q_\eta,t-\tau)F_\eta(Q,\tau)d\Omega(Q) \\ &= \int_0^t d\tau\int_{\Omega_I} u_i(P,Q_\eta,t-\tau)\{f_\eta(Q,\tau)+F_\eta^u(Q,t)+F_\eta^V(Q,\tau)\}d\Omega(Q) \\ &= \int_0^t d\tau\int_{\Omega_I} u_i(P,Q_\eta,t-\tau)f\ 'Q,\tau)d\Omega(Q) \end{aligned} \tag{99}$$

$$
\begin{aligned}
&+ \int_{\Omega_I} \mathrm{d}\Omega(Q) \int_0^t u_i(P, Q_\eta, t-\tau)\rho u_{0\eta}(Q)\delta_{,\tau}(\tau - 0^+)\mathrm{d}\tau \\
&+ \int_{\Omega_I} \mathrm{d}\Omega(Q) \int_0^t u_i(P, Q_\eta, t-\tau)\rho V_{0\eta}(Q)\delta(\tau - 0^+)\mathrm{d}\tau \\
&= \int_0^t \mathrm{d}\tau \int_{\Omega_I} u_i(P, Q_\eta, t-\tau) f_\eta(Q,\tau)\mathrm{d}\Omega(Q) \\
&+ \int_{\Omega_I} \{u_i(P, Q_\eta, t-0^+)\rho V_{0\eta}(Q) + u_{i,t}(P, Q_\eta, t-0^+)\rho u_{0\eta}(Q)\}\mathrm{d}\Omega(Q)
\end{aligned}
$$

Similarly, the stress term depending upon $F_\eta(Q,t)$ in eqn (95) can be expressed as follows (denoted by $S_{ij}(P,t)$).

$$
\begin{aligned}
S_{ij}(P,t) &= \int_0^t \mathrm{d}\tau \int_{\Omega_I} \sigma_{ij}(P, Q_\eta, t-\tau) F_\eta(Q,\tau)\mathrm{d}\Omega(Q) \qquad (100) \\
&= \int_0^t \mathrm{d}\tau \int_{\Omega_I} \sigma_{ij}(P, Q_\eta, t-\tau) f_\eta(Q,\tau)\mathrm{d}\Omega(Q) \\
&+ \int_{\Omega_I} \{\sigma_{ij}(P, Q_\eta, t-0^+)\rho V_{0\eta}(Q) + \sigma_{ij,t}(P, Q_\eta, t-0^+)\rho u_{0\eta}(Q)\}\mathrm{d}\Omega(Q)
\end{aligned}
$$

Consequently, the elastodynamic displacement field and stress field in the region Ω_I can be expressed by superposing all the actions (action of the given body force, action of the body force equivalent to the initial conditions, actions of the point forces and the discrepancies to be distributed along the imaginary boundary) as follows.

$$
\begin{aligned}
u_i(P,t) &= \int_0^t \mathrm{d}\tau \int_\Gamma u_i(P, Q_\eta, t-\tau)\phi_\eta(Q,\tau)\mathrm{d}\Gamma(Q) \qquad (101) \\
&+ \int_0^t \mathrm{d}\tau \int_\Gamma u_i^*(P, Q_\eta^n, t-\tau)\psi_\eta(Q,\tau)\mathrm{d}\Gamma(Q) + R_i(P,t), \quad (P \in \Omega_I, t > 0)
\end{aligned}
$$

$$
\begin{aligned}
\sigma_{ij}(P,t) &= \int_0^t \mathrm{d}\tau \int_\Gamma \sigma_{ij}(P, Q_\eta, t-\tau)\phi_\eta(Q,\tau)\mathrm{d}\Gamma(Q) \qquad (102) \\
&+ \int_0^t \mathrm{d}\tau \int_\Gamma \sigma_{ij}^*(P, Q_\eta^n, t-\tau)\psi_\eta(Q,\tau)\mathrm{d}\Gamma(Q) + S_{ij}(P,t)\,, \quad (P \in \Omega_I, t > 0)
\end{aligned}
$$

Equations (101) and (102) are the most general fundamental equations for expressing the elastodynamic field of the region in question based on the principle of superposition*. If the densities of point forces and the discrepancies to be distributed along the imaginary boundary are known, the elastodynamic field in the region can be calculated from eqns (101) and (102). One of the two unknowns $\phi_\eta(Q,t)$ and $\psi_\eta(Q,t)$, in eqns (101) and (102) can be specified arbitrarily and the remaining unknown is determined from the boundary condition equation, which is obtained from eqn (101)

* The body force $f_\eta^E(Q,t)$ can also be distributed in auxiliary region similarly as in the case of elastostatic problems. However, for simplicity $f_\eta^E(Q,t)$ is put to be 0 here.

and (102) by moving point P on the boundary. The details of numerical procedures are similar to those in the case of the elastostatic problems.

References

1. Nisitani, H. Two-dimensional problem solved using a digital computer, *Journal of the Japan Society of Mechanical Engineers*, JSME, 1967, **70**, 627-635.

2. Nisitani, H. The two-dimensional stress problem solved using an electric digital computer, *Bulletin of the JSME*, 1968, **11**(43), 14-23.

3. Nisitani, H. *Proceedings of the 1st Conference on Numerical Methods in Fracture Mechanics*, A.L. Luxmore & D.R.J. Own (ed), 1978, Swansea, UK, pp. 67-80.

4. Nisitani, H. Stress analysis of notch problems by body force method, *Chapter 1, Mechanics of Fracture 5*, G.C. Sih (ed), 1978, Sijthoff & Noordhoff, pp. 1-68.

5. Nisitani, H. & Murakami, Y. Stress intensity factors of an elliptical crack or a semi-elliptical crack subject to tension, *International Journal of Fracture*, 1974, **10**(3), 353-368.

6. Murakami, Y. & Nisitani, H. *Proceedings of International Conference on Fracture Mechanics Technology*, Vol. II, 1977, pp. 1171-1185.

7. Nisitani, H. & Isida, M. Simple procedure for calculating K_I of a notch with a crack of arbitrary size and its application to non-propagating fracture crack, *Proc. Joint Conference on Exp. Mechanics*, (SESA & JSME), 1982.

8. Nisitani, H., Murakami, Y. & Chen, D.-H. Development of body force method and its application to crack problems, *Proc. of ICF International Symposium on Fracture Mechanics*, 1983.

9. Isida, M., Chen, D.-H. & Nisitani, H. Plane problems of an arbitrary array of cracks emanating from the edge of an elliptical hole, *Engng. Frac. Mech.*, 1985, **21**(5).

10. Nisitani, H. & Noda, T. Tension of a cylindrical bar having an infinite row of circumferential cracks, *Engng. Frac. Mech.*, 1984, **20**(4).

11. Chen, D.-H. *Proceedings of the International Conference on Role of Fracture Mechanics in Modern Technology (Fukuoka)*, 1978, pp. 153.

12. Nisitani, H. & Chen, D.-H. Elastic-plastic analysis of an infinite plate with an elliptic hole by body force method, *Trans. Japan Soc. Mech. Engrs*, 31-465, pp. 1471.

13. Nisitani, H. & Chen, D.-H. Elastic-plastic analysis of a crack by the body force method (Tension of an infinite plate with a crack), *Trans. Japan Soc. Mech. Engrs*, 52-474, pp. 544.

14. Chen, D.-H., Nisitani, H. & Noguchi, H. *Bulletin of the JSME*, 1990, **33**(2), 193-202.

15. Love, A.E.H. *Mathematical Theory of Elasticity*, 4th edition, Cambridge, 1927, pp. 183.

16. Brebbia, C.A. *The Boundary Element Method for Engineers*, Pentech Press, 1978.

17. Nisitani, H. & Chen, D.-H. *Body Force Method*, Baifukan Publications, Tokyo, 1988.

18. Brebbia, C.A. *Progress in Boundary Element Methods*, Vol. 1, Pentech Press, 1981.

19. Isida, M. Private communication, 1965.

20. Noguchi, H., Nisitani, H., Gotou, H. & Mori, K. *Trans. Japan Soc. Engrs*, 1987, 53-488, 820.

21. Isida, M. Analysis of stress intensity factors of plate with arbitrary array cracks and bent cracks, *Trans Japan, Soc. Mech. Engrs*, 1978, **44**, 1122-1133.

22. Nisitani, H., Chen, D.-H. & Mori, K. Stress intensity factors for an internal semi-elliptical surface crack in cylindrical pressure vessels, *Advances in Fracture and Fatigue for the 1990s Volume II*, G. Yagawa & G.M. Wilkowski (ed), 1989, pp. 43.

23. Nisitani, H., Saimoto, A. & Noguchi, H. Versatile method of analysis of two-dimensional elastic problem by body force method, *Proceedings of the First International Conference on Computer Aided Assessment and Control Localized Damage*, 1990, pp. 33.

24. Isida, M. & Sato, R. Analysis of finite regions containing a circular hole or inclusion with applications to their periodic array, *Trans Japan Soc. Mech. Engrs*, 1984, **50**, 1619-1627.

25. Kitagawa, Y. & Yuuki, R. Stress intensity factors for branched cracked in a two-dimensional stress state, *Trans Japan Soc. Mech. Engrs*, **41**, 1641.

26. Isida, M. & Noguchi, H. Formulae of stress intensity factors of branched cracks in plane problems, *Trans Japan Soc. Mech. Engrs*, 1983, **49**, 469-479.

27. Noda, N., Nisitani, H. & Fukasako, I. Tension and in-plane bending of a strip with a 60° V-shaped single edge notch, *Trans. Japan Soc. Mech. Engrs*, 1986, **52**, 1066-1072.

28. Nisitani, H. & Mori, K. Influence of supporting conditions on stress intensity factors for single-edge-cracked specimens under bending, Technical Report of the Kyushu University, 1985, **58**, pp. 751-755.

29. Gross, B. & Srawly, J.E. Stress intensity factor for single-edge notch specimens in bending or combined bending and tension by boundary collocation of a stress function, NASA Technical Report, D-2603, 1965.

30. Nisitani, H. & Noda, N. Stress concentration of a strip with double edge notches under tension or in-plane bending, *Eng. Frac. Mech.*, 1986, **6**, 1051-1065.

31. Benthem, J.P. & Koiter, W. Asymptotic approximation to cracked problems, *Method of Analysis of Cracked Problems*, G.C. Shi (ed), Sijthoff & Noordhoff, 1972.

32. Nisitani, H. & Noda, N. Stress concentration of a cylindrical bar with a V-shaped circumferential groove under torsion, tension or bending, *Eng. Frac. Mech.*,. 1984, **20**(5), 743-766.

33. Raju, I.S. & Newnam. J.C., Jr. Stress intensity factors for internal and external surface cracks in cylindrical vessels, *ASME Journal of Pressure Vessel Technology*, November 1982, **104**, 293-298.

34. Isida, M., Noguchi, H. & Yoshida, T. Tension and bending of finite thickness plates with a semi-elliptical surface crack, *Int. Journal of Fracture*, 1984, **26**(3), 157-188.

35. Kobayashi, A.S., Emery, A.F., Polvanich, N. & Love, W.J. Inner and outer surface cracks in internally pressurized cylinders, *ASME Journal of Pressure Vessel Technology*, Feb. 1977, 83-89.

36. Hellot, J., Labbens, R.C. & Pellissiertanon, A. Semi-elliptical cracks in the meridional plate or a cylinder subjected to stress gradients, *Fracture Mechanics*, ASTM STP 67, American Society for Testing and Materials, 1979, 431-446.

37. McGowan, J.J. and Raymund, M. Stress intensity factor solutions for internal longitudinal semi-elliptical surface flaws in a cylinder under arbitrary loadings, *Fracture Mechanics*, ASTM STP 67, American Society for Testing and Materials, 1979, 365-380.

38. Benthem, J.P. & Koiter, W.T. *Mechanics of Fracture 1*, Noordhoff, 1973, pp. 131-178.

39. Yamada, Y., Yoshimura, N. & Sakurai, T. Plastic stress-strain matrix and its application for the solution of elastic-plastic problems by the finite element method. *Int. J. Mech. Soc.*, **10**, 343.

40. Marcal, P.V. & King, I.P. Elasto-plastic analysis of two-dimensional stress systems by the finite element method, *Int. J. Mech. Soc.*, **9**, 143.

Chapter 5

Special crack-tip elements for 3D problems

M.L. Luchi & S. Rizzuti

Department of Mechanical Engineering, University of Calabria 87030, Arcavacata di Rende (Cosenza), Italy

Introduction

Three-dimensional linear elastic fracture mechanics was early recognized as one of the fields where the boundary element method can most successfully compete with the finite element method. The first applications of BEM to the solution of 3-dimensional elastic crack problems do, in fact, date back to 1971.[1] Since low order elements, either constant[1] or linear[2,3] were employed in those earliest analyses, a very fine discretization was required in the crack-front region in order that the $\sqrt{r}$ displacement and the $1/\sqrt{r}$ traction variations (r distance from the front) could be acceptably approximated.

Later, the successful introduction of quarter-point isoparametric elements in FEM fracture mechanics analysis.[4,5] suggested the use of quarter-point boundary elements along the crack-front[6,7] as an alternative to standard eight-node quadratic elements.[8] However, while quarter-point finite elements both represent the $\sqrt{r}$ variation and introduce a $1/\sqrt{r}$ singularity in the stress field, the use of quarter-point elements in the boundary element method, in which displacements and tractions are separately approximated, does only enable the displacement behaviour to be properly represented. On the other hand, in order that the main advantages of the method with respect to FEM, i.e. simpler discretization and higher accuracy, can be fully attained it is clearly necessary that the crack-front elements be capable of modeling the singularity present in the traction field.

Although the above arguments were fully developed by Cruse as early as 1977,[9] traction-singular boundary elements for three-dimensional crack analysis, and the associated numerical procedures, were only thoroughly discussed in 1983.[10]

The formulation of special crack-front elements by means of proper displacement and traction interpolation functions is illustrated in detail in the first section of the

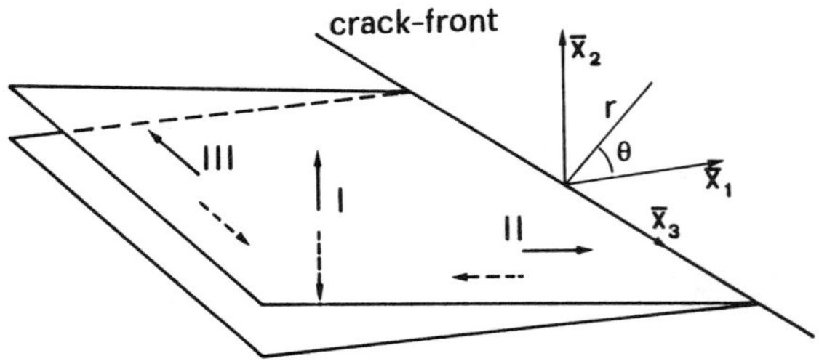

Figure 1: Fracture modes I, II and III

present work.

In order that singular integrals extending over special elements can be accurately evaluated, particular integration procedures must be used to remove the higher order singularity resulting from the use of singular traction interpolation functions. Two different integration procedures, proposed respectively in Ref. 11 and 12, are given in Section 2.

Besides yielding higher accuracy, the use of traction-singular elements presents another significant advantage. As shown in Section 3, the traction value along the crack-front can in fact be interpreted as a stress intensity factor. The SIF values can thus be directly obtained as primary unknowns, eliminating the need of assuming a particular planar state in the crack-front region.

Some examples, taken from the literature, of three-dimensional crack analyses carried out by using traction-singular elements are reported in Section 4.

1 Special element formulation

It is well-known that when inelastic effects at the crack-tip are disregarded, the displacements u_1, u_2 and u_3 in directions $\overline{x}_1$, $\overline{x}_2$ and $\overline{x}_3$, respectively parallel to the normal, binormal and tangent to the crack edge (see Fig. 1), can be expressed in terms of the polar co-ordinates r and θ (measured on a plane normal to the crack-front with respect to an origin where the plane is cut by the front) as

$$u_1 = \frac{1+\nu}{4E}\sqrt{\frac{2r}{\pi}}\left\{K_I[(2\kappa-1)\cos\frac{1}{2}\theta - \cos\frac{3}{2}\theta] + K_{II}[(2\kappa+3)\sin\frac{1}{2}\theta + \sin\frac{3}{2}\theta]\right\} \quad (1)$$

$$u_2 = \frac{1+\nu}{4E}\sqrt{\frac{2r}{\pi}}\left\{K_I[(2\kappa+1)\sin\frac{1}{2}\theta - \sin\frac{3}{2}\theta] + K_{II}[(2\kappa-3)\cos\frac{1}{2}\theta + \cos\frac{3}{2}\theta]\right\}$$

$$u_3 = \frac{2(1+\nu)}{E}\sqrt{\frac{2r}{\pi}}K_{III}\sin\frac{1}{2}\theta$$

where E is Young's modulus, ν Poisson's ratio and K_I, K_{II} and K_{III} respectively, represent the stress intensity factors in the opening, sliding and tearing modes, i.e. in the modes of deformation associated with symmetric opening, in-plane and anti-plane shear. Stresses σ_{ij} can be accordingly given as

$$\sigma_{11} = \frac{1}{\sqrt{2\pi r}}\left\{K_I\cos\frac{1}{2}\theta[1-\sin\frac{1}{2}\theta\sin\frac{3}{2}\theta] - K_{II}\sin\frac{1}{2}\theta[2+\cos\frac{1}{2}\theta\cos\frac{3}{2}\theta]\right\} \quad (2)$$

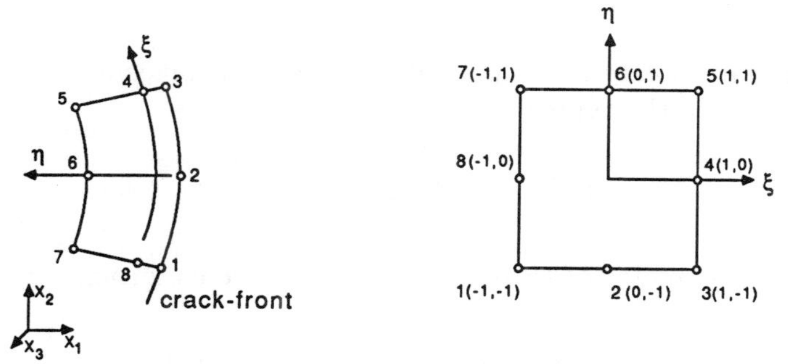

Figure 2: Mapping of the traction-singular quarter-point element

$$
\begin{aligned}
\sigma_{22} &= \frac{1}{\sqrt{2\pi r}}\left\{K_I \cos\frac{1}{2}\theta[1+\sin\frac{1}{2}\theta\sin\frac{3}{2}\theta] + K_{II}\sin\frac{1}{2}\theta\cos\frac{1}{2}\theta\cos\frac{3}{2}\theta]\right\} \\
\sigma_{33} &= \frac{3-\nu-\kappa-\nu\kappa}{2\nu}(\sigma_{11}+\sigma_{22}) \\
\sigma_{12} &= \frac{1}{\sqrt{2\pi r}}\left\{K_I \sin\frac{1}{2}\theta\cos\frac{1}{2}\theta\cos\frac{3}{2}\theta + K_{II}\cos\frac{1}{2}\theta[1-\sin\frac{1}{2}\theta\sin\frac{3}{2}\theta]\right\} \\
\sigma_{23} &= \frac{1}{\sqrt{2\pi r}}K_{III}\cos\frac{1}{2}\theta \\
\sigma_{31} &= -\frac{1}{\sqrt{2\pi r}}K_{III}\sin\frac{1}{2}\theta
\end{aligned}
$$

where $\kappa = 3 - 4\nu$ in plane strain and $\kappa = (3-\nu)/(1+\nu)$ in plane stress.

The asymptotic relations (1) and (2) do actually represent the leading terms of the infinite power series which describe the displacement and stress fields

$$
\begin{aligned}
u_i(r,\theta) &= \alpha_i^1(\theta)r^{\frac{1}{2}} + \alpha_i^2(\theta)r + \alpha_i^3(\theta)r^{\frac{3}{2}} + \ldots \\
\sigma_{ij}(r,\theta) &= \beta_{ij}^1(\theta)r^{-\frac{1}{2}} + \beta_{ij}^2(\theta)r^0 + \beta_{ij}^3(\theta)r^{\frac{1}{2}} + \ldots
\end{aligned}
\tag{3}
$$

and can therefore approximate displacements and stresses only within a small region around the crack-front ($r << a$, where a is the size of the defect).

It is evident from the above considerations that, in order to accurately model the elastic behaviour, displacement and traction variations corresponding to the first terms of the respective power series should be allowed in the elements adjacent to the crack-front.

As in two dimensions,[13] displacement variations proportional to $r^{\frac{1}{2}}$ and r and traction variations proportional to $r^{-\frac{1}{2}}$ and $r^{\frac{1}{2}}$ can be represented by suitably modifying the traction interpolation functions in quarter-point elements. Quarter-point isoparametric elements (see Fig. 2, where the side $\eta = -1$ lies on the crack-front), where quadratic shape functions ϕ^s

$$
\begin{aligned}
\phi^s &= \frac{1}{4}(1+\xi\xi_s)(1+\eta\eta_s)(\xi\xi_s+\eta\eta_s-1) \quad \text{for } s=1,3,5,7 \\
\phi^s &= \frac{1}{2}(1-\xi^2)(1+\eta\eta_s) \quad \text{for } s=2,6
\end{aligned}
\tag{4}
$$

$$\phi_s = \frac{1}{2}(1+\xi\xi_s)(1-\eta^2) \quad \text{for } s = 4, 8$$

are employed to interpolate the nodal values of both displacement and traction, can represent variations of the type

$$\begin{matrix} u_i \\ t_i \end{matrix} = a_i^1 + a_i^2\sqrt{r} + a_i^3 r \tag{5}$$

Provided that the geometry of the element satisfies the condition

$$x_i^6 - x_i^2 = \frac{1}{2}[(x_i^1 + x_i^3) - (x_i^5 + x_i^7)] \tag{6}$$

the distance r from the front along a line $x =$ const. is in fact proportional to $(1+\eta)^2$. In order to model the singular behaviour of tractions

$$t_i = \frac{b_i^1}{\sqrt{r}} + b_i^2 + b_i^3\sqrt{r} \tag{7}$$

proper traction interpolation functions must consequently be defined. Suitable functions can be straightforwardly derived from the quadratic shape functions by dividing each of them by $(1+\eta)$ and by multiplying ϕ^5, ϕ^6 and ϕ^7 by 2 so that they are equal to unity at nodes 5, 6 and 7, respectively.

$$\begin{aligned}
\phi_t^1 &= \frac{1}{2}(1-\xi)[-\frac{\xi}{1+\eta} - (1-\frac{1}{2}\xi) + \frac{1}{2}(1+\eta)] \\
\phi_t^2 &= (1-\xi^2)[\frac{1}{1+\eta} - \frac{1}{2}] \\
\phi_t^3 &= \frac{1}{2}(1+\xi)[\frac{\xi}{1+\eta} - (1+\frac{1}{2}\xi) + \frac{1}{2}(1+\eta)] \\
\phi_t^4 &= \frac{1}{2}(1+\xi)(1-\eta) \\
\phi_t^5 &= \frac{1}{2}(1+\xi)[(\xi-2) + \frac{1}{2}(1+\eta)] \\
\phi_t^6 &= (1-\xi^2) \\
\phi_t^7 &= \frac{1}{2}(1-\xi)[-(\xi+2) + \frac{1}{2}(1+\eta)] \\
\phi_t^8 &= \frac{1}{2}(1-\xi)(1-\eta)
\end{aligned} \tag{8}$$

Quoting the title of Ref. 4, it can thus be affirmed that 'crack-tip boundary elements are necessary'; i.e. that, in contrast with the finite element method, an accurate boundary element crack analysis necessarily requires that special elements be included in the boundary element library. The above traction-singular element formulation appears particularly straightforward mainly because it derives from the familiar quarter-point elements; however the use of different crack-front elements where the side nodes are left in their usual mid-side position and both displacement and traction interpolation functions differ from the quadratic shape functions does not present any

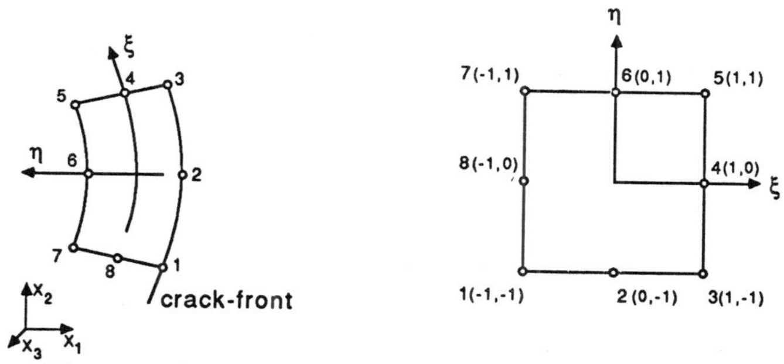

Figure 3: Mapping of the traction-singular element

particular disadvantage. On the contrary, as will be shown in the following section, this latter approach facilitates the development of numerical procedures which enable the singular integrals extending over a special element to be accurately evaluated.

With reference to Fig. 3, where the side $\eta = -1$ lies once again on the crack-front, proper displacement interpolation functions can be defined as

$$
\begin{aligned}
\phi_{\mathrm{d}}^{1} &= \frac{1}{2}(1-\xi)[-\xi-(1-\frac{\sqrt{2}}{2}\xi)\sqrt{1+\eta}+\frac{\sqrt{2}}{2}(1+\eta)] \qquad (9)\\
\phi_{\mathrm{d}}^{2} &= (1-\xi^{2})[1-\frac{\sqrt{2}}{2}\sqrt{1+\eta}]\\
\phi_{\mathrm{d}}^{3} &= \frac{1}{2}(1+\xi)[\xi-(1+\frac{\sqrt{2}}{2}\xi)\sqrt{1+\eta}+\frac{\sqrt{2}}{2}(1+\eta)]\\
\phi_{\mathrm{d}}^{4} &= \frac{1}{2}(1+\xi)[(2+\sqrt{2})\sqrt{1+\eta}-(1+\sqrt{2})(1+\eta)]\\
\phi_{\mathrm{d}}^{5} &= \frac{1}{4}(1+\xi)[\sqrt{2}(\xi-2-\sqrt{2})\sqrt{1+\eta}+(2+\sqrt{2})(1+\eta)]\\
\phi_{\mathrm{d}}^{6} &= \frac{\sqrt{2}}{2}(1-\xi^{2})\sqrt{1+\eta}\\
\phi_{\mathrm{d}}^{7} &= \frac{1}{4}(1-\xi)[-\sqrt{2}(\xi+2+\sqrt{2})\sqrt{1+\eta}+(2+\sqrt{2})(1+\eta)]\\
\phi_{\mathrm{d}}^{8} &= \frac{1}{2}(1-\xi)[(2+\sqrt{2})\sqrt{1+\eta}-(1+\sqrt{2})(1+\eta)]
\end{aligned}
$$

Traction interpolation functions, ϕ_{t}^{s} can be derived from ϕ_{d}^{s} by dividing each of them by $\sqrt{(1+\eta)}$ and by multiplying ϕ_{d}^{5}, ϕ_{d}^{6} and ϕ_{d}^{7} by $\sqrt{2}$ and are given as

$$
\begin{aligned}
\phi_{\mathrm{t}}^{1} &= \frac{1}{2}(1-\xi)[-\frac{\xi}{\sqrt{1+\eta}}-(1-\frac{\sqrt{2}}{2}\xi)+\frac{\sqrt{2}}{2}\sqrt{1+\eta}] \qquad (10)\\
\phi_{\mathrm{t}}^{2} &= (1-\xi^{2})[\frac{\xi}{\sqrt{1+\eta}}-\frac{\sqrt{2}}{2}]\\
\phi_{\mathrm{t}}^{3} &= \frac{1}{2}(1+\xi)[\frac{\xi}{\sqrt{1+\eta}}-(1+\frac{\sqrt{2}}{2}\xi)+\frac{\sqrt{2}}{2}\sqrt{1+\eta}]
\end{aligned}
$$

$$
\begin{aligned}
\phi_t^4 &= \frac{1}{2}(1+\xi)[(2+\sqrt{2})-(1+\sqrt{2})\sqrt{1+\eta}] \\
\phi_t^5 &= \frac{1}{4}(1+\xi)[2(\xi-2-\sqrt{2}))+2(1+\sqrt{2})\sqrt{1+\eta}] \\
\phi_t^6 &= (1-\xi^2) \\
\phi_t^7 &= \frac{1}{4}(1+\xi)[-2(\xi+2+\sqrt{2})+2(1+\sqrt{2})\sqrt{1+\eta}] \\
\phi_t^8 &= \frac{1}{2}(1-\xi)[(2+\sqrt{2})-(1+\sqrt{2})\sqrt{1+\eta}]
\end{aligned}
$$

It can be easily shown that the above interpolation functions represent displacement and traction variations, as given by (5) and (7) respectively, since the distance r along a line ξ = const. varies now linearly with $(1+\eta)$, where η = -1 corresponds to the crack-front.

For the sake of completeness, it is worth recalling that different interpolation functions can be defined[14] which are capable of representing more terms of the power series (3).

When modeling a surface crack six-node elements are generally employed around the crack-tip on the free surface. Inter-element compatibility obviously requires that the special elements formulated by (9) and (10) be only used along the crack-front when the six-node elements on the free surface are either standard quadratic elements or elements (analogous to those formulated by Blackburn[15] in the finite element method) where special interpolation functions are defined to represent the displacement variation given in (5). It is evident that the same variation (5) can be obtained by using six-node quarter-point elements; in this case quarter-point elements must also be used along the crack-front. Since the traction-singular quarter-point elements described in the first part of the section present some drawbacks where the numerical integration is concerned (see next section), quarter-point elements have been proposed[10] over which the same integration procedures illustrated in Section 2 can be employed. These special traction-singular elements differ from the standard ones in that the side nodes in the ξ, η plane ($\eta = 0$) are also moved to the 1/4 position ($\eta = -0.5$, where $\eta = -1$ lies on the crack-front). Since the relation between $(1+\eta)$ and the distance from the crack-front along a line ξ = const. is thus once again linear, in order to represent the variations defined in (5) and (7), proper displacement and traction interpolation functions must be employed with the same dependence on $(1+\eta)$ as in (9) and (10). The derivation of the two sets of functions is given in Appendix A.

It is quite obvious that traction-singular elements are to be employed in the region externally adjacent to the crack-front, whereas the elements internally adjacent to the front, lying on the crack surface, must be capable to correctly represent the displacement variation only.

2 Procedures for numerical integration

For the sake of clarity, let us recall that the boundary integral equation (BIE) for an elastic homogeneous isotropic body in the absence of body forces can be written as

$$c_{ij}(P)u_j(P) + \int_S T_{ij}(P,Q)u_j(Q)\mathrm{d}S = \int_S U_{ij}(P,Q)t_j(Q)\mathrm{d}S \tag{11}$$

where u_j and t_j are displacement and traction boundary values, U_{ij} and T_{ij} are Kelvin's point force solutions for three-dimensional elastostatics, P and Q respectively denote source point and integration point on the boundary and c_{ij} is a coefficient dependent on the local boundary geometry at point P. Integration is performed over the entire boundary S of the body.

Let the surface be divided into m elements and on each r-node element the geometry, displacement and tractions be represented in terms of the respective nodal values x_i^{s}, u_i^{s} and $t_i^{\mathrm{s}}(s \in \{1,\ldots,r\})$ as

$$\begin{aligned} x_i(\xi,\eta) &= \phi^{\mathrm{s}}(\xi,\eta)x_i^{\mathrm{s}} \\ u_i(\xi,\eta) &= \phi_{\mathrm{d}}^{\mathrm{s}}(\xi,\eta)u_i^{\mathrm{s}} \\ t_i(\xi,\eta) &= \phi_{\mathrm{t}}^{\mathrm{s}}(\xi,\eta)t_i^{\mathrm{s}} \end{aligned} \tag{12}$$

where ϕ^{s}, $\phi_{\mathrm{d}}^{\mathrm{s}}$ and $\phi_{\mathrm{t}}^{\mathrm{s}}$ are functions of the intrinsic co-ordinates ξ and η. Integral eqn (11) can consequently be discretized as

$$c_{ij}(P^a)u_j(P^a) + \sum_{l=1}^{m}{}^{\infty}\sum_{s=1}^{r} H_{ij}^{als}u_j^{ls} = \sum_{l=1}^{m}{}^{\infty}\sum_{s=1}^{r} G_{ij}^{als}t_j^{ls} \tag{13}$$

where the surface point P has been taken to coincide with node α of the discretization. Coefficients H_{ij}^{als} and G_{ij}^{als} are given by the integrals

$$G_{ij}^{als} = \int_{-1}^{1} U_{ij}(P^a, Q^l(\xi,\eta))\phi_{\mathrm{t}}^{\mathrm{s}}J^l(\xi,\eta)\mathrm{d}\xi\mathrm{d}\eta \tag{14a}$$

$$H_{ij}^{als} = \int_{-1}^{1} T_{ij}(P^a, Q^l(\xi,\eta))\phi_{\mathrm{d}}^{\mathrm{s}}J^l(\xi,\eta)\mathrm{d}\xi\mathrm{d}\eta \tag{14b}$$

where $J^l(\xi,\eta)$ is the Jacobian $\mathrm{d}S/\mathrm{d}\xi\mathrm{d}\eta$, which are generally evaluated numerically by Gaussian quadrature. It is well-known that the existence of a singularity in the U_{ij} and T_{ij} kernels of BIE (11),

$$U_{ij} = O(\frac{1}{R}) \tag{15}$$

$$T_{ij} = O(\frac{1}{R^2}) \tag{16}$$

where R is the distance from P to Q, involves evaluating a singular integral whenever the source point P^a belongs to the element l over which the integral extends. The singular integral can nevertheless be accurately evaluated by standard quadrature formulae, since special steps can be taken to remove the singularity. The use of singular traction interpolation functions not only produces a singularity in the integrand in

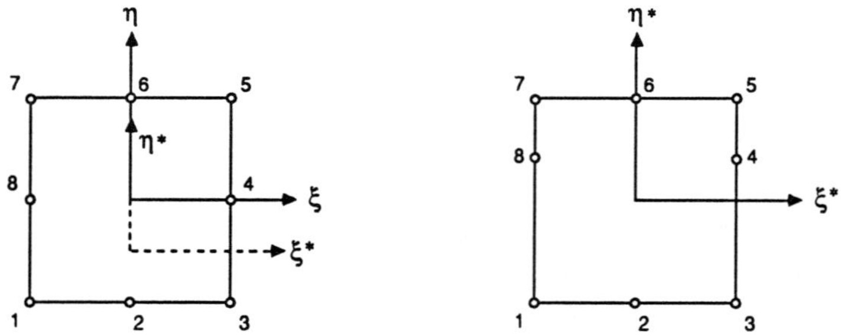

Figure 4: Mapping from the ξ, η plane to the ξ^*, η^* plane

(14a) even when the source point P^a lies outside the element but it also increases the order of the pre-existing singularity in the case where P^a belongs to l. The integrals do however all exist and appropriate co-ordinate transformations can be used to map the integration domain $(-1 \leq \xi \leq 1; -1 \leq \eta \leq 1)$ onto a different domain over which the integrand is bounded.

Let us distinguish two different cases according to whether or not the source node α belongs to the special element l.

In the case where node a does not belong to element l, the integration procedure is quite simple. Since only the singularity introduced in the integrand in (14a) by the traction interpolation functions $\phi_{\rm t}^{\rm s}$ needs to be removed, it is sufficient to carry out the integration in the ξ^*, η^* plane (see Fig. 4), where

$$\xi = \xi^*$$
$$(1+\eta) = \frac{1}{2}(1+\eta*)^2 \tag{17}$$

within the limits $-1 \leq \xi^* \leq 1$ and $-1 \leq \eta^* \leq 1$. The product of the singular shape functions $\phi_{\rm t}^{\rm s}(s \in \{1,2,3\})$, both as given by (10) and by (A2)

$$\phi_{\rm t}^{\rm s} = O(\frac{1}{\sqrt{1+\eta}}) = O(\frac{1}{(1+\eta^*)}) \tag{18}$$

and the Jacobian

$$\frac{d\xi d\eta}{d\xi^* d\eta^*} = (1+\eta^*) \tag{19}$$

is in fact bounded. It should be noted that in evaluating integral (14.b) the displacement interpolation functions $\phi_{\rm d}^{\rm s}$ (both as given by (9) and by (A1)) assume a quadratic form when expressed in terms of the co-ordinates ξ^* and η^*.

It is worth pointing out that when the traction-singular quarter-point elements represented in Fig. 2 are employed, transformation of the integration domain is unnecessary since the singularity in the shape functions (8), $\phi_{\rm t}^{\rm s}$ $(s \in \{1,2,3\})$

$$\phi_{\rm t}^{\rm s} = O(\frac{1}{(1+\eta)}) \tag{20}$$

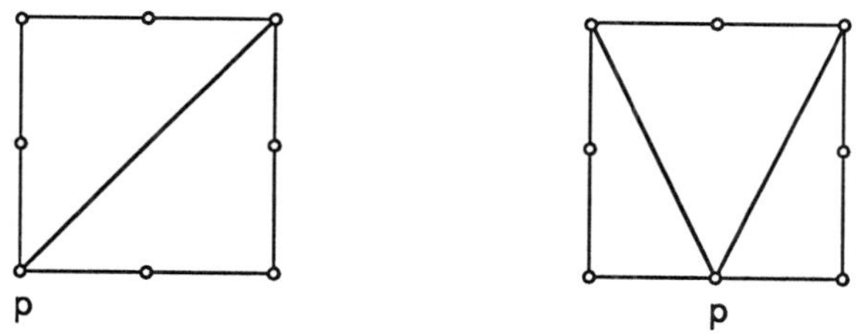

Figure 5: Subdivision into triangles of the domain in the ξ, η plane

is removed by the Jacobian $\mathrm{d}S/\mathrm{d}\xi\mathrm{d}\eta$

$$\frac{\mathrm{d}S}{\mathrm{d}\xi\mathrm{d}\eta} = O(1+\eta) \tag{21}$$

As shown in Fig. 2, the mapping from x_i to ξ, η brings in fact the quarter-points to the middle of the side as does the mapping from ξ, η to ξ^*, η^* and therefore from x_i to ξ^*, η^*.

In the case where the source point P^a belongs to the special element l, both the kernel and interpolation functions present a singularity within the integration domain. Devising a sequence of co-ordinates mappings, capable of transforming the singular integral to an integral over a smooth function, appears in this case a less trivial task.

Let us assume that a standard boundary element code for non-singular field analysis is modified to include traction-singular elements in the element library. A procedure will most certainly be implemented in the code to remove the kernel singularity, since this is present even when integrals (14) extend over a non-singular element (say, an eight-node isoparametric element). Most of the procedures involve subdividing the domain in the ξ, η plane onto which element l is transformed by the isoparametric mapping into two triangles (three when P^a is a side-node) with a common vertex at p, image of P^a in the ξ, η plane, as shown in Fig. 5; each triangle is in turn mapped onto a different domain over which the integrand is bounded.

A possible mapping simply derives from considering each triangle to be a degenerate quadrangle with one side collapsed onto the singular point. If, for the sake of simplicity, we consider only the case where the source point P^a coincides with the corner node of intrinsic co-ordinates $\xi = \eta = -1$ as in Fig. 6, the mapping of triangles α, β into square domains in the ξ', η' plane ($-1 \leq \xi' \leq 1; -1 \leq \eta' \leq 1$) can be described, on triangle α, by

$$\begin{aligned}(1+\xi) &= (1+\eta') \\ (1+\eta) &= \frac{1}{2}(1-\xi')(1+\eta')\end{aligned} \tag{22}$$

and, on triangle β, by

$$\begin{aligned}(1+\xi) &= \frac{1}{2}(1+\xi')(1+\eta') \\ (1+\eta) &= (1+\eta')\end{aligned} \tag{23}$$

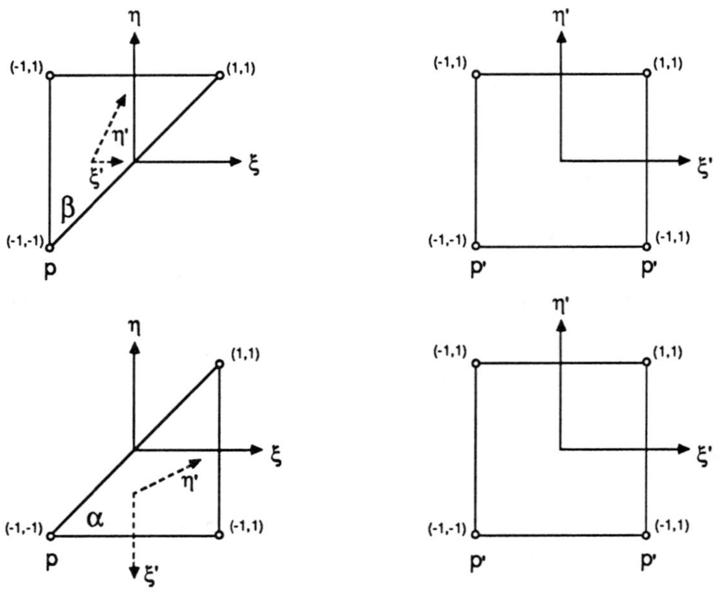

Figure 6: Mapping of triangles α and β onto square domains in the ξ', η' plane

where the side $\eta' = -1$ corresponds to the collapsed side. The Jacobian $d\xi d\eta / d\xi' d\eta'$ of both transformations is given by

$$\frac{d\xi d\eta}{d\xi' d\eta'} = \frac{1}{2}(1 + \eta') \tag{24}$$

The square of the distance R from P^a can be written in terms of the local co-ordinates ξ and η as

$$R^2 = c_1(1+\xi)^2 + c_2(1+\eta)^2 + c_3(1+\xi)(1+\eta) + \dots \qquad c_1 > 0, c_2 > 0 \tag{25}$$

plus higher order terms. Substituting relations (22) or (23), whichever applies to the triangle considered, into relation (25) to express R^2 in terms of the co-ordinates ξ' and η' yields in both cases

$$R^2 = O((1+\eta')^2) \tag{26}$$

The singularity in kernel U_{ij} in (14a)

$$U_{ij} = O(\frac{1}{R}) = O(\frac{1}{(1+\eta')}) \tag{27}$$

is thus removed by the Jacobian (24). Where integral (14b) is concerned, the higher order singularity in kernel T_{ij}

$$T_{ij} = O(\frac{1}{R^2}) = O(\frac{1}{(1+\eta')^2}) \tag{28}$$

also cancels out in the product with the Jacobian (24) and the quadratic shape function $\phi^s(\xi,\eta)$, since this latter vanishes, as $(1+\eta') \to 0$, at least with the order

$$\phi^s = O((1+\eta')) \tag{29}$$

whenever $\beta(l,s) > \alpha$ (where $\beta(l,s)$ defines the nodal numbering). The strong singularity arising in (14b) when $\beta(l,s) = a$ need not in fact to be considered since it is well-known that computation of coefficient H_j^{aa} can always be avoided by applying the rigid body motion condition.

The above arguments can be obviously extended to the case where P^a is a mid-side node and the element is subdivided into three triangles.

Let us now suppose that the special elements we want to introduce are the traction-singular quarter-point elements represented in Fig. 2. Since, as previously shown by relations (20) and (21), the singularity introduced by the traction interpolation functions cancels out in the product with $J(\xi,\eta)$, only the singularity due to the kernel U_{ij} is present within the integration domain in the ξ,η plane. It does therefore appear reasonable to expect that the procedure previously described, i.e. the subdivision of the domain into triangles and the subsequent mapping of each triangle onto a square domain, can once again completely remove the singularity. Consider again the case where point p has intrinsic co-ordinates $\xi = \eta = -1$. As a consequence of the mapping whereby quarter-points are taken to the middle of the side, the square of the distance R can now be expressed in terms of local co-ordinates as

$$R^2 = c_1(1+\xi)^2 + \frac{c_2}{4}(1+\eta)^4 + \ldots \tag{30}$$

When relations (22), which describe the mapping of triangle a onto a square domain, are substituted into (30), the variation of R^2 can once again be represented as in (26). Since all the arguments developed above for the isoparametric quadratic elements are still valid, the singularities present in both integral (14a) and (14b) are completely removed. When, however triangle β is considered, substitution of relations (23) into (30) yields

$$R^2 = (1+\eta')^2[\frac{c_1}{4}(1+\xi')^2 + \frac{c_2}{4}(1+\eta')^2 + \ldots] \tag{31}$$

The product of kernel U_{ij} by the Jacobian (24) is consequently unbounded at the point $\xi' = \eta' = -1$. Similarly, when integral (14b) is considered, the product of the singular kernel T_{ij} by the Jacobian (24) and the quadratic displacement interpolation function ϕ^s_d, vanishing as in (29), presents a singularity at the same point p'.

By extending the above arguments to the case where P^a coincides with the side node of intrinsic co-ordinates $\xi = 0, \eta = -1$, it can thus be affirmed that the usual procedure employed to remove kernel singularities, is no longer sufficient in the particular case where the source point P^a not only belongs to a traction-singular quarter-point element but it also lies on the crack-front.

It is quite evident that the same conclusions could be drawn for the other types of traction-singular elements (both the mid-side node elements defined by the interpolation functions (9) and (10) and the quarter-point elements described in Appendix A) in the case where the subdivision into triangles is operated in the ξ^*,η^* plane onto

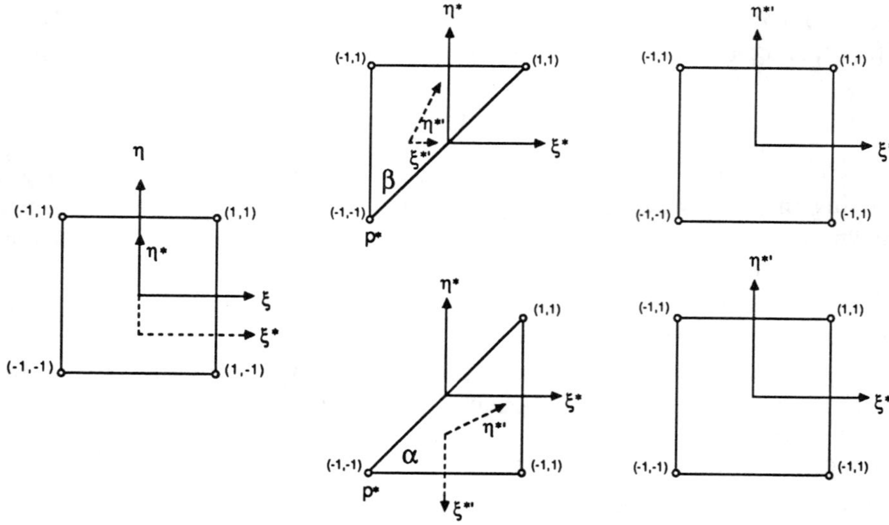

Figure 7: Mapping from the ξ, η plane to the ξ^*, η^* plane; subdivision into triangles of the domain in the ξ^*, η^* plane and mapping of each triangle onto a square domain in the $\xi^{*'}, \eta^{*'}$ plane

which the element has been previously mapped to remove the traction field singularity, as in Fig. 7. Relations (24), (29), (30) and (31), and the related arguments, are in fact still valid provided that ξ^*, η^* substitute for ξ, η and $\xi^{*\prime}, \eta^{*\prime}$ for ξ', η'.
In order to obtain an integration domain over which the integrand is bounded it is necessary to 'invert' the sequence of the co-ordinate mappings represented in Fig. 7, that is to operate the subdivision into triangles in the ξ, η plane and to further map the square in the ξ', η' plane, onto which triangle β has been mapped, onto a square domain in the ξ'^*, η'^* plane ($-1 \leq \xi'^* \leq 1; -1 \leq \eta'^* \leq 1$) according to

$$\xi' = \xi'^*$$

$$(1 + \eta') = \frac{1}{2}(1 + \eta'^*)^2 \tag{32}$$

as illustrated in Fig. 8. By substituting relations (32) into (26) the variation of the square of R can now be given as

$$R^2 = O((1 + \eta'^*)^4) \tag{33}$$

The kernel singularities hence become

$$U_{ij} = O(\frac{1}{R}) = O(\frac{1}{(1 + \eta'^*)^2}) \tag{34}$$

$$T_{ij} = O(\frac{1}{R^2}) = O(\frac{1}{(1 + \eta'^*)^4}) \tag{35}$$

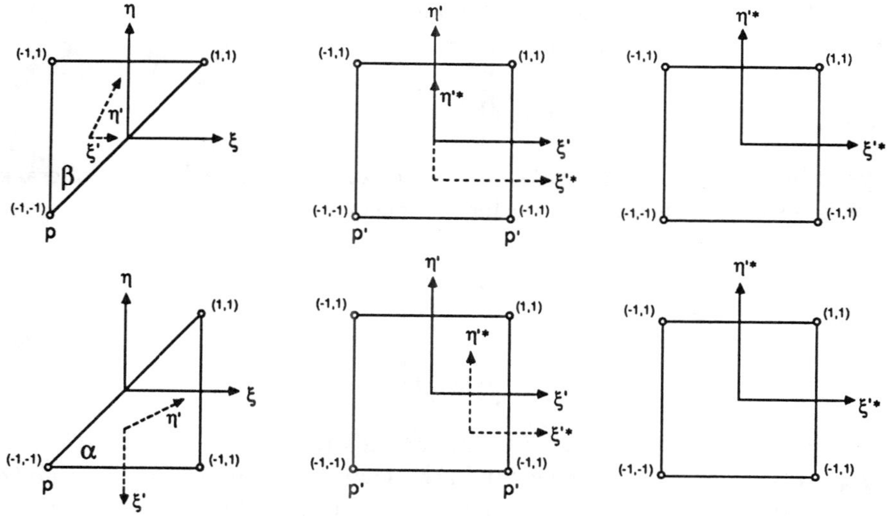

Figure 8: Mapping of each triangle onto a square domain in the ξ', η' plane and subsequent mapping from the ξ', η' plane to the ξ'^*, η'^* plane

A $(1+\eta'^*)^{-3}$ singularity is therefore present in the products of the kernel and interpolation function both in (14a) where the traction interpolation function $\phi_t^s(\xi,\eta)(s \in \{1,2,3\})$ is singular as

$$\phi_t^s = O(\frac{1}{\sqrt{1+\eta}}) = O(\frac{1}{\sqrt{1+\eta'}}) = O(\frac{1}{(1+\eta'^*)}) \tag{36}$$

and in (14b) where the displacement interpolation function $\phi_d^s(\xi,\eta)(b(l,s) > a)$ vanishes with $(1+\eta'^*)$ at least as

$$\phi_d^s = O((1+\eta'^*)) \tag{37}$$

The singularity is thus removed by the Jacobian

$$\frac{d\xi d\eta}{d\xi'^* d\eta'^*} = \frac{d\xi d\eta}{d\xi' d\eta'}\frac{d\xi' d\eta'}{d\xi'^* d\eta'^*} = \frac{1}{2}(1+\eta')(1+\eta'^*) = \frac{1}{4}(1+\eta'^*)^3 \tag{38}$$

When integrating on triangle a (see Fig. 8), we must consider that the mapping is now defined by the co-ordinate transformations (22); the traction field singularity present within the integration domain in the ξ', η' plane is therefore

$$\phi_t^s = O(\frac{1}{\sqrt{1+\eta}}) = O(\frac{1}{\sqrt{(1-\xi')(1+\eta')}}), \qquad (s \in \{1,2,3\}) \tag{39}$$

The domain in the ξ', η' plane must consequently be transformed according to

$$\begin{aligned}(1-\xi') &= \frac{1}{2}(1-\xi'^*)^2 \\ (1+\eta') &= \frac{1}{2}(1+\eta'^*)^2\end{aligned} \tag{40}$$

By extending the above argument to the case where P^a corresponds to the mid-side node ($\xi = 0,\ \eta = -1$) or to the corner node ($\xi = 1,\ \eta = -1$), it can be generalized that the co-ordinate transformation given by

$$(1+\eta') = \frac{1}{2}(1+\eta'^*)^2 \tag{41}$$

is always to be carried out and that, when a side of the triangle lies on the crack-front, a further transformation must be performed according to

$$(1+\xi') = \frac{1}{2}(1+\xi'^*)^2 \tag{42}$$

or to

$$(1-\xi') = \tfrac{1}{2}(1-\xi'^*)^2$$

depending on whether the crack-front corresponds to $\xi' = -1$ or to $\xi' = 1$.

It is quite obvious that the inversion in the sequence of co-ordinate transformations which leads to the correct integration procedure cannot be applied to the traction quarter-point elements in Fig. 2. The quadratic transformation which takes quarter-points to the middle of the side is in fact intrinsic to the formulation of this type of elements and it does therefore necessarily precede any subdivision.

An alternative integration procedure, involving a combination of cartesian and polar co-ordinate transformations as well as singular element subdivision, has been proposed in Ref. 12. The procedure derives from the different mapping employed to remove kernel singularities on each of the triangles into which the non-singular element is, as previously, subdivided. Limiting the discussion to the case where the source point P^a coincides with the side node $\xi = 0,\ \eta = -1$, a polar co-ordinate transformation with pole in the singular point p

$$\begin{aligned} \xi &= 2\rho\cos\vartheta \\ \eta &= 2\rho\sin\vartheta - 1 \end{aligned} \tag{43}$$

is carried out on each of the three triangles (a, b and g in Fig. 9). Integration is hence performed in the ρ, ϑ plane within the limits

$$\begin{array}{llll} (\alpha) & 0 \le \vartheta \le \arctan 2 & 0 \le \rho \le \dfrac{1}{2}\sec\vartheta & \\ (\beta) & \arctan 2 \le \vartheta \le \pi - \arctan 2 & 0 \le \rho \le \operatorname{cosc}\vartheta & \\ (\gamma) & \pi - \arctan 2 \le \vartheta \le \pi & 0 \le \rho \le -\dfrac{1}{2}\sec\vartheta & \end{array} \tag{44}$$

Since the square of the distance R can be given as

$$R^2 = c_1\xi^2 + c_2(1+\eta)^2 + \ldots = O(\rho^2) \tag{45}$$

considering for brevity integral (14a) only, it can be easily shown that the singularity in kernel U_{ij}

$$U_{ij} = O(\frac{1}{R}) = O(\frac{1}{\rho}) \tag{46}$$

is removed by the Jacobian of transformation (43)

$$\frac{d\xi d\eta}{d\rho d\vartheta} = 4\rho \tag{47}$$

When integral (14a) extends over a special element, the contribution to the singularity deriving from the presence of singular traction interpolation functions

$$\phi_t^s = O(\frac{1}{\sqrt{1+\eta}}) = O(\frac{1}{\sqrt{\rho \sin\vartheta}}) \qquad (s \in \{1,2,3\}) \tag{48}$$

must be taken into account. Since the integrand is singular both at $\rho = 0$, and at $\vartheta = 0$ or $\vartheta = \pi$, a further transformation is needed

$$\begin{aligned} \rho &= s^2 \\ \sin\vartheta &= \sin^2 t \end{aligned} \tag{49}$$

When integral (14a) is carried out in the s, t plane within the limits

$$\begin{array}{llll} (\alpha) & 0 \le t \le \arcsin\sqrt{2/\sqrt{5}} & 0 \le s \le (\frac{\sec t}{2\sqrt{1+\sin^2 t}})^{1/2} & (50) \\ (\beta) & \arcsin\sqrt{2/\sqrt{5}} \le t \le \pi - \arcsin\sqrt{2/\sqrt{5}} & 0 \le s \le \mathrm{cosc} t & \\ (\gamma) & \pi - \arcsin\sqrt{2/\sqrt{5}} \le t \le \pi & 0 \le s \le (\frac{-\sec t}{2\sqrt{1+\sin^2 t}})^{1/2} & \end{array}$$

the traction field singularity

$$\phi_t^s = O(\frac{1}{s \sin t}) \tag{51}$$

cancels out in the product with the Jacobian

$$\frac{d\rho d\vartheta}{ds dt} = \frac{4s \sin t}{\sqrt{1+\sin^2 t}} \tag{52}$$

Note that the co-ordinate transformation described from the second of (49) is only necessary for the domains obtained by mapping triangles α and γ, or, more generally, triangles with one of the sides lying on the crack-front.

3 Computation of stress intensity factors

Whenever the elasticity solution is available, stress intensity factors can be evaluated at any point on the crack-front from the asymptotic behaviour either of stresses or displacements. Considering that at $\theta = 180°$ and $\theta = 0°$ the interaction between the first and second mode uncouples in expressions (1) and (2), respectively, it is possible to evaluate K_I, K_{II} and K_{III} from the crack surface displacement ($\theta = 180°$) as

$$\begin{aligned} K_I &= \frac{E}{(1+\nu)(1+\kappa)} \lim_{r\to 0} \sqrt{\frac{\pi}{2r}}\ (u_{2_{\theta=180°}} - u_{2_{\theta=-180°}}) \\ K_{II} &= \frac{E}{(1+\nu)(1+\kappa)} \lim_{r\to 0} \sqrt{\frac{\pi}{2r}}\ (u_{1_{\theta=180°}} - u_{1_{\theta=-180°}}) \\ K_{III} &= \frac{E}{2(1+\nu)} \lim_{r\to 0} \sqrt{\frac{\pi}{2r}}\ (u_{3_{\theta=180°}} - u_{3_{\theta=-180°}}) \end{aligned} \tag{53}$$

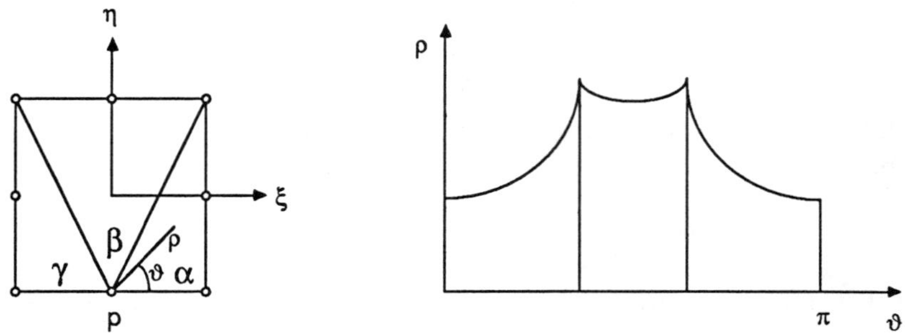

Figure 9: Mapping of triangles α, β and γ onto integration domains in the ρ, θ plane

and from the stresses in the plane of the crack ($\theta = 0$) as

$$\begin{aligned} K_I &= \lim_{r \to 0} \sqrt{2\pi r}\ \sigma_{22_{\theta=0^\circ}} \\ K_{II} &= \lim_{r \to 0} \sqrt{2\pi r}\ \sigma_{12_{\theta=0^\circ}} \\ K_{III} &= \lim_{r \to 0} \sqrt{2\pi r}\ \sigma_{23_{\theta=0^\circ}} \end{aligned} \tag{54}$$

When an approximate solution is found by a numerical method such as FEM or BEM, an extrapolation procedure can be employed. Special elements which correctly represent the variation of displacements in the vicinity of the crack-front generally enable accurate values of the stress intensity factors to be extrapolated from two nodal values only.

Observing that when only one mode is present the coefficient of the r-term is zero in the series describing the displacement behaviour around the crack, the mode I stress intensity factor at a point on the crack-front which coincides with a corner node C (see Fig. 10) may be linearly extrapolated from the values computed for displacement at the mid-side and corner nodes (B and A, respectively) as

$$K_I = H \frac{2\sqrt{2}\ u_2^{\prime B} - u_2^{\prime A}}{\sqrt{L_{\rm i}}} \tag{55}$$

where $H = \sqrt{\pi/2}\ \frac{E'}{2}$ (with $E' = E$ in plane stress and $E' = E/(1-\nu^2)$ in plane strain), $L_{\rm i}$ is the length of the side of the element internally adjacent to the front and the prime indicates that the relative displacements with respect to the crack-tip C must in general be used in the extrapolation procedure. Whenever elements (either singular or not) are employed with the side-node B located at the 1/4 position, the extrapolation formula modifies as

$$K_I = H \frac{8u_2^{\prime B} - u_2^{\prime A}}{3\sqrt{L_{\rm i}}} \tag{56}$$

Expression (55), or (56), where u_1 and u_3 replace u_2, can be used to evaluate K_{II} and K_{III}, respectively.

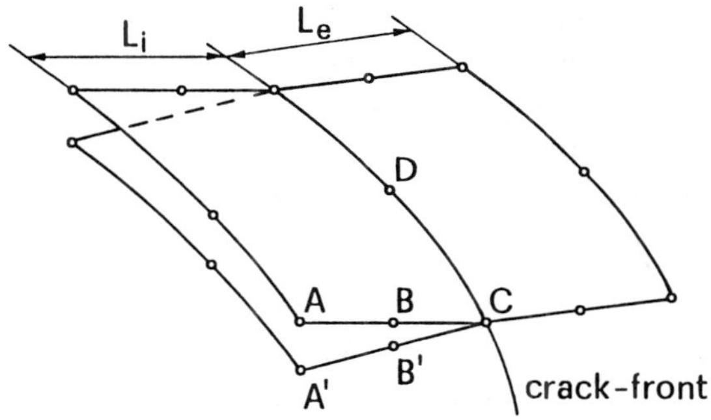

Figure 10: Crack-front elements

In a mixed mode problem, when both modes I and II are present, K_I and K_{II} can be evaluated as

$$\begin{aligned} K_I &= H\frac{[2\sqrt{2}(u_2^B - u_2^{B'}) - (u_2^A - u_2^{A'})]}{\sqrt{L_\mathrm{i}}} \\ K_{II} &= H\frac{[2\sqrt{2}(u_1^B - u_1^{B'}) - (u_1^A - u_1^{A'})]}{\sqrt{L_\mathrm{i}}} \end{aligned} \tag{57}$$

where nodes A' and B', coinciding geometrically with A and B, belong to the other crack surface. A linear extrapolation is in fact still feasible since the coefficient of the linear term in the eigenfunction expansion of $u_{1_\theta} = 180° - u_{1_\theta} = -180°$ and $u_{2_\theta} = 180° - u_{2_\theta} = -180°$ is once again null. When quarter-point elements are employed, expressions (57) must be replaced by

$$\begin{aligned} K_I &= H\frac{[8(u_2^B - u_2^{B'}) - (u_2^A - u_2^{A'})]}{3\sqrt{L_\mathrm{i}}} \\ K_{II} &= H\frac{[8(u_1^B - u_1^{B'}) - (u_1^A - u_1^{A'})]}{3\sqrt{L_\mathrm{i}}} \end{aligned} \tag{58}$$

The use of traction-singular elements enables stress intensity factors to be evaluated from tractions in a more direct way. Recalling that

$$t_i = t_i^\mathrm{s}\phi_i^\mathrm{s}(\xi,\eta) = \frac{b_i'^1(\xi)}{\sqrt{1+\eta}} + b_i'^2(\xi) + b_i'^3(\xi)\sqrt{1+\eta} \tag{59}$$

where the function $b_i'^1$ interpolates along the crack-front the nodal values computed from traction t_i^s ($s \in \{1,2,3\}$), and that the distance r from the crack-front measured along a line $\xi =$ const. can be expressed as

$$r = (1+\eta)\frac{L_\mathrm{e}}{2} \tag{60}$$

where L_e is the size of the element along the same line, it follows that

$$b_i'^1(\xi) = \lim_{\eta\to-1} t_i\sqrt{1+\eta} = \lim_{r\to0} t_i\sqrt{\frac{2r}{L_\mathrm{e}}} \tag{61}$$

When the traction-singular element lies on the plane of the crack ($\theta = 0°$), if for simplicity we assume that the rectangular co-ordinate system (x_1, x_2, x_3) adopted in the boundary element analysis coincides with the reference system $(\overline{x}_1,\ \overline{x}_2,\ \overline{x}_3)$ defined along the crack-front and represented in Fig. 1, the stress components σ_{12}, σ_{22} and σ_{23} respectively coincide with t_1, t_2 and t_3. A comparison between relations (61) and (54) clearly shows that the stress intensity factors are given, but for a constant, by the nodal values of traction

$$\begin{aligned} K_I &= \sqrt{\pi L_e}\ t_2 \\ K_{II} &= \sqrt{\pi L_e}\ t_1 \\ K_{III} &= \sqrt{\pi L_e}\ t_3 \end{aligned} \tag{62}$$

at any point on the crack-front which coincides either with a corner node or a mid-side node (C and D, respectively, in Fig. 10). Whatever the reference system adopted in the analysis, the above formulae can still be employed provided that t_1, t_2 and t_3 are replaced by $\overline{t}_1$, $\overline{t}_2$ and $\overline{t}_3$, where these latter represent the nodal components of traction in the $\overline{x}_1$, $\overline{x}_2$ and $\overline{x}_3$ directions, respectively.

While all the above arguments do also apply to the quarter-point traction-singular element described in Appendix A, in the case of the quarter-point traction-singular element represented in Fig. 2, where the traction interpolation functions are singular as

$$\phi_t^s = O(\frac{1}{1+\eta}) \qquad (s \in \{1,2,3\}) \tag{63}$$

and a quadratic relation holds between the distance from the front measured along a line $\xi =$ const. and $(1+\eta)$

$$r = (1+\eta)^2 \frac{L_e}{4} \tag{64}$$

relations (61) becomes

$$b_i'^1 = \lim_{\eta \to -1} t_i(1+\eta) = \lim_{r \to 0} t_i\ 2\sqrt{\frac{r}{L_e}} \tag{65}$$

Stress intensity factors can hence be still evaluated by formulae (62) except for the constant which is now equal to $\sqrt{\pi L_e / 2}$.

When, on the other hand, the traction-singular element lies on the plane $\theta = \overline{\theta}(\overline{\theta} > 0)$, i.e. when the interface in a multi-domain representation does not lie on the plane of the crack, it is more convenient to derive SIFs from the stress components relative to the cylindrical co-ordinate system r, θ, z (where z coincides with $\overline{x}_3$)). The asymptotic expressions for the three cylindrical components of stress concerning the calculation can be easily obtained from (2) and are given as

$$\begin{aligned} \sigma_{\theta\theta} &= \frac{1}{\sqrt{2\pi r}} \left[K_I \left(\frac{3}{4}\cos\frac{1}{2}\theta + \frac{1}{4}\cos\frac{3}{2}\theta\right)\right] - \left[K_{II}\left(\frac{3}{4}\sin\frac{1}{2}\theta + \frac{3}{4}\sin\frac{3}{2}\theta\right)\right] \\ \sigma_{r\theta} &= \frac{1}{\sqrt{2\pi r}} \left[K_I \left(\frac{1}{4}\sin\frac{1}{2}\theta + \frac{1}{4}\sin\frac{3}{2}\theta\right)\right] + \left[K_{II}\left(\frac{1}{4}\cos\frac{1}{2}\theta + \frac{3}{4}\cos\frac{3}{2}\theta\right)\right] \\ \sigma_{\theta z} &= \frac{1}{\sqrt{2\pi r}} K_{III} \cos\frac{1}{2}\theta \end{aligned} \tag{66}$$

Relations, analogous to (54), can be derived between SIFs and the limit values of the cylindrical stress components

$$\begin{aligned}
\lim_{r\to 0}\sqrt{2\pi r}\ \sigma_{\theta\theta} &= K_I\left(\frac{3}{4}\cos\frac{1}{2}\bar{\theta}+\frac{1}{4}\cos\frac{3}{2}\bar{\theta}\right)-K_{II}\left(\frac{3}{4}\sin\frac{1}{2}\bar{\theta}+\frac{3}{4}\sin\frac{3}{2}\bar{\theta}\right) \quad (67)\\
\lim_{r\to 0}\sqrt{2\pi r}\sigma_{r\theta} &= K_I\left(\frac{1}{4}\sin\frac{1}{2}\bar{\theta}+\frac{1}{4}\sin\frac{3}{2}\bar{\theta}\right)+K_{II}\left(\frac{1}{4}\cos\frac{1}{2}\bar{\theta}+\frac{3}{4}\cos\frac{3}{2}\bar{\theta}\right)\\
\lim_{r\to 0}\sqrt{2\pi r}\sigma_{\theta z} &= K_{III}\cos\frac{1}{2}\bar{\theta}
\end{aligned}$$

and, recalling relations (61), between SIFs and the nodal values of the components of traction in the directions defined by the cylindrical co-ordinate system, t_r t_θ and t_z,

$$\begin{aligned}
\sqrt{\pi L_e}\ t_\theta &= K_I\left(\frac{3}{4}\cos\frac{1}{2}\bar{\theta}+\frac{1}{4}\cos\frac{3}{2}\bar{\theta}\right)-K_{II}\left(\frac{3}{4}\sin\frac{1}{2}\bar{\theta}+\frac{3}{4}\sin\frac{3}{2}\bar{\theta}\right) \quad (68)\\
\sqrt{\pi L_e}\ t_r &= K_I\left(\frac{1}{4}\sin\frac{1}{2}\bar{\theta}+\frac{1}{4}\sin\frac{3}{2}\bar{\theta}\right)+K_{II}\left(\frac{1}{4}\cos\frac{1}{2}\bar{\theta}+\frac{3}{4}\cos\frac{3}{2}\bar{\theta}\right)\\
\sqrt{\pi L_e}\ t_z &= K_{III}\cos\frac{1}{2}\bar{\theta}
\end{aligned}$$

Modes I and II do, however, no longer uncouple and a linear system of two equations in two unknowns must be solved to compute the mixed-mode stress intensity factors. The expression for SIFs which can be derived by solving the system are given in Appendix B.

Note, finally, that all the formulae for evaluating SIFs given in the present work have been derived assuming that the sides of the elements adjacent to the crack-front are normal to the front. In some cases (e.g. in the analysis of curved crack-fronts) it may be however convenient, or even necessary, to discretize the crack-front region in such a way that the side does not lie on the normal to the front but forms with it an angle Φ. This can be corrected by dividing the SIF values extrapolated from displacements by $\sqrt{\cos\Phi}$ and by multiplying the values derived from tractions by $\sqrt{\cos\Phi}$.[16]

4 Numerical examples

In order to evidence the good accuracy which can be attained by using traction-singular boundary elements, the present section reports the results obtained from the analyses of a few simple cases for which the analytical solution is known. All the problems considered involve a regularly-shaped crack (either penny-shaped or elliptical) embedded in an infinite body.

The first analysis concerns a penny-shaped crack of radius a in an infinite body subjected to a uniform tensile stress σ, normal to the crack faces, at infinity. The infinite body has been approximated by a finite cylinder of radius $5a$. Because of symmetry, only one fourth of the cylinder surface was discretized with no elements on the planes of symmetry; the mesh, consisting of sixteen elements, is shown in Fig. 11. In Fig. 12 the percentage difference is plotted between the exact solution

$$K_I = 2\sigma\sqrt{a/\pi} \quad (69)$$

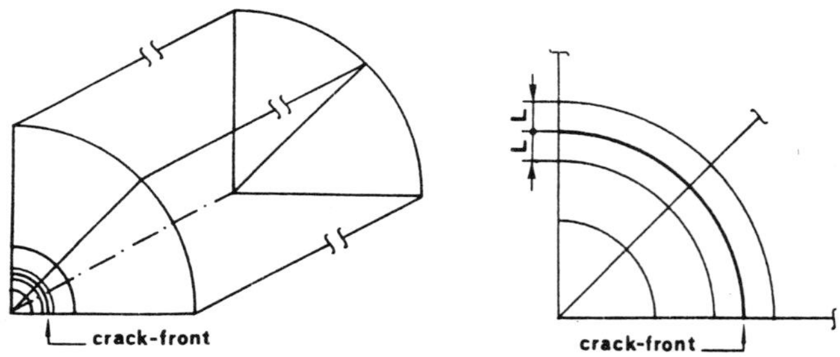

Figure 11: Boundary element discretization of one fourth of the cylinder surface

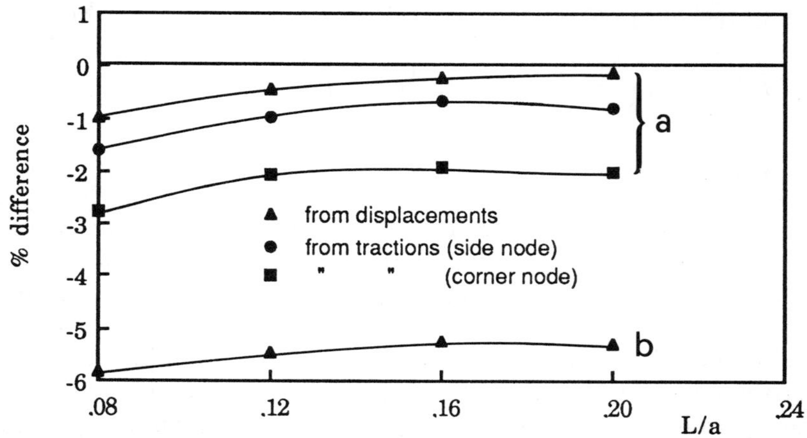

Figure 12: Percent error in the value of K_I calculated for the penny-shaped crack under uniform tension by using: a) traction-singular elements; b) non-singular quarter-point elements

and the results, reported in Ref. 10, obtained for different values of the ratio between the radial size of the special elements ($L_e = L_i = L$) and the crack radius. A comparison can be drawn between the values of K_I extrapolated from the nodal displacements obtained from two different analyses. In the first analysis the special elements defined by the interpolation functions (9) and (10) were employed (functions (9) were used both as displacement and traction interpolation functions in the non-singular elements lying on the crack surface), whereas in the second analysis standard (non-singular) quarter-point elements were used both internally and externally to the crack-front. Where the results of the first analysis are concerned, the SIF values derived from the traction values calculated at the mid-side and corner nodes of the traction-singular elements are also given.

Similar results were obtained by Jia, Shippy & Rizzo[12] who analyzed the same case using a multi-domain approach which involves subdividing the infinite body into two subregions separated by a closed surface containing one of the crack faces. The mesh of the cylindrical surface which represents the common boundary of the two

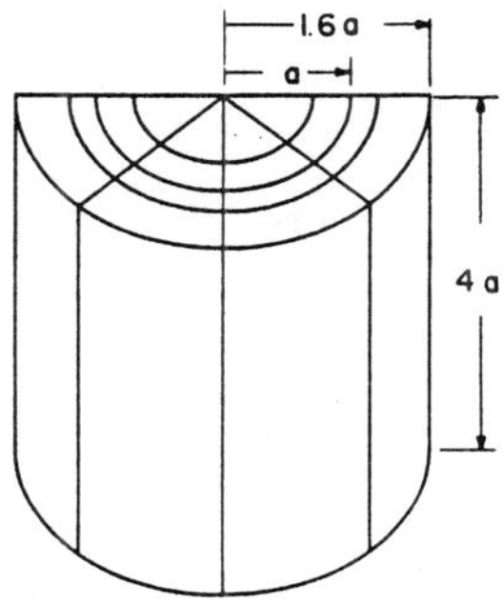

Figure 13: Boundary element discretization of one half of the cylindrical surface which separates the two regions in the multi-domain representation[12]

Table 1: Normalized K_I calculated for a penny-shaped crack under uniform tension (exact value of $K_I = 1.1284$)

Mesh	No of elements	K_I BIE	Error (%)
1	24	1.1056	-2.0
2	28	1.1097	-1.7
3	36	1.1150	-1.2
4	42	1.1192	-0.8

regions is represented in Fig. 13; half of the surface was discretized (with no elements on the symmetry plane) for a total of 24 elements. The special elements externally and internally adjacent to the crack-front were respectively 0.3 and 0.2 times the crack radius a. Since the particular approach adopted does not enable a remote loading to be directly represented, a uniform pressure was applied to the crack surfaces. In Table 1 the value of K_I, normalized by $\sigma\sqrt{a}$, obtained directly from the corner nodes of traction-singular elements, is compared with the exact value. In the table the mesh shown in Fig. 13 is referred to as mesh 1; mesh 2 is obtained from mesh 1 by adding a layer of elements to the top portion of the mesh thus increasing the radius of the cylinder from $1.6a$ to $2.5a$; meshes 3 and 4 differ from 1 and 2, respectively, in that they contain six circumferential segments rather than four.

Two further problems were analyzed by Jia and co-workers which involve an identical geometry and different boundary conditions: a remote torque and a remote shear load, respectively.

In the first case a torsional shear stress, linearly varying from 0 to τ in the radial direction, was assumed to act on the crack surfaces. The whole cylindrical surface was discretized; the same meshes 1-4 (consisting here of a double number of elements) were employed. A comparison between the numerical value of K_{III}, once again derived directly from the traction value at a corner node, and the exact solution

$$K_{III} = \frac{4\tau}{3}\sqrt{a/\pi} \tag{70}$$

Table 2: Normalized K_{III} calculated for a penny-shaped crack under torsion (exact value of $K_{III} = 0.7523$)

Mesh	No of elements	K_{III} BIE	Error (%)
1	48	0.7261	-3.5
2	56	0.7272	-3.4
3	72	0.7339	-2.4
4	84	0.7351	-2.3

Table 3: Percent error in the value of K_{II} calculated for a penny-shaped crack under in-plane shear load

θ	Mesh 1	Mesh 2	Mesh 3	Mesh 4
0.0	-1.9	-0.5	-1.7	-0.5
15.0			-1.2	-0.1
22.5	+0.7	+2.0		
30.0			-1.7	-0.5
45.0	-1.9	-0.5	-1.2	0.0
60.0			-1.7	-0.5
67.5	+0.7	+2.0		
75.0			-1.2	0.0
90.0				

is reported in Table 2 where the SIF values are normalized by $\tau\sqrt{a}$.

When analyzing a penny-shaped crack embedded in an infinite body subjected to shear load at infinity, a uniform shear stress t was applied to the crack surfaces. Because of symmetry half of the cylinder was once again discretized. The mixed-mode stress intensity factors, K_{II} and K_{III},

$$K_{II} = \frac{4\tau\sqrt{a/\pi}}{2-\nu}\cos\theta \tag{71}$$

$$K_{III} = \frac{-4(1-\nu)\tau\sqrt{a/\pi}}{2-\nu}\sin\theta$$

where θ is the polar angle measured from the direction of the applied stress, are compared with the numerical values directly obtained from both corner and mid-side nodes of traction-singular elements in Tables 3 and 4, respectively, where the percent relative errors with respect to the exact values of K_{II} and K_{III} are given.

Another 3-D crack problem whose closed form solution is known is represented by an elliptical crack in an infinite body under uniform normal tensile stress, σ, at infinity.

$$K_I = \frac{\sigma\sqrt{\pi b}}{E(m)}(\sin^2\varphi + \frac{b^2}{a^2}\cos^2\varphi)^{1/4} \tag{72}$$

Table 4: Percent error in the value of K_{III} calculated for a penny-shaped crack under in-plane shear load

θ	Mesh 1	Mesh 2	Mesh 3	Mesh 4
0.0				
15.0			-2.8	-1.6
22.5	-3.7	-2.4		
30.0			-2.9	-1.7
45.0	-4.2	-3.0	-2.8	-1.6
60.0			-2.9	-1.7
67.5	-3.7	-2.4		
75.0			-2.8	-1.6
90.0	-4.2	-2.9	-2.9	-1.7

where a and b are respectively the major and minor axis of the ellipse, φ is the elliptical angle and $E(m)$ is an elliptical integral of the second kind

$$E(m) = \int_0^{\pi/2} (1 - m\sin^2\varphi)^{1/2} \mathrm{d}\varphi \tag{73}$$

with $m = 1 - (b/a)^2$.

The analysis, reported in Ref. 11, has been carried out for a ratio of the minor to the major axis equal to 0.5. The crack was considered to be embedded in a cylinder with an approximately elliptical cross-section whose axes were taken to be 6 and 8 times the minor axis of the crack. Because of symmetry only one fourth of the surface of the cylinder needed to be discretized; the mesh, consisting of 24 elements, is shown in Fig. 14. The lengths L_e and L_i of the sides (normal to the front) of the special elements externally and internally adjacent to the crack-front were $0.16a$ and $0.32a$, respectively. In Table 5 the percent errors in the calculated K_I are reported for those values of the elliptical angle which define the position of corner and mid-side nodes along the front.

Traction-singular elements have been employed in more complex fracture mechanics analyses. The interaction between two circular coplanar cracks has been investigated in Ref. 11; the same reference presents the results obtained for a semicircular inner surface crack in an internally pressurized thick-walled cylinder. The variation of K_I along the crack-front (both straight and curved) of CT and three-point bend specimens is analyzed in Refs 16 & 17, respectively. The solution for a penny-shaped crack embedded in a circular cylinder of infinite length and finite diameter is given in Ref. 12.

References

1. Cruse, T.A. & Vanburen, W. Three-dimensional elastic stress analysis of a fracture specimen with an edge crack, *Int. J. Fract. Mech.*, 1971, **17**, 1-15.

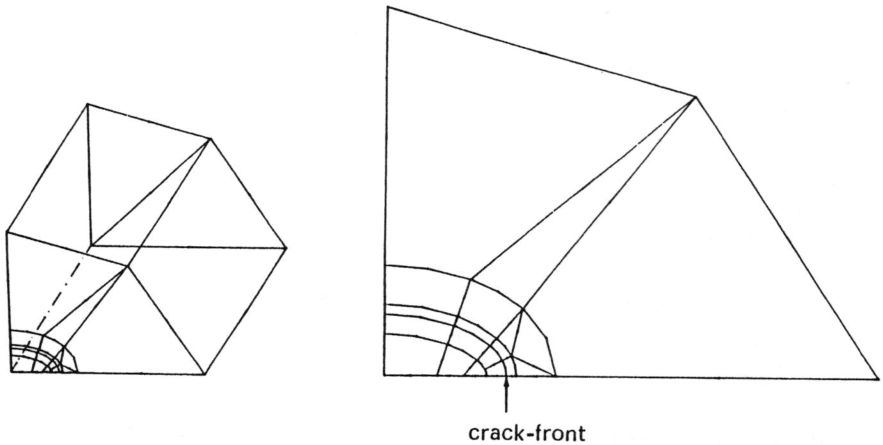

Figure 14: Boundary element mesh for the elliptical crack problem

Table 5: Percent error in the value of K_I calculated for the elliptical crack problem

φ	K_I from displacements at corner nodes	K_I from tractions at corner nodes	K_I from tractions at mid-side nodes
0.0	+0.14	-1.77	
7.68			-0.87
15.0	+0.04	-1.26	
22.86			-0.76
30.0	-0.01	-1.52	
43.35			-0.28
55.0	+0.24	-0.17	
73.11			+0.20
90.0	+0.56	+0.67	

2. Cruse, T.A. Boundary-integral equation method for three-dimensional elastic fracture mechanics analysis, AFOSR-TR-75-0813, May 1975.

3. Cruse, T.A., Osias, J.R. & Wilson, R.B. Boundary-integral equation method for elastic fracture mechanics analysis, AFOSR-TR-76-0878, May 1976.

4. Henshell, R.D. & Shaw, K.G. Crack-tip finite elements are unnecessary, *Int. J. Numer. Methods Eng.*, 1975, **9**, 495-507.

5. Barsoum, R.S. On the use of isoparametric finite elements in linear fracture mechanics, *Int. J. Numer. Methods Eng.*, 1976, **10**, 25-37.

6. Cruse, T.A. & Wilson, R.B. Advances application of boundary-integral equation methods, *Nucl. Eng. Des.*, 1978, **46**, 223-234.

7. Lange, D. 3-D Fracture analysis using the boundary integral equation method, in *Numerical Methods in Fracture Mechanics*, A.R. Luxmoore & D.R.J. Owen (ed), Proceedings of the First Int. Conf. on Numerical Methods in Fracture Mechanics, Swansea, West Glamorgan, UK, 1978, pp. 115-127.

8. Tan, C.L. & Fenner, R.T. Elastic fracture mechanics analysis by the boundary integral equation method, *Proc. R. Soc. London*, A269, 1979, pp. 243-260.

9. Cruse, T.A. & Wilson, R.B. Boundary-integral method for elastic fracture mechanics, AFOSR-TR-78-0355, November 1977.

10. Luchi, M.L. & Poggialini, A. Computation of 3-dimensional stress intensity factors using special boundary elements, in *Boundary Elements*, C.A. Brebbia, T. Futagami & M. Tanaka (ed), Proceedings of the Fifth Int. Conf. on BEM , Hiroshima, Japan, Springer-Verlag, 1983, pp. 461-470.

11. Luchi, M.L. & Rizzuti, S. Boundary elements for three-dimensional elastic crack analysis, *Int. J. Numer. Methods Eng.*, 1987, **24**, 2253-2271.

12. Jia, Z.H., Shippy, D.J. & Rizzo, F.J. Three-dimensional crack analysis using singular boundary elements, *Int. J. Numer. Methods Eng.*, 1989, **28**, 2257-2273.

13. Blandford, G.E., Ingraffea, A.R. & Ligget, J.A. Two-dimensional stress intensity factor computation using the boundary element method, *Int. J. Numer. Methods Eng.*, 1981, **17**, 387-404.

14. Tanaka, M. & Itoh, H. New crack elements applied to boundary element computation of stress intensity factors in 3-D elastostatics, *Role of Fracture Mechanics in Nuclear Technology*, G.C. Sih, H. Nisitani & T. Ishihara (ed), Elsevier, Amsterdam, 1987, pp. 835-847.

15. Blackburn, W.S. Calculation of stress intensity factors at a crack-tip using finite elements, in *The Mathematics of Finite Elements and Applications*, J.R. Whiteman (ed), Academic Press, 1973, pp. 327-336.

16. Luchi, M.L. & Rizzuti, S. Boundary element analysis of *CT* specimen with straight and curved crack-front, *Int. Journ. Fract.*, 1987, **34**, 23-40.

17. Luchi, M.L. & Rizzuti, S. Calcolo del fattore di intensitá degli sforzi lungo il fronte del difetto di un provino a flessione su tre punti, *Proc. of XV National Conference of AIAS, Pisa, Italia*, ETS Editrice, 1987, pp. 1-14.

Appendix A

As pointed out in the final part of Section 1, traction-singular quarter-point boundary elements have been proposed[10] in which the side-nodes in the ξ, η plane are also moved to the 1/4 position (see Fig. A1, where the side $\eta = -1$ corresponds to the crack-front).

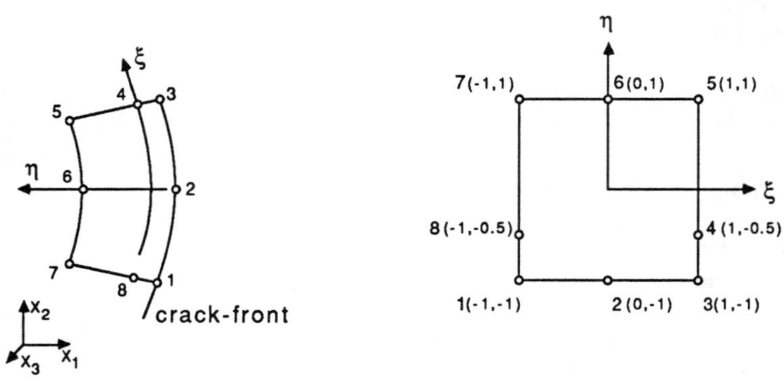

Figure A1: Mapping of the special traction-singular quarter-point element

Displacement and traction variations, as given by (9) and (10) respectively, are also in this case represented by means of proper interpolation functions. Appropriate functions can be straightforwardly derived by noticing that they have the same form as that assumed by the displacement and traction interpolation functions employed in the traction-singular quarter-point boundary elements represented in Fig. 2, following a transformation which brings the side-nodes from the middle of the side to the 1/4 position, i.e. an inverse transformation to that described by relations (17). Assuming that the quadratic shape functions (4) and the singular traction interpolation functions (8) are defined in the ξ^*, η^* plane, it is thus sufficient to introduce the relation

$$1 + \eta^* = \tfrac{1}{2}(1 + \eta)^2 \tag{A1}$$

into the expressions of $\phi_s(\xi^*, \eta^*)$ and $\phi_t^s(\xi^*, \eta^*)$, given respectively by (4) and (8) where ξ^* and η^* replace ξ and η, to derive displacement interpolation functions, $\phi_d^s(\xi, \eta)$, as

$$\phi_d^1 = \frac{1}{2}(1 - \xi)[-\xi - \sqrt{2}(1 - \frac{1}{2}\xi)\sqrt{1 + \eta} + (1 + \eta)]$$

$$\phi_d^2 = (1 - \xi^2)[1 - \frac{\sqrt{2}}{2}\sqrt{1 + \eta}]$$

$$\phi_d^3 = \frac{1}{2}(1 + \xi)[\xi - \sqrt{2}(1 + \frac{1}{2}\xi)\sqrt{1 + \eta} + (1 + \eta)]$$

$$\phi_d^4 = (1 + \xi)[\sqrt{2}\sqrt{1 + \eta} - (1 + \eta)] \tag{A2}$$

$$\phi_{\rm d}^5 = \frac{1}{2}(1+\xi)[\frac{\sqrt{2}}{2}(\xi-2)\sqrt{1+\eta}+(1+\eta)]$$

$$\phi_{\rm d}^6 = \frac{\sqrt{2}}{2}(1-\xi^2)\sqrt{1+\eta}$$

$$\phi_{\rm d}^7 = \frac{1}{2}(1-\xi)[-\frac{\sqrt{2}}{2}(\xi+2)\sqrt{1+\eta}+(1+\eta)]$$

$$\phi_{\rm d}^8 = (1-\xi)[\sqrt{2}\sqrt{1+\eta}-(1+\eta)]$$

and traction interpolation functions, $\phi_{\rm t}^{\rm s}(\xi,\eta)$, as:

$$\phi_{\rm t}^1 = \tfrac{1}{2}(1-\xi)[-\frac{\xi}{\sqrt{1+\eta}}-\sqrt{2}(1-\frac{1}{2}\xi)+\sqrt{1+\eta}]$$

$$\phi_{\rm t}^2 = (1-\xi^2)[\frac{1}{\sqrt{1+\eta}}-\frac{\sqrt{2}}{2}]$$

$$\phi_{\rm t}^3 = \frac{1}{2}(1+\xi)[\frac{\xi}{\sqrt{1+\eta}}-(\sqrt{2}(1+\frac{1}{2}\xi)+\sqrt{1+\eta}]$$

$$\phi_{\rm t}^4 = (1+\xi)[1-\frac{\sqrt{2}}{2}\sqrt{1+\eta}] \tag{A3}$$

$$\phi_{\rm t}^5 = \frac{1}{2}(1+\xi)[(\xi-2)+\sqrt{2}\sqrt{1+\eta}]$$

$$\phi_{\rm t}^6 = (1-\xi^2)$$

$$\phi_{\rm t}^7 = \frac{1}{2}(1-\xi)[-(\xi+2)+\sqrt{2}\sqrt{1+\eta}]$$

$$\phi_{\rm t}^8 = (1-\xi)[1-\frac{\sqrt{2}}{2}\sqrt{1+\eta}]$$

Note that the singular traction interpolation functions $\phi_{\rm t}^{\rm s}$ can be derived by dividing displacement interpolation functions $\phi_{\rm d}^{\rm s}$ by the term $\sqrt{1+\eta}$; $\phi_{\rm d}^4$ and $\phi_{\rm d}^8$ are further divided by $\sqrt{2}$ and $\phi_{\rm d}^5$; $\phi_{\rm d}^6$, $\phi_{\rm d}^7$ multiplyed by $\sqrt{2}$ in order to be unity at the respective nodes.

Appendix B

Equations (68), which relate the mixed mode stress intensity factors, K_I and K_{II}, to the cylindrical components of traction in the case where the traction-singular elements externally adjacent to the crack-front lie on a plane which forms an angle $\overline{\theta}(\overline{\theta}>0)$ with the plane of the crack, can be written in a more compact form as

$$\begin{aligned}\sqrt{\pi L_{\rm e}}\ t_\theta &= K_I f_1(\overline{\theta})+K_{II}f_2(\overline{\theta})\\ \sqrt{\pi L_{\rm e}}\ t_r &= K_I f_3(\overline{\theta})+K_{II}f_4(\overline{\theta})\end{aligned} \tag{B1}$$

where

$$
\begin{aligned}
f_1(\theta) &= \cos\frac{\theta}{2}(1-\sin^2\frac{\theta}{2}) \\
f_2(\theta) &= -3\sin\frac{\theta}{2}(1-\sin^2\frac{\theta}{2}) \\
f_3(\theta) &= \sin\frac{\theta}{2}(1-\sin^2\frac{\theta}{2}) \\
f_4(\theta) &= \cos\frac{\theta}{2}(1-3\sin^2\frac{\theta}{2})
\end{aligned}
\qquad \text{(B2)}
$$

Solving for the unknowns K_I and K_{II} yields

$$
\begin{aligned}
K_I &= \sqrt{\pi L_e}\;\frac{t_\theta f_4(\overline{\theta}) - t_r f_2(\overline{\theta})}{f_4(\overline{\theta})f_1(\overline{\theta}) - f_2(\overline{\theta})f_3(\overline{\theta})} \\
K_{II} &= \sqrt{\pi L_e}\;\frac{t_r f_1(\overline{\theta}) - t_\theta f_3(\overline{\theta})}{f_4(\overline{\theta})f_1(\overline{\theta}) - f_2(\overline{\theta})f_3(\overline{\theta})}
\end{aligned}
\qquad \text{(B3)}
$$

By substituting expressions (B2) for the terms $f_1(\overline{\theta})$, $f_2(\overline{\theta})$, $f_3(\overline{\theta})$ and $f_4(\overline{\theta})$, we obtain, after some manipulation

$$
\begin{aligned}
K_I &= \sqrt{\pi L_e}\;\frac{t_\theta(1-3\sin^2\frac{\overline{\theta}}{2}) + t_r(3\sin\frac{\overline{\theta}}{2}\cos\frac{\overline{\theta}}{2})}{\cos^3\frac{\overline{\theta}}{2}} \\
K_{II} &= \sqrt{\pi L_e}\;\frac{t_r\cos\frac{\overline{\theta}}{2} - t_\theta\sin\frac{\overline{\theta}}{2}}{\cos^2\frac{\overline{\theta}}{2}}
\end{aligned}
\qquad \text{(B4)}
$$

Chapter 6

Stress intensity factor weight functions for cracks in 3-D finite geometries

R. Bains,[a] M.H. Aliabadi[a] & D.P. Rooke[b]

[a] *Wessex Institute of Technology/University of Portsmouth, Ashurst Lodge, Ashurst, Southampton, SO4 2AA, UK*

[b] *Defence Research Agency, Aerospace Division, RAE Farnborough, Hants GU14 7TU, UK*

1 Introduction

All structures contain cracks to some extent, mainly as a result of the manufacturing processes or due to localized in-service damage. These cracks act as sites of stress elevation; subsequent growth due to service loads could lead to a degradation in the structural strength. If these structures are part of an aircraft, the presence of a crack in a critical component could have catastrophic consequences. The severity of a crack in a structure that is subjected to load has been characterised by a single parameter, the stress intensity factor (SIF). This is one of the most important parameters used in fracture mechanics to determine crack growth. A knowledge of this parameter allows accurate determination, through the fatigue laws, of the life expectancy of engineering structures.

In general for three-dimensional crack problems, the geometry and loading encountered is such that the stress intensity factor cannot be solved for analytically. The problem is further complicated since this parameter is known to be a function of position on the crack front. The only economical means of solving linear elastic problems is through the use of the boundary element method (BEM).[1] For crack problems in which the geometry is fixed and the applied loads vary, the weight function method of determining stress intensity factors has proved to be efficient and versatile, see Bains.[2]

The weight function concept of determining stress intensity factors (SIF) within the realm of classical elasticity for isotropic bodies was first introduced by Bueckner[3] for two-dimensional crack analysis, under states of plane strain. He showed that these weight functions once obtained for a given configuration are unique universal functions for that particular crack configuration and body geometry. Once the weight functions associated with a specific crack-tip in a given body are known, the stress intensity factor for that crack-tip can be evaluated directly for any surface and body force loadings.[3] Bueckner[3] identified the weight functions as proportional to the displacements of his so-called fundamental field. The fundamental field may be considered to be generated through the application of a pair of concentrated forces on the crack at the tip in a body in equilibrium with zero body forces and surface tractions. The displacements of this field which satisfy Navier's equation have a singularity at the crack tip which normally would not be admissible as a physical field, since it gives rise to unbounded strain energy of deformation at the load-point. Thus its existence is outside the class of displacement fields corresponding to bounded energy upon which proof of uniqueness in linear elasticity theory is based. In contrast any elastic field arising from applied loadings are referred to as a regular field if the displacements are bounded everywhere and the stresses obey the usual asymptotic relations at the crack-tip or along the whole crack front in a three-dimensional problem. For two-dimensional crack problems the fundamental field formulation gives rise to singularities of order $\rho^{-\frac{3}{2}}$ and $\rho^{-\frac{1}{2}}$ in the stresses and displacements respectively, where ρ denotes radial distance measured from the crack tip. The fundamental fields in the two-dimensional analysis were derived within the framework of analytic function theory as presented by Muskhelishvili.[4] Through the application of Betti's reciprocal work theorem[5] for two identical geometrical states, Bueckner showed that the stress intensity factor is directly proportional to the integral, over the boundary, of the product of the traction field associated with one geometrical state and the weight function associated with the second geometrical state, in the absence of body forces. The weight function theory, for two-dimensional crack problems was further developed by Bueckner in Ref. 6. He has also derived in closed form weight functions for the notched bar[7] and for the configuration of collinear plane strain cracks.[8]

Early extensions of the weight function theory using fundamental fields to three-dimensional crack problems were restricted purely to mode I deformation[9,10] and crack face displacements only were obtained. In Ref. 9, the edge crack in an elastic half-space was formulated in terms of integro-differential equations and a relationship between crack pressure and crack opening displacement was ascertained. Fundamental fields and weight functions presented in Ref. 10 for the penny-shaped crack were derived in terms of a Boussinesq-Papkovitch potential; while for the elliptical crack, crack fields are derived from knowledge of Dyson's formulas;[11,12] these are simply another representation of the Boussinesq-Papkovitch potentials as harmonic potentials of a single layer. The most significant advances as regards weight functions and fundamental fields in three dimensions were made by Bueckner,[13] who derived in closed form, without recourse to integral transforms of any kind, weight functions for all three fracture modes for the penny-shaped and the half-plane crack in infinite elastic bodies. He constructed the fundamental fields for all three modes from the knowledge of a single Boussinesq-Papkovitch potential. Through Betti's reciprocity theorem,[5] Bueckner formulated a relationship for the stress intensity factor for any mode β(=I,II or III) as

$$K_\beta = \int_\Gamma \mathbf{T}\cdot\mathbf{W}_\beta d\Gamma + \int_V \mathbf{F}\cdot\mathbf{W}_\beta dV \tag{1}$$

where $\cdot$ denotes the dot product, $\mathbf{T}$ and $\mathbf{F}$ are the prescribed surface tractions on the boundary Γ of V, and the body force per unit volume respectively, of one geometrical state; $\mathbf{W}_\beta$ are proportional to the displacement fields in the β mode, induced by the presence of concentrated forces acting on the crack faces near the crack front, in the same geometrical configuration for which $\mathbf{T}$ and $\mathbf{F}$ are prescribed. The vector fields $\mathbf{W}_\beta$ are associated with a specific point on the crack front where K_β is required. Note that the quantities $\mathbf{T}$ and $\mathbf{F}$ are completely independent of $\mathbf{W}_\beta$ and that once $\mathbf{W}_\beta$ is available for a given geometry, stress intensity factors can be found through eqn (1) for any surface traction and body force vectors, hence $\mathbf{W}_\beta$ is a universal characteristic of an elastic body and is commonly referred to as the weight function.

From knowledge of mode I Boussinesq-Papkovitch potentials Bueckner[13] systematically constructed closed form weight functions for mode II and III deformations, of the penny-shaped and half-plane cracks. In Ref. 13, the Boussinesq-Papkovitch potential which generates the fundamental fields gives rise to singularities in the stresses and displacements of order $\rho^{-\frac{5}{2}}$ and $\rho^{-\frac{3}{2}}$ respectively, where ρ denotes radial distance measured from the location of the point loads on the crack front. In addition, the formulation also generates the standard stress singularity at all points on the crack front other than the location of the point loads. This standard stress singularity of order $\rho^{-\frac{1}{2}}$ is normally encountered in traction free crack problems which are formulated through linear elasticity concepts.

Gao,[14] using a similar solution strategy to that developed in Ref. 13 derived weight functions for the embedded and surface circular cracks in infinite elastic solids. The crack face weight functions presented in Ref. 14 are defined as the intensity factors induced by a pair of equal, oppositely sensed unit point forces acting on the upper and lower crack faces. Bueckner's method is extended to determine weight functions for an external circular crack, formed as two elastic half-spaces joined over a circular connected region. In Ref. 15 Bueckner presents a review of the many aspects of two- and three-dimensional weight functions.

Rice[16] developed his two-dimensional weight function concepts through a consideration of the strain energy release and derivative of displacement fields with respect to crack length. He showed[16] that if the displacement fields $\mathbf{u}^{(1)}$ and stress intensity factor $K^{(1)}$ are known as functions of crack length a, for any symmetrical load system acting on a linear elastic body in plane strain, then the stress intensity factor $K^{(2)}$ for any other symmetrical load system whatsoever on the same body can be directly determined, i.e.

$$K^{(2)} = \frac{E'}{2K^{(1)}}\left[\int_\Gamma \mathbf{T}^{(2)}\cdot\frac{\partial \mathbf{u}^{(1)}}{\partial a} d\Gamma + \int_A \mathbf{F}^{(2)}\cdot\frac{\partial \mathbf{u}^{(1)}}{\partial a} dA\right] \tag{2}$$

where E' is the elastic modulus: for an isotropic material it is $E/(1-\nu^2)$ for plane strain and E for generalised plane stress. The vectors $\mathbf{T}^{(2)}$ and $\mathbf{F}^{(2)}$ are the prescribed surface tractions on the boundary Γ and body force within region A, respectively, for which the stress intensity factor $K^{(2)}$ is desired. Since the stress intensity factor for load system 2 is independent of the particular load system represented by 1, the function

$$h^{(1)} = \frac{E'}{2K^{(1)}} \frac{\partial \mathbf{u}^{(1)}}{\partial a} \tag{3}$$

was referred to as the weight function. In the Appendix of Ref. 16, Rice further shows that the three-dimensional theory of weight functions is based on displacement field variation associated, to first order, with arbitrary variation in the position of the crack front. The difficulty of this approach is that three-dimensional solutions for any load system 1 will not be known with such complete generality that the first order variation can be determined for completely arbitrary variations along the crack front. Rice[17] has presented results for the first order variation of an elastic displacement field associated with the arbitrary incremental planar advance of the front of a half-plane crack in a loaded elastic full space, discussed the relation for such results to a three-dimensional weight function theory, and derived an expression for the distribution of the mode I stress intensity factor for a slightly curved crack front. Gao & Rice[18] extended the work in Ref. 17 to determine also the shear-mode stress intensity factors along the front of a generally loaded half-plane crack when that front is slightly perturbed from a straight line. Application of this procedure to various crack configurations and loading modes can be found in Refs 19, 20 and 21. The crack front perturbation of Ref. 16 was also adopted by Vainshtok & Varfolomeyev[22] to determine weight functions in three dimensions. Rice[23,24] has further developed the weight function method to analyse the interaction between elastic cracks and transformation strains. Weight functions have also been applied to determine the elastic field of a three-dimensional dislocation loop.[25,26,27]

Weight function determination for finite cracked bodies using analytical techniques is generally extremely difficult, if not impossible with present mathematical techniques, and some form of numerical solution scheme is usually employed. The two most commonly used techniques for generating numerical weight functions in both two- and three-dimensional crack analysis are the FEM and BEM.

2 Review of numerical weight functions

The fundamental field concept of determining weight functions gives rise to unbounded energy of deformation in a finite area surrounding the crack tip. Application of the standard finite element procedure to determine stress intensity factors using fundamental fields would thus fail since this numerical solution technique is based upon the minimization of potential energy of the cracked body. Paris & McMeeking[28] and Paris, McMeeking & Tada[29] overcame this difficulty by removing a circle of material centered at the crack tip, and solved the resulting auxiliary boundary value problem for this geometry using standard FEM analysis. In Refs 28 and 29, the weight functions for the given crack geometry in an infinite solid were prescribed as displacement boundary conditions on the surface of the circular cavity while zero traction was prescribed on the remaining part of the bounding surface to the body. Basically the circular cavity acts to regularise the problem by removing the unboundedness of the strain energy at the crack tip. Thus any numerical solution technique which is based on the principle of minimum potential energy can be used to solve for the weight functions in the auxiliary problem. The accuracy of the solution obtained by Refs 28 and 29 showed sensitivity to the radius of the circular cavity at the tip. Cartwright & Rooke[30] showed using the cavity idea of Refs 28 and 29 that the weight functions could be obtained more accurately and efficiently with less degrees of freedom using a boundary element method than using the FEM for the same problem. Aliabadi, Rooke & Cartwright[31] further extended the work of Ref. 30 to both mode I and II deformations. By imposing traction boundary conditions on the cavity boundary Aliabadi & Rooke[32] obtained more reliable results.

All the analyses based on the cavity concept have an inherent drawback in that the radius of the cavity, on which the singular fields are imposed, must be selected before a solution can be computed. Reducing the cavity to zero radius, to remove the uncertainty in the choice of the radius, means that singular tractions and displacements fields must be directly modelled; this would require extreme mesh refinement at the crack tip. It is worth mentioning that unbounded strain energy at the tip does not pose any problems for the BEM since this numerical tool is not based on the principle of minimum potential energy. Despite its drawback, the cavity concept has been used to solve a large number of two-dimensional problems.[1] This idea has also been extended to three-dimensional cracked solids. Rooke, Cartwright & Aliabadi[33] developed the technique for rectilinear planar crack fronts. The weight functions obtained by Aliabadi, Rooke & Cartwright[34] have been used to generate stress intensity factors for a straight-fronted edge crack in a bar of rectangular cross-section subjected to various mode I loadings. Later, Bains, Aliabadi & Rooke[35] showed that the fields obtained by Ref. 33 are valid only in certain restricted regions of the complex plane.

Aliabadi, Rooke & Cartwright[36] introduced an alternative technique that gives efficient and accurate stress intensity factor solutions by the direct removal of the highly singular terms in the crack tip fields. This technique, which is based on regularization of the problem, bypasses the requirement of highly refined meshes around the crack tip and/or special techniques often incorporated for representing the singularity and calculating stress intensity factors. Aliabadi[37] further extended the work of Ref. 36 to two-dimensional mixed mode crack problems. For two-dimensional analysis the technique of Refs 36 and 37 has been extensively used and with considerable success.[1,2,39] The validity of this technique in three-dimensional crack problems was demonstrated by Bains *et. al.*[35,40] they obtained stress intensity factors for a planar rectilinear crack front in a bar of rectangular cross-section subjected to various mode I loadings. Furthermore, the technique has also been extended to penny-shaped cracks in finite cylindrical bars, see Bains, Aliabadi & Rooke[41] and to elliptical cracks in square cross-sectional bars, see Bains[42] for mode I deformation of the crack.

In an independent formulation, Parks & Kamenetzky[43] and Vanderglas[44] adopted the displacement derivative formulation[16] and computed the mode I weight functions with the aid of eqn (3) through a finite element procedure which employed the virtual crack extension technique[45,46] For mode I, this is an efficient technique but it still suffers from two drawbacks. The first of which concerns the differencing scheme employed in certain parts of the finite element mesh for computing the stiffness derivative. Thus all the disadvantages associated with numerical differentiation would apply to such computation. Secondly, as discussed in Refs 43 and 44, there are limitations associated with the method used in the calculation of the displacement derivatives with respect to the crack tip coordinates because of the special perturbation techniques used in simulating the virtual crack extension, for nodal points near the crack tip. Despite these drawbacks, Parks & Kamenetzky[43] produced solutions which were comparable with Ref. 29 and furthermore outlined a generalisation of their numerical procedure to the three-dimensional crack configurations by following the three-dimensional weight function theory development of Rice.[16] For three-dimensional problems this technique introduces additional complexities, in that the direction, as well as the magnitude, of the perturbation needs to be specified at various points along the crack front. Numerically this implies the shifting of nodes on the crack front individually and can be computationally expensive.

Sham[47] presented a new finite element procedure based on Bueckner's fundamental field concept for two- and three-dimensional weight function determination; this presentation is an alternative to the Parks & Kamentzky[43] virtual crack extension method. This new procedure incorporates a variational principle for the determination of singular fields in finite bodies with unbounded strain energy. In his approach Sha[47] considers a small region surrounding the point on the crack front where a pair of concentrated forces are applied and where the stress intensity factor is required. The singular part of the displacement field in this small region of the body is approximated by the singular displacements due to the same concentrated forces being applied in an infinite body. He then needs to determine the regular field in this small region, which results from the presence of the finite boundaries. This analysis uses 'quarter-point' cubic elements in the vicinity of the crack front to model the singular stress behaviour. Sham & Zhou[48] applied this technique to a finite element analysis of the stress intensity factors, for circular and elliptical cracks in isotropic, finite bodies with prescribed displacement and traction boundary conditions. Stress intensity factors were only given by Sham *et. al.*,[48] for uniform tensile loads and solutions for the elliptical crack were generated through knowledge of fundamental fields for the circular crack. This technique has also been extended to two-dimensional cracks in orthotropic materials by Sham & Zhou.[49]

Recently, Nikishkov & Piminov[50] have proposed a boundary element procedure, for computing weight functions, which is analogous to the finite element approach of Sham.[47] This technique can be used in the determination of weight functions for each of the three modes of fracture for symmetrical as well as nonsymmetrical crack geometries. The drawback of this technique is that it would require considerable effort in terms of implementation into a 3-D boundary element code.

Cruse & Besuner[51] and Besuner[52] have extended Rice's formulation[16] to determine stress intensity factors for fatigue crack propagation. In Refs 51 and 52 the authors obtained, using the boundary element method, average stress intensity factors for each perturbation of the crack front. The instantaneous value at a specific point on the crack front and the average value along the whole crack front are not exactly equivalent for most three-dimensional problems since the instantaneous stress intensity factor along the crack front generally varies with position. Further, this technique requires many iterations to obtain a single stress intensity factor solution and is thus computationally expensive.

Another technique using Rice's derivation which is similar to Refs 51 and 52 is due to Labbens, Heliot & Pellissier-Tannon.[53] This technique[53] is an extension of the approximate polynomial distribution technique as proposed by Grandt.[54] In Ref. 53 the computing time required to solve a semi-elliptical crack and a penny-shaped crack were too long for engineering applications and as a consequence they developed[55] a similar technique using boundary element analysis. Here polynomial influence functions were defined to correspond to the terms of a polynomial expansion of the stress field acting on the crack faces; these influence functions also depended on the radii and depth of the semi-elliptical crack. Numerical crack-face weight functions were obtained after five computer runs, one for each term in the polynomial; furthermore stress intensity factors were obtained from extrapolation of the local crack tip field.

When crack opening displacement fields for a reference system are not available, they can be approximately determined, see Morawietz, Mattheck & Munz[56] and Fett, Mattheck & Munz.[57] Saff & Sanger[58] proposed another approximate technique for the evaluation of weight functions in three dimensions. The analysis of Ref. 58 is based on the slice synthesis technique formulated by Fujimoto.[59] Zhao, Wu & Yan[60,61] have applied this technique to obtain weight functions for the embedded and surface elliptical cracks in plates. Other weight function applications to crack problem can be found in Refs 62 and 63.

In an interesting work, Oore & Burns[64] developed an integral for calculating the opening mode stress intensity factor, at any point on the front of an irregular flat crack embedded in an infinite elastic solid and subjected to an arbitrary normal stress field. Using a line integral they incorporated into their analysis the effect of crack front geometry and location of the point force loading in the geometry. Crack face weight functions are finally arrived at by dimensional analysis arguments. The final solution for the stress intensity factor is obtained by integration of the crack face weight function and the applied load over one of the crack face areas. Stress intensity factor results obtained using the analysis of Ref. 64 are compared with experimental results obtained by Khattab, Burns, Pick & Thompson.[65] On average, Ref. 65 states that the integral and stress-freezing photoelasticity estimates agree to within 5%.

In general, evaluation of the stress intensity factor by means of the weight function concept is more favorable than that of any other technique because of the load independent characteristics of weight functions. The key advantage of the weight function method is that once the weight functions are predetermined explicitly for a given geometry, accurate stress intensity factors for any loading conditions for the same geometry can be obtained easily and inexpensively. Therefore the use of the weight function concept can obviate the need for repeated computer computations of the stress intensity factors for a given geometry under different loading conditions.

A knowledge of fundamental fields in closed form for various crack front configurations is essential for the development of weight functions and hence the generation of stress intensity factors. To date closed form fundamental fields, applicable within the whole elastic domain are available only for certain idealised crack front configurations such as the half-plane and the penny-shaped.

3 Fundamental fields formulation

Mode I, II & III SIF formulation: rectilinear crack front

For a given geometrical configuration, see Fig. 1, under the influence of two independent loading conditions, Betti's reciprocal work theorem[5] states that the work of the external applied tractions of the first state through the displacements of the second

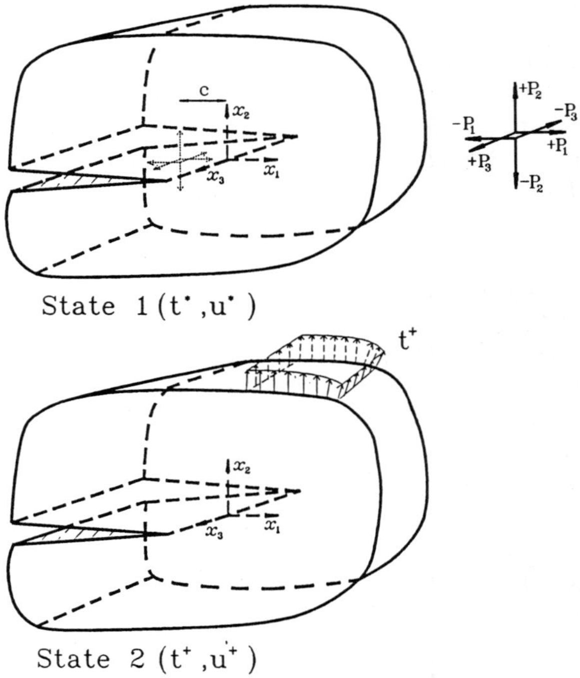

Figure 1: Two identical geometrical configurations: different loading states

is equal to the work of the applied tractions of the second through the displacements of the first state. Mathematically it can be stated as,

$$\int_{\Gamma} \mathbf{u}^{+} \cdot \mathbf{t}^{*} d\Gamma = \int_{\Gamma} \mathbf{t}^{+} \cdot \mathbf{u}^{*} d\Gamma \tag{4}$$

where $\mathbf{t}^+$, $\mathbf{u}^+$ are the self-equilibrated traction and displacement fields, respectively, corresponding to one state of loading and $\mathbf{t}^*$, $\mathbf{u}^*$ are those corresponding to another independent state of loading. The boundary of the solid is denoted by Γ. The choice of $\mathbf{t}^+$, $\mathbf{u}^+$ and $\mathbf{t}^*$, $\mathbf{u}^*$ are arbitrary except that they must satisfy equilibrium, compatibility and the boundary conditions. The three-dimensional weight function formulation presented below follows that given by Rooke, Cartwright & Aliabadi.[66]

States 1 and 2 in Fig. 1, both represent a crack in a three-dimensional solid. The boundary of the solid in state 2 is subjected to self-equilibrated tractions $\mathbf{t}^+$; this is the problem for which the desired solution is required. The displacement vector $\mathbf{u}^+$

for this state can be written as,

$$\mathbf{u}^{+} = \sum_{j=1}^{3} u_j \hat{x}_j \tag{5}$$

where $\hat{x}_j$ are the unit vector components in the orthogonal coordinate system ($j = 1, 2, 3$). From Sih & Liebowitz[67] the leading terms in the general expansion of displacement components close to the crack tip can be written in the following form:

$$u_1 = \frac{(1+\nu)}{E}\sqrt{\left(\frac{2r}{\pi}\right)}\left[K_I \cos\frac{\theta}{2}\left((1-2\nu)+\sin^2\frac{\theta}{2}\right)+ K_{II}\sin\frac{\theta}{2}\left(2(1-\nu)+\cos^2\frac{\theta}{2}\right)\right] \tag{6}$$

$$u_2 = \frac{(1+\nu)}{E}\sqrt{\left(\frac{2r}{\pi}\right)}\left[K_I \sin\frac{\theta}{2}\left(2(1-\nu)-\cos^2\frac{\theta}{2}\right)- K_{II}\cos\frac{\theta}{2}\left((1-2\nu)-\sin^2\frac{\theta}{2}\right)\right] \tag{7}$$

and

$$u_3 = \frac{2(1+\nu)}{E}\sqrt{\left(\frac{2r}{\pi}\right)} K_{III}\sin\frac{\theta}{2} \tag{8}$$

The coordinates (r, θ) are cylindrical polar coordinates in the $\hat{x}_1$, $\hat{x}_2$ plane, centred on the crack front at x_3=0. The three modes of fracture are represented by the stress intensity factors K_N (N =I, II, III), E and ν are the elastic modulus and Poisson's ratio respectively. On the crack surfaces (i.e. $\theta = \pm\pi$) eqns (6), (7) and (8) reduce to

$$u_1 = \pm\frac{2(1-\nu^2)}{E}\sqrt{\frac{2|x_1|}{\pi}} K_{II} \tag{9}$$

$$u_2 = \pm\frac{2(1-\nu^2)}{E}\sqrt{\frac{2|x_1|}{\pi}} K_{I} \tag{10}$$

and

$$u_3 = \pm\frac{2(1+\nu)}{E}\sqrt{\frac{2|x_1|}{\pi}} K_{III} \tag{11}$$

In state 1 the crack is subjected to pairs of concentrated symmetrical forces acting on the crack surfaces near the crack front as shown in Fig. 1. The point forces on the crack surfaces generate displacement and traction fields $\mathbf{u}^*$, $\mathbf{t}^*$ respectively. The traction vector $\mathbf{t}^*$ can be represented in terms of the point force vector $\mathbf{P}$ with the aid of the Dirac delta function. If the point forces act on both faces ($x_2 = 0$) at $x_1 = -c$, and $x_3 = 0$ the traction vector $\mathbf{t}^*$ can be written for either face as

$$\mathbf{t}^* = \delta(x_1 + c)\delta(x_3)\mathbf{P} \tag{12}$$

where $\delta(....)$ denotes the Dirac delta function. The point force vector $\mathbf{P}$ can be expressed in terms of components P_j $(j = 1,2,3)$, in each of the three orthogonal coordinate directions x_j as

$$\mathbf{P} = \sum_{j=1}^{3} P_j \hat{x}_j \tag{13}$$

Substituting eqns (5) and (12) in eqn (4) gives, for small c,

$$P_1 K_{II} + P_2 K_I + \frac{P_3 K_{III}}{(1-\nu)} = \frac{E\sqrt{\pi}}{2\sqrt{2}(1-\nu^2)\sqrt{c}} \int_\Gamma \mathbf{t}^+ \cdot \mathbf{u}^* \mathrm{d}\Gamma \tag{14}$$

This relationship can be used to find any stress intensity factor K_N (N =I, II, III) by allowing the corresponding P_j $(j = 1,2,3)$ to be nonzero. The displacements corresponding to $\mathbf{u}^*$ will be those on the surface, Γ, corresponding to loading from the nonzero P_j. For mode I deformation $P_1 = P_3 = 0$ and eqn (14) reduces to,

$$P_2 K_I = \frac{E\sqrt{\pi}}{2\sqrt{2}(1-\nu^2)\sqrt{c}} \int_\Gamma \mathbf{t}^+ \cdot \mathbf{u}^* \mathrm{d}\Gamma \tag{15}$$

where $\mathbf{u}^*$ is the three-dimensional weight function and corresponds to the displacement on Γ due to the presence of the point forces on the crack faces and $\mathbf{t}^+$ is the applied traction vector, for which the stress intensity factor is required. The components of $\mathbf{u}^*$ are called weight functions, since the integral in eqn (15) represents a weighted average of the applied tractions over the surface Γ. The weight functions $\mathbf{u}^*$ are independent of the tractions $\mathbf{t}^+$, thus the stress intensity factor can be obtained for any loading once $\mathbf{u}^*$ is known.

The above equations were derived for the point $x_3 = 0$ on the crack front. The same equations would be obtained for any other point on the crack front but the stress intensity factors K_N (N =I, II, III) would be functions of x_3.

3.2 Mode I fundamental fields: rectilinear crack front

The presence of the localised forces acting on the crack surfaces will generate displacement and stress fields in the domain. The displacement fields for a planar, rectilinear crack subjected to equal and opposite localised forces on the crack surface at a distance c from the crack front, (see Fig. 2), have been expressed by Kassir & Sih[68] in terms of Papkovitch-Neuber potentials,[69] viz

$$u_j = 4(1-\nu) f_j - \frac{\partial F}{\partial x_j} \tag{16}$$

The Papkovitch-Neuber potential f_j generates displacement components u_j, which satisfy Navier's equation with zero body force, i.e.

$$\nabla^2 u_j + \frac{1}{(1-2\nu)} \nabla e = 0 \qquad (j = 1,2,3.) \tag{17}$$

where ∇ is the del operator and ∇^2 the Laplacian operator, i.e.

$$\nabla^2 = \sum_{j=1}^{3} \frac{\partial^2}{\partial x_j^2} \tag{18}$$

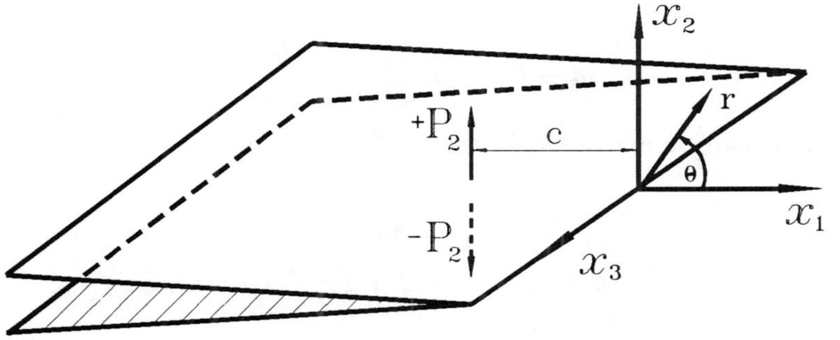

Figure 2: Concentrated symmetrical forces on planar crack geometry

and the dilatation e is given by

$$e = \sum_{j=1}^{3} \frac{\partial u_j}{\partial x_j} \tag{19}$$

In eqn (16), F is a linear function of the harmonic functions $f_n(x_1, x_2, x_3)$, $n = 0, 1, 2, 3$, given by

$$F = f_0 + \sum_{n=1}^{3} x_n f_n \tag{20}$$

For the case when the applied loads cause deformations symmetric with respect to variables x_2 and x_3, the harmonic functions f_0, f_1 and f_2 must be even in x_3 and f_3 odd in x_3. For this particular problem the solution can be expressed[68] in terms of one harmonic function $f(x_1, x_2, x_3)$, as

$$f_0 = -(1-2\nu) \int_{\infty}^{x_2} f(x_1, \xi, x_3) \mathrm{d}\xi \tag{21}$$

$$f_2 = -f(x_1, x_2, x_3) = -f \tag{22}$$

and

$$f_1 = f_3 = 0 \tag{23}$$

If I_o is defined as

$$I_o = \int_{\infty}^{x_2} f(x_1, \xi, x_3) \mathrm{d}\xi = \int_{\infty}^{x_2} f \mathrm{d}\xi \tag{24}$$

then

$$f_0 = -(1-2\nu) I_o \tag{25}$$

and the displacement fields can now be expressed as

$$u_1 = (1-2\nu) \frac{\partial I_o}{\partial x_1} + x_2 \frac{\partial f}{\partial x_1} \tag{26}$$

$$u_2 = 2(\nu - 1)f + x_2 \frac{\partial f}{\partial x_2} \tag{27}$$

and

$$u_3 = (1 - 2\nu)\frac{\partial I_o}{\partial x_3} + x_2 \frac{\partial f}{\partial x_3} \tag{28}$$

The stress fields can be expressed in terms of strains from Hooke's law. For normal stresses

$$\frac{\sigma_{ii}}{2\mu} = \frac{\nu e}{(1 - 2\nu)} + e_{ii} \tag{29}$$

where μ is the shear modulus and $i, j = 1, 2, 3$; for shear stresses,

$$\frac{\sigma_{ij}}{2\mu} = e_{ij} \qquad (i \neq j) \tag{30}$$

where

$$e_{ij} = \frac{1}{2}\left(\frac{\partial u_i}{\partial x_j} + \frac{\partial u_j}{\partial x_i}\right)$$

After differentiation of the displacement fields and substitution into Hooke's law, the normal stresses become

$$\frac{\sigma_{11}}{2\mu} = -2\nu \frac{\partial f}{\partial x_2} + (1 - 2\nu)\frac{\partial^2 I_o}{\partial x_1^2} - x_2 \frac{\partial^2 f}{\partial x_1^2} \tag{31}$$

$$\frac{\sigma_{22}}{2\mu} = -\frac{\partial f}{\partial x_2} + x_2 \frac{\partial^2 f}{\partial x_2^2} \tag{32}$$

and

$$\frac{\sigma_{33}}{2\mu} = -2\nu \frac{\partial f}{\partial x_2} + (1 - 2\nu)\frac{\partial^2 I_o}{\partial x_3^2} + x_2 \frac{\partial^2 f}{\partial x_3^2} \tag{33}$$

the shear stresses become

$$\frac{\sigma_{12}}{2\mu} = x_2 \frac{\partial^2 f}{\partial x_1 \partial x_2} \tag{34}$$

$$\frac{\sigma_{23}}{2\mu} = x_2 \frac{\partial^2 f}{\partial x_3 \partial x_2} \tag{35}$$

and

$$\frac{\sigma_{31}}{2\mu} = x_2 \frac{\partial^2 f}{\partial x_1 \partial x_3} + (1 - 2\nu)\frac{\partial^2 I_o}{\partial x_1 \partial x_3} \tag{36}$$

For point forces near the tip of a straight-fronted crack, see Fig. 2, Kassir & Sih have shown[68] that the harmonic function $f(x_1, x_2, x_3)$ can be expressed as

$$f = \frac{-P}{2\mu\pi^2\rho} \arctan\left[\frac{2\sqrt{(cr)}\sin\frac{\theta}{2}}{\rho}\right] \tag{37}$$

where

$$\rho^2 = \sum_{j=1}^{3} x_j^2$$

$$r^2 = \sum_{j=1}^{2} x_j^2$$

and

$$\theta = \arctan\left[\frac{x_2}{x_1}\right]$$

Taking the limit of eqn (37) as the point force approaches the crack front, we obtain

$$f = \frac{-R_I\sqrt{r}}{\mu\rho^2}\sin\frac{\theta}{2} = \frac{-R_I\sqrt{r-x_1}}{\sqrt{2}\mu\rho^2} \tag{38}$$

where

$$R_I = \lim_{c\to 0}\left(\frac{P\sqrt{c}}{\pi^2}\right)$$

The derivatives of the harmonic function f required for the displacement and stress fields above can be found in Ref. 2. Rooke & Aliabadi[70] defined the derivatives of the integral I_o in terms of integrals $I(m,n)$, given by

$$I(m,n) = \frac{r^{(2-m)}R_I}{\sqrt{2}\mu}\int_0^1 \frac{S^{(m+n-\frac{1}{2})}\mathrm{d}S}{\sqrt{(r+x_1S)}(r^2+x_3^2S^2)^{(\frac{n+2}{2})}} \tag{39}$$

those derivatives required are given in Ref. 2. General analytical solutions to this integral are algebraically complex and so $I(m,n)$ was evaluated numerically by Gauss-Chebyshev quadrature.[71] In terms of these integrals the displacement fields take the following form:

$$u_1 = -(1-2\nu)\left[\frac{1}{2}I(1,0) + 2x_1I(0,2)\right] + x_2\frac{\partial f}{\partial x_1} \tag{40}$$

$$u_2 = 2(\nu-1)f + x_2\frac{\partial f}{\partial x_2} \tag{41}$$

and

$$u_3 = -(1-2\nu)\left[2x_3I(0,2)\right] + x_2\frac{\partial f}{\partial x_3} \tag{42}$$

It must be pointed out that the displacement fields u_1, u_2 and u_3, given by expressions (40), (41) and (42) respectively, are different from those given by Rooke *et. al.*[66,70] who obtained I_o and its derivatives from the work of Grobner *et. al.*[72] Their fields have limited application because their integration is confined to a restricted region in the complex plane.

The stress fields in terms of f and $I(m,n)$ can be expressed as

$$\frac{\sigma_{11}}{2\mu} = -2\nu\frac{\partial f}{\partial x_2} + (1-2\nu)\left[\frac{1}{4}I(2,0) + 8x_1^2I(0,4) + 2x_1I(1,2) - \right.$$

$$\left. 2I(0,2) + \frac{x_1}{2}I(3,0)\right] + x_2\frac{\partial^2 f}{\partial x_1^2} \tag{43}$$

$$\frac{\sigma_{22}}{2\mu} = -\frac{\partial f}{\partial x_2} + x_2 \frac{\partial^2 f}{\partial x_2^2} \tag{44}$$

$$\frac{\sigma_{33}}{2\mu} = -2\nu \frac{\partial f}{\partial x_2} + (1-2\nu)\left[8x_3^2 I(0,4) - 2I(0,2)\right] + x_2 \frac{\partial^2 f}{\partial x_3^2} \tag{45}$$

$$\frac{\sigma_{12}}{2\mu} = x_2 \frac{\partial^2 f}{\partial x_1 \partial x_2} \tag{46}$$

$$\frac{\sigma_{23}}{2\mu} = x_2 \frac{\partial^2 f}{\partial x_3 \partial x_2} \tag{47}$$

and

$$\frac{\sigma_{13}}{2\mu} = (1-2\nu)\left[x_3 I(1,2) + 8x_1 x_3 I(0,4)\right] + x_2 \frac{\partial^2 f}{\partial x_1 \partial x_3} \tag{48}$$

The above displacement and stress fields are in agreement with the work of Bueckner[13] and Meade & Keer.[73] The point force at the crack tip generates displacements of order $\rho^{-\frac{3}{2}}$ and stresses of order $\rho^{-\frac{5}{2}}$, whereas for standard traction-free cracks the displacements are of order $\rho^{\frac{1}{2}}$ and stresses are of order $\rho^{-\frac{1}{2}}$.

3.3 Mode I fundamental fields: penny-shaped crack front

Consider an isotropic, homogeneous elastic body of infinite extent, weakened by a penny-shaped crack. Let a fixed cartesian coordinate system $(x_j,\ j = 1,2,3)$ be attached to this configuration such that the coordinate origin is coincident with the centre of the crack, (see Fig. 3). The radius of the crack is a. It is also necessary to introduce a cylindrical coordinate system (r, θ, x_3), since this is more suitable for this problem, Fig. 3. Let Ω be a crack in the plane $x_3 = 0$. Ω^+ and Ω^- will be used to denote its faces on the upper and lower x_3-half space respectively and $\Omega = \Omega^+ \cup \Omega^-$. The periphery of the crack is denoted by $\mho$.

The elastic fields of mode I, generated by the presence of the collinear unit point forces P, acting on Ω^+ and Ω^-, see Fig. 3, can be described by a Boussinesq-Papkovitch potential $G(x_1, x_2, x_3)$. Such a potential G is harmonic, i.e.

$$\nabla^2 G = \sum_{j=1}^{3} G_{,x_j x_j} = 0 \tag{49}$$

where the following convention is used,

$$f_{,r} = \frac{\partial f}{\partial r}$$

In cylindrical coordinates (r, θ, x_3), eqn (49) becomes

$$\nabla^2 G = G_{,rr} + \frac{G_{,\theta\theta}}{r^2} + \frac{G_{,r}}{r} + G_{,x_3 x_3} = 0 \tag{50}$$

This potential gives the displacement fields as

$$u = -x_3 G_{,r x_3} - (1-2\nu) G_{,r} \tag{51}$$

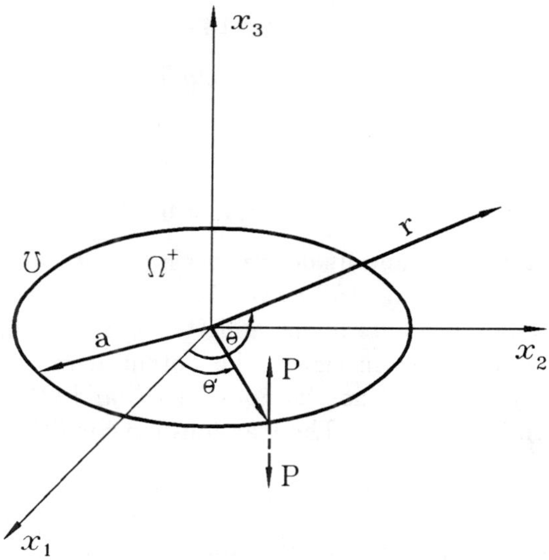

Figure 3: Coordinate systems for the penny-shaped crack configuration

$$v = -x_3 \frac{G_{,\theta x_3}}{r} - (1-2\nu)\frac{G_{,\theta}}{r} \tag{52}$$

and

$$w = -x_3 G_{,x_3 x_3} + 2(1-\nu) G_{,x_3} \tag{53}$$

and the stress fields as

$$\sigma_{rr} = -2\mu \left[\left(x_3 G_{,rr}\right)_{,x_3} + 2\nu \left(\frac{G_{,\theta\theta}}{r^2} + \frac{G_{,r}}{r} \right) \right] \tag{54}$$

$$\sigma_{\theta\theta} = -2\mu \left[\left(x_3 \frac{G_{,\theta\theta}}{r^2} + x_3 \frac{G_{,r}}{r} \right)_{,x_3} + 2\nu G_{,rr} \right] \tag{55}$$

$$\sigma_{x_3 x_3} = -2\mu \left[x_3 G_{,x_3 x_3 x_3} - G_{,x_3 x_3} \right] \tag{56}$$

$$\tau_{\theta x_3} = -2\mu x_3 \frac{G_{,\theta x_3 x_3}}{r} \tag{57}$$

$$\tau_{r x_3} = -2\mu x_3 G_{,r x_3 x_3} \tag{58}$$

and

$$\tau_{r\theta} = -2\mu \left[x_3 \left(\frac{G_{,\theta x_3}}{r} \right)_{,r} + (1-2\nu) \left(\frac{G_{,\theta}}{r} \right)_{,r} \right] \tag{59}$$

where u, v, w are the cylindrical components of the displacement vector and σ_{rr}, $\sigma_{\theta\theta}$, $\sigma_{x_3 x_3}$, $\tau_{\theta x_3}$, $\tau_{r x_3}$, $\tau_{r\theta}$ the cylindrical stress components. Material properties are denoted by the shear modulus μ and Poisson's ratio ν. These fields are without body forces. If G and its derivatives up to the 3rd order are continuous, then on Ω^+ and Ω^- we have

$$w = 2(1-\nu)G_{,x_3} \tag{60}$$

$$\sigma_{x_3x_3} = 2\mu G_{,x_3x_3} \tag{61}$$

$$\tau_{\theta x_3} = 0 \tag{62}$$

and

$$\tau_{rx_3} = 0 \tag{63}$$

The boundary values of G and its derivatives may differ on opposite faces of the crack; for mode I, this potential has to be even in x_3. This implies that $\sigma_{x_3x_3}$ has the same value on corresponding points of Ω^+ and Ω^-, while the displacement w is the same in magnitude but opposite in sign. The potential must have the property that it generates no body force and no tractions on Ω^+ and Ω^- and the stresses must tend to zero as $|x_3|$ tends to infinity. The traction-free condition on the crack faces implies that

$$G_{,x_3x_3} = 0 \tag{64}$$

In order to derive the explicit form of G and its derivatives, it is necessary to introduce oblate spheroidal coordinates s, t (Bueckner[13]). The oblate spheroidal coordinates s, t are convenient for the description of circular cracks. These new coordinates are related to the cylindrical coordinates by

$$r = a \cosh s \cos t \tag{65}$$

and

$$x_3 = a \sinh s \sin t \tag{66}$$

where

$$0 \leq s < \infty \ , \quad -\frac{\pi}{2} \leq t \leq \frac{\pi}{2}$$

Within this range, $s = 0$ on Ω. Also, t is positive for $x_3 > 0$ and negative for $x_3 < 0$; $t = 0$ for $x_3 = 0$ and $r > a$. Bueckner[13] decomposed G according to

$$G = \sum_{n=0}^{\infty} G_n(x_1, x_2, x_3) = \sum_{n=0}^{\infty} q^n H_n(s) \tag{67}$$

with

$$q = \frac{r}{a} e^{i(\theta - \theta')} \tag{68}$$

where $i = \sqrt{-1}$, n is a non negative integer and θ' is the angular location, with respect to x_1, of the point force on $\mho$. The harmonic condition on the decomposed functions is satisfied if solutions to the following differential equation exist.

$$\frac{\mathrm{d}^2}{\mathrm{d}s^2} H_n(s) + (1+2n) \tanh s \frac{\mathrm{d}}{\mathrm{d}s} H_n(s) = 0 \tag{69}$$

In Ref. 13, the author found that the differential eqn (69) has the following solution,

$$H_n(s) = c_o \int_s^\infty \frac{d\varrho}{\cosh^{2n+1}\varrho} \tag{70}$$

with c_o defined as

$$c_o = -\frac{\ell^{\frac{5}{2}}}{\sqrt{2a}(1-\nu)} = -\frac{1}{\sqrt{2a}(1-\nu)}$$

where c_o has the dimension of the square of length, since ℓ is of unit length. In order to determine the potential, it is necessary to express G as

$$G = G_o + \sum_{n=1}^{\infty} G_n \tag{71}$$

where

$$G_o = H_o(s) = c_o \int_s^\infty \frac{d\varrho}{\cosh\varrho} = c_o \left[\frac{\pi}{2} - \arctan(\sinh s)\right] \tag{72}$$

Providing eqn (71) is uniformly convergent, then the harmonic function can be expressed[13] as,

$$G = -c_o \left[\frac{1}{\sqrt{q-1}} \log\left(\frac{\sinh s - \sqrt{q-1}}{\sinh s + \sqrt{q-1}}\right) + \frac{\pi}{2} - \arctan(\sinh s)\right] \tag{73}$$

subject to

$$\Im\left(\sqrt{q-1}\right) \geq 0 \tag{74}$$

and

$$-\pi \leq \Im(\log(....)) < 0 \tag{75}$$

Here $\Im$ refers to the imaginary part of a complex expression. At this point is must be emphasized that the harmonic potential G represented by eqn (73) involves complex variables. Only the real parts of the final results are implied for various real quantities such as the weight functions. By the appropriate differentiations of G and upon substitution into the displacement and stress equations the complete solution to the problem of two collinear unit point forces acting on a penny-shaped crack in an infinite elastic body is obtained. The required derivatives of G, for the computation of the fundamental fields can be found in Refs 2 and 48.

With reference to Fig. 4, the crack face weight functions are given as

$$u^{cf} = \frac{\pi(1-2\nu)}{(1-\nu)2r\sqrt{2d}} \left[\frac{a}{d}\cos\left(\frac{3\lambda}{2}\right) - \cos\left(\frac{\lambda}{2}\right)\right] \tag{76}$$

$$v^{cf} = \frac{\pi(1-2\nu)}{(1-\nu)2r\sqrt{2d}} \left[\sin\left(\frac{\lambda}{2}\right) - \frac{a}{d}\sin\left(\frac{3\lambda}{2}\right)\right] \tag{77}$$

and

$$w^{cf} = \frac{\sqrt{2(a^2-r^2)}}{\sqrt{a}d^2} \tag{78}$$

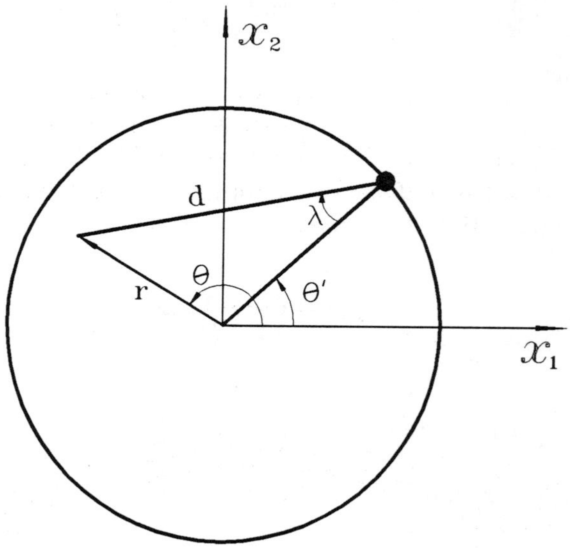

Figure 4: Definition of λ and d

they are in agreement with the work of Ref. 13, where

$$d^2 \equiv r^2 + a^2 - 2ra\cos(\theta - \theta') \tag{79}$$

and λ is the angle as shown in Fig. 4. The fundamental fields generated by G have the property that at θ equal to θ' there are singularities of order $\rho^{-\frac{5}{2}}$ and $\rho^{-\frac{3}{2}}$ in the stresses and displacements respectively, where ρ is the distance measured from the point $(a, \theta', 0)$ on the crack front, $\mho$. Furthermore, at all points on $\mho$, excluding the location of the point force (i.e. θ equal to θ'), the field generates a standard stress singularity of order $(r-a)^{-\frac{1}{2}}$ on the crack plane. The displacement field is bounded at all points along $\mho$ with the exception of θ equal to θ'. Because of the singular nature of the fundamental field (i.e. unbounded stresses and displacements) it gives rise to unbounded strain energy in the vicinity of the point θ equal to θ'.

3.4 Mode I SIF formulation: penny-shaped crack front

Stress intensity factors for a penny-shaped (traction-free) crack, subjected to mode I loading, can be determined by applying Betti's reciprocal work theorem[5] to a regular and a fundamental field. The regular field (in absence of body forces) generates stress intensity factors which can be expressed in terms of a Fourier series along the crack-front,[13] namely,

$$K(\theta') = K_o + \sum_{n=1}^{\infty} K_n(\theta') \tag{80}$$

where K_o and $K_n(\theta')$ are defined according to

$$K(\theta') = \Re\left(\frac{1}{2\sqrt{\pi}}\int_0^{2\pi} K(\theta)\mathrm{d}\theta + \frac{1}{\sqrt{\pi}}\sum_{n=1}^{\infty}\int_0^{2\pi} K(\theta)e^{ni(\theta-\theta')}\mathrm{d}\theta\right) \tag{81}$$

In eqn (81), $\Re$ denotes the real part of a complex expression. The fundamental fields generated by the potential G_n can be used to determine $K(\theta')$. For this, consider a spherical region V' of radius r_s containing the crack. Replace the crack front by a torus of radius r_w and boundary Γ_w, see Fig. 5. To aid the analysis it is necessary

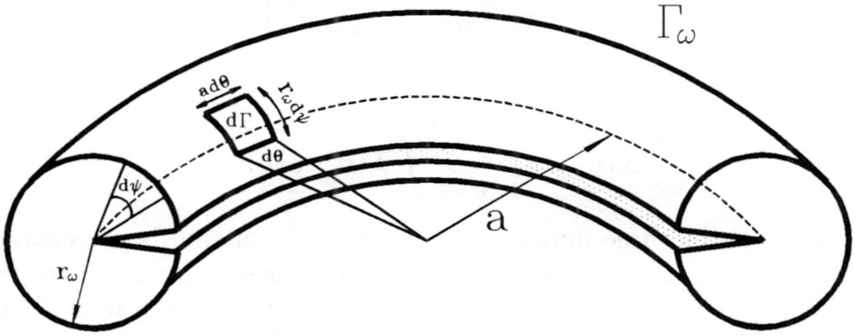

Figure 5: Torus surrounding crack front

to decompose the spherical region, V' into Γ_w and Γ_R, where Γ_R denotes the total boundary excluding Γ_w. On boundaries Γ_w and Γ_R the regular field generates tractions $\mathbf{t}^+$ and displacements $\mathbf{u}^+$, while the potential G_n generates displacements $\Re(\mathbf{u}_n^*)$ and tractions $\Re(\mathbf{t}_n^*)$. Application of Betti's reciprocal work theorem[5] to the regular and fundamental fields gives

$$\int_{\Gamma_w}\left[\mathbf{u}^+\cdot\Re(\mathbf{t}_n^*) - \Re(\mathbf{u}_n^*)\cdot\mathbf{t}^+\right]\mathrm{d}\Gamma = \int_{\Gamma_R}\left[\Re(\mathbf{u}_n^*)\cdot\mathbf{t}^+ - \mathbf{u}^+\cdot\Re(\mathbf{t}_n^*)\right]\mathrm{d}\Gamma \tag{82}$$

Defining an element of area $\mathrm{d}\Gamma = ar_w\mathrm{d}\varphi\mathrm{d}\theta$ on the torus (see Fig. 5) enables the integral over Γ_w to be expressed as

$$\int_{\Gamma_w}\left[\mathbf{u}^+\cdot\Re(\mathbf{t}_n^*) - \Re(\mathbf{u}_n^*)\cdot\mathbf{t}^+\right]\mathrm{d}\Gamma = \int_0^{2\pi}\int_0^{2\pi}\left[\mathbf{u}^+\cdot\Re(\mathbf{t}_n^*) - \Re(\mathbf{u}_n^*)\cdot\mathbf{t}^+\right] ar_w\mathrm{d}\varphi\mathrm{d}\theta \tag{83}$$

In the limit as $r_w \to 0$, Bueckner[13] showed that

$$\begin{aligned}
&\lim_{r_w\to 0}\left(\int_0^{2\pi}\int_0^{2\pi}\left[\mathbf{u}^+\cdot\Re(\mathbf{t}_n^*) - \Re(\mathbf{u}_n^*)\cdot\mathbf{t}^+\right] ar_w\mathrm{d}\varphi\mathrm{d}\theta\right) \\
&= \int_0^{2\pi}\lim_{r_w\to 0}\left(\int_0^{2\pi}\left[\mathbf{u}^+\cdot\Re(\mathbf{t}_n^*) - \Re(\mathbf{u}_n^*)\cdot\mathbf{t}^+\right] ar_w\mathrm{d}\varphi\right)\mathrm{d}\theta \\
&= \pi\sqrt{2}\int_0^{2\pi} K(\theta)e^{ni(\theta-\theta')}\mathrm{d}\theta
\end{aligned} \tag{84}$$

Thus for traction-free crack faces (i.e. $\Re(\mathbf{t}_n^*) = 0$) and as $r_s \to \infty$, the integral over Γ_R becomes

$$\pi\sqrt{2}\int_0^{2\pi} K(\theta)e^{ni(\theta-\theta')}\mathrm{d}\theta = \int_{\Gamma_R} \Re(\mathbf{u}_n^*)\cdot\mathbf{t}^+\mathrm{d}\Gamma \tag{85}$$

Substituting (85) into (81) gives

$$K(\theta') = \frac{1}{2\pi\sqrt{2\pi}}\left[\int_{\Gamma_R} \Re(\mathbf{u}_o^*)\cdot\mathbf{t}^+\mathrm{d}\Gamma + 2\int_{\Gamma_R}\left(\sum_{n=1}^{\infty}\Re(\mathbf{u}_n^*)\right)\cdot\mathbf{t}^+\mathrm{d}\Gamma\right] \tag{86}$$

By denoting the fundamental displacement field as

$$\Re(\mathbf{u}^*) = \Re\left(\mathbf{u}_o^* + 2\sum_{n=1}^{\infty}\mathbf{u}_n^*\right) \tag{87}$$

eqn (86) can be rewritten as

$$K(\theta') = \frac{1}{2\pi\sqrt{2\pi}}\int_{\Gamma} \Re(\mathbf{u}^*)\cdot\mathbf{t}^+\mathrm{d}\Gamma \tag{88}$$

In eqn (88) $\Re(\mathbf{u}^*)$ is the three-dimensional weight function and is proportional to the displacements on the boundary Γ of the elastic body, arising from the presence of the point forces at angular location θ' on the crack front; and $\mathbf{t}^+$ is the applied traction vector, for which the stress intensity factor is required. The weight functions $\Re(\mathbf{u}^*)$ are independent of the tractions $\mathbf{t}^+$, thus the stress intensity factor $K(\theta')$ can be obtained for any loading on Γ once $\Re(\mathbf{u}^*)$ is known. The weight function $\Re(\mathbf{u}^*)$ can be generated for a particular cracked geometry through numerical solution techniques such as the boundary element method (BEM).

4 Numerical formulation of BEM

The boundary element method (BEM) essentially involves the transformation of the governing field equations in the domain, to an integral equation over the boundary.[1] This boundary integral equation relating displacements u_j and tractions $t_j(=\sigma_{ij}n_i)$ on the surface Γ of an elastic homogeneous isotropic body in the absence of body forces can be expressed as

$$C_{ij}(\mathbf{x}')u_j(\mathbf{x}') + \int_{\Gamma} T_{ij}(\mathbf{x},\mathbf{x}')u_j(\mathbf{x})\mathrm{d}\Gamma(\mathbf{x}) = \int_{\Gamma} U_{ij}(\mathbf{x},\mathbf{x}')t_j(\mathbf{x})\mathrm{d}\Gamma(\mathbf{x}) \quad (i,j = 1,2,3) \tag{89}$$

where $\mathbf{x}'$ and $\mathbf{x}$ are points on the surface Γ respectively, see Fig. 6. The tensors $U_{ij}(\mathbf{x},\mathbf{x}')$ and $T_{ij}(\mathbf{x},\mathbf{x}')$ are known fundamental displacement and traction solutions respectively for a point force in an infinite domain (i.e. Kelvin's solutions). The fundamental solutions are given by

$$\begin{aligned} T_{ij}(\mathbf{x},\mathbf{x}') &= \frac{-1}{8\pi(1-\nu)\mathbf{R}^2}\left[\frac{\partial\mathbf{R}}{\partial\mathbf{n}}\left((1-2\nu)\delta_{ij} + 3\frac{\partial\mathbf{R}}{\partial x_i}\frac{\partial\mathbf{R}}{\partial x_j}\right)\right. \\ &+ \left.(1-2\nu)\left(\frac{\partial\mathbf{R}}{\partial x_j}n_i - \frac{\partial\mathbf{R}}{\partial x_i}n_j\right)\right] \end{aligned} \tag{90}$$

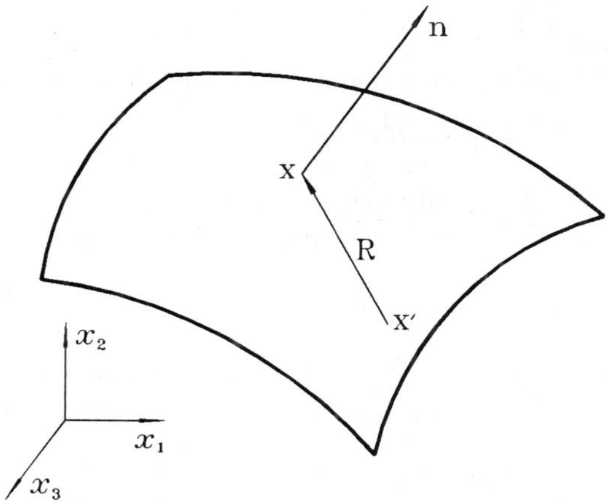

Figure 6: Definition of points $\mathbf{x}'$ and $\mathbf{x}$ on Γ.

and

$$U_{ij}(\mathbf{x},\mathbf{x}') = \frac{(1+\nu)}{8\pi E(1-\nu)\mathbf{R}}\left[(3-4\nu)\delta_{ij} + \frac{\partial \mathbf{R}}{\partial x_i}\frac{\partial \mathbf{R}}{\partial x_j}\right] \tag{91}$$

where $\mathbf{R}$ is the distance from $\mathbf{x}$ to $\mathbf{x}'$, $\mathbf{n}$ is the unit outward normal and n_i represents its components in the orthogonal coordinate system, δ_{ij} is the Kronecker delta function, E the elasticity modulus. The constant C_{ij} may be determined from rigid body considerations.[1]

When the surface is divided into N elements with $\mathcal{M}$ nodes each, the geometry, displacements, and tractions in each element are represented in terms of the nodal values $\mathbf{x}^\alpha$, $\mathbf{u}^\alpha$ and $\mathbf{t}^\alpha$ as

$$\mathbf{x}(\xi,\eta) = \sum_{\alpha=1}^{\mathcal{M}} N^\alpha(\xi,\eta)\mathbf{x}^\alpha \tag{92}$$

$$\mathbf{u}(\xi,\eta) = \sum_{\alpha=1}^{\mathcal{M}} N^\alpha(\xi,\eta)\mathbf{u}^\alpha \tag{93}$$

and

$$\mathbf{t}(\xi,\eta) = \sum_{\alpha=1}^{\mathcal{M}} N^\alpha(\xi,\eta)\mathbf{t}^\alpha \tag{94}$$

where $N^\alpha(\xi,\eta)$ are shape functions of the intrinsic coordinates ξ,η, such that $(-1 \le \xi,\eta \le 1)$. With this shape function representation the boundary integral eqn (89) can be written in a discretized form as

$$C_{ij}(\mathbf{x}^\beta)u_j(\mathbf{x}^\beta) + \sum_{m=1}^{N}\sum_{\alpha=1}^{\mathcal{M}} u_j^\alpha \int_{-1}^{+1}\int_{-1}^{+1} T_{ij}(\mathbf{x}(\xi,\eta),\mathbf{x}^\beta)N^\alpha(\xi,\eta)J^m(\xi,\eta)\mathrm{d}\xi\mathrm{d}\eta =$$
$$\sum_{m=1}^{N}\sum_{\alpha=1}^{\mathcal{M}} t_j^\alpha \int_{-1}^{+1}\int_{-1}^{+1} U_{ij}(\mathbf{x}(\xi,\eta),\mathbf{x}^\beta)N^\alpha(\xi,\eta)J^m(\xi,\eta)\mathrm{d}\xi\mathrm{d}\eta \quad (95)$$

where $\mathbf{x}^\beta$ is the nodal collocation point and $J^m(\xi,\eta)$ is the Jacobian of the transformation from the global coordinate system (x_1, x_2, x_3) to the local coordinate system (ξ,η) for the mth element. Equation (95) can thus be solved once the boundary conditions have been imposed. After substitution of the prescribed boundary conditions, the resulting system of algebraic equations can be written as,

$$[A]\{\mathbf{b}\} = \{\mathrm{d}\} \quad (96)$$

where vector $\{\mathbf{b}\}$ contains the unknown boundary displacements or tractions, $[A]$ is the coefficient matrix which is usually non-symmetric and densely populated and $\{\mathrm{d}\}$ is a vector containing the prescribed boundary conditions and the integrals of the appropriate displacement or traction kernels.

5 Regularization procedure

The problem associated with the modelling of the singularity in the fundamental fields can be avoided through a regularization procedure which is applied before the problem is solved numerically. This procedure which will be referred to as the subtraction of fundamental fields (SFF), is an extension of the technique presented for the two-dimensional analysis.[74] Furthermore, the difficulties associated with modelling the fundamental fields on the cylindrical cavity of radius r_o, together with the uncertainties in the actual choice of the value of r_o, can be avoided by using the SFF technique. Moreover the cylindrical cavity concept would be extremely difficult to extend to crack fronts of arbitrary curvature. In this new technique the singularities are not modelled by boundary conditions on a cylindrical cavity surface but are removed entirely from the numerical problem.

The difference between the current technique and the regularization procedure used by Sham *et. al.*,[47,48] is that they require accurate determination of the regular fields in a small region surrounding the point of application of the pair of concentrated forces at the crack front; it is therefore necessary to define the size of this arbitrary region and investigate its influence on the results. Whereas in the present approach, the regularization is applied over the whole problem region and detailed modelling in the vicinity of the crack front is not required. Further, in the work of Refs 47 and 48 they obtain their weight functions through a finite element analysis of the problem; thus they require discretisation of the entire problem domain. For linear elastic fracture mechanics problems the boundary element method of analysis has been established as a more efficient and accurate method relative to the FEM.

In the direct modelling of fundamental fields both the singularity at the point force location as well as the standard stress singularity at all other points along the crack front must be simulated with the discretisation through the use of elements; it is thus expected that large numbers of elements, with considerable mesh refinement around the singular points would be required to give adequate accuracy in the weight functions. Coupling of the subtraction of fundamental fields technique with special crack front elements such has traction-singular quarter-point elements enables firstly the removal of the highly singular displacements and stresses at the application of the point loads, and secondly the simulation with the special crack front elements of the standard stress singularity. This coupling allows the problem to be modelled extremely efficiently and consequently gives rise to accurate weight functions and hence accurate stress intensity factors for arbitrary loadings on the boundary of the problem.

5.1 Subtraction of fundamental fields (SFF)

The subtraction of fundamental fields (SFF) technique circumvents the direct modelling of the singular behaviour of the fundamental fields through a modification of the displacement and traction fields of the finite geometry for which the weight functions are sought. To fully appreciate this consider Fig. 7. Suppose that the desired

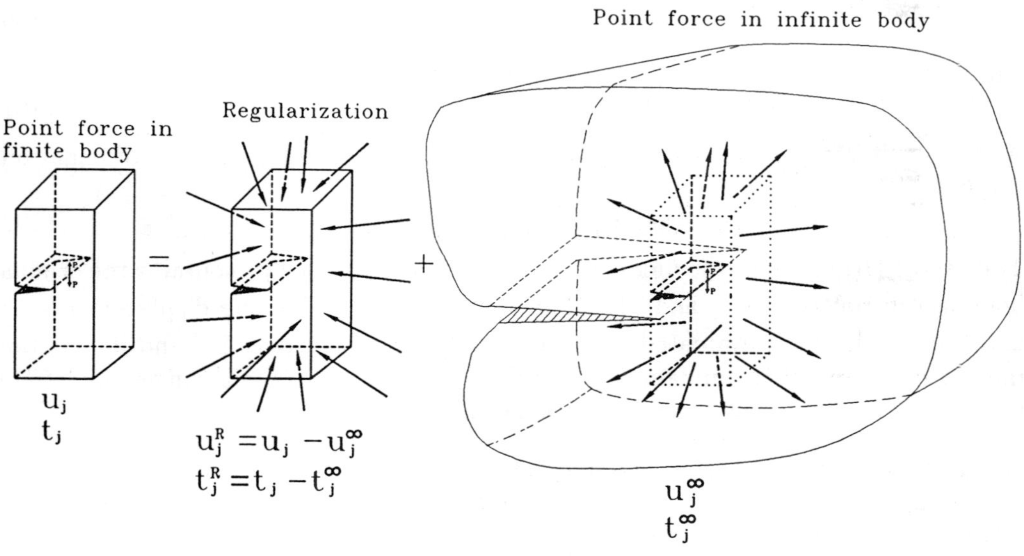

Figure 7: Regularization of the problem domain

weight-function solution of a problem are the displacements and tractions $u_j = u_j^*$ and $t_j = t_j^*$ respectively, arising from the presence of concentrated forces acting at the crack front in a finite geometry. Bearing in mind that the elastic fields arising from the presence of concentrated forces on the crack front in an infinite body are available, see Sections 3.2 and 3.3, then through a superposition principle the elastic fields of the desired problem can be modified by subtracting out the fundamental fields that occur (at the site of the finite-geometry surfaces) due to the presence of the point force in the infinite body. This removes the principal singularities at the crack-tip, and the resulting regularised boundary value problem can be solved accurately using BEM. The desired final solution can then be obtained by addition of the known fundamental fields back to the solution obtained from the BEM.

Mathematically the fields of the desired problem, see Fig. 7, can be written as

$$u_j = (u_j - u_j^{\infty}) + u_j^{\infty} = u_j^R + u_j^{\infty} \tag{97}$$

and

$$t_j = (t_j - t_j^{\infty}) + t_j^{\infty} = t_j^R + t_j^{\infty} \tag{98}$$

where u_j^{∞} and t_j^{∞} are the known fundamental fields (occurring at the site of the finite-geometry surfaces); they are obtained from eqns (40)-(48). Substitution for u_j and t_j into the three-dimensional boundary integral equation gives,

$$C_{ij}(\mathbf{x}')u_j^R(\mathbf{x}') + \int_{\Gamma} T_{ij}(\mathbf{x}',\mathbf{x})u_j^R(\mathbf{x})d\Gamma(\mathbf{x}) = \int_{\Gamma} U_{ij}(\mathbf{x}',\mathbf{x})t_j^R(\mathbf{x})d\Gamma(\mathbf{x}) \tag{99}$$

Only the field quantities u_j^R and t_j^R appear in eqn (99), since u_j^{∞} and t_j^{∞} satisfy the governing equations of elasticity exactly. Equation (99) is now solved, subject to the following new boundary conditions:

$$\overline{u}_j^R = \overline{u}_j - \overline{u}_j^{\infty} \tag{100}$$

and

$$\overline{t}_j^R = \overline{t}_j - \overline{t}_j^{\infty} \tag{101}$$

After substitution of the new boundary conditions the resulting system of equations can be written as

$$[A]\{\mathbf{b}\} = [C]\{\mathbf{y}\} - [C]\{\mathbf{g}\} \tag{102}$$

The vector $\{\mathbf{b}\}$ contains the unknowns u_j^R and t_j^R; the vector $\{\mathbf{y}\}$ contains the original boundary conditions $\overline{u}_j$, and $\overline{t}_j$. The vector $\{\mathbf{g}\}$ contains the displacements $\overline{u}_j^{\infty}$ and the tractions $\overline{t}_j^{\infty}$, evaluated at the boundary, while matrices $[A]$ and $[C]$ contain integrals of the Kelvin fundamental solutions T_{ij} and U_{ij}. A rearrangement of eqn (102) leads to the following system of equations,

$$[A]\{\mathbf{b}\} = \{\mathbf{f}\} \tag{103}$$

where

$$\{\mathbf{f}\} = [C]\{\mathbf{y}\} - [C]\{\mathbf{g}\}$$

Equation (103) has the same form as the system of equations originated by the use of standard BEM, therefore this system of equations can be solved by the BEM code using standard techniques. In the analysis that follows the subtraction of fundamental fields technique is used to remove displacement and stress singularities of order $\rho^{-\frac{3}{2}}$ and $\rho^{-\frac{5}{2}}$, respectively, ρ denotes distance measured from the location of the point forces on the crack front. Hence this technique reduces a highly singular set of equations to the more normal singular set that exist in standard crack problems.

5.2 Square root traction singularity modelling

The subtraction of fundamental fields technique removes the higher order singularities in the displacement and traction fields at the point of application of the loads $\pm$P. The remaining boundary value problem still contains a traction singularity around the crack front. This traction singularity of order $\rho^{-\frac{1}{2}}$ is the one normally encountered in standard crack problems. Thus in addition to the use of SFF, accurate modelling of this singularity should give rise to accurate weight functions on the boundaries of the problem domain. The standard interpolation functions employed in the BEM numerical solution technique are polynomials and consequently the behaviour of the remaining singular fields near the crack front can never be correctly modelled, even with mesh refinement.

The use of special crack front elements, see Bains,[2] circumvents this problem since these elements contain the traction singularity. These special crack front elements were first employed in the FEM by Barsoum[75] and Henshell & Shaw.[76] The most commonly used special crack front elements employed in BEM are the displacement quarter-point (DQP) and the traction-singular quarter-point (TSQP) elements; possibly due to the ease of implementation into the standard code. Basically a $\rho^{\frac{1}{2}}$ variation of both displacements and tractions is introduced by the repositioning of the midside nodes to the quarter-position in the elements adjacent to the crack front. These quarter-point elements in the global coordinates (x_1, x_2, x_3) are mapped to local elements in intrinsic coordinates (ξ, η) as shown in Fig. 8. To simulate the traction

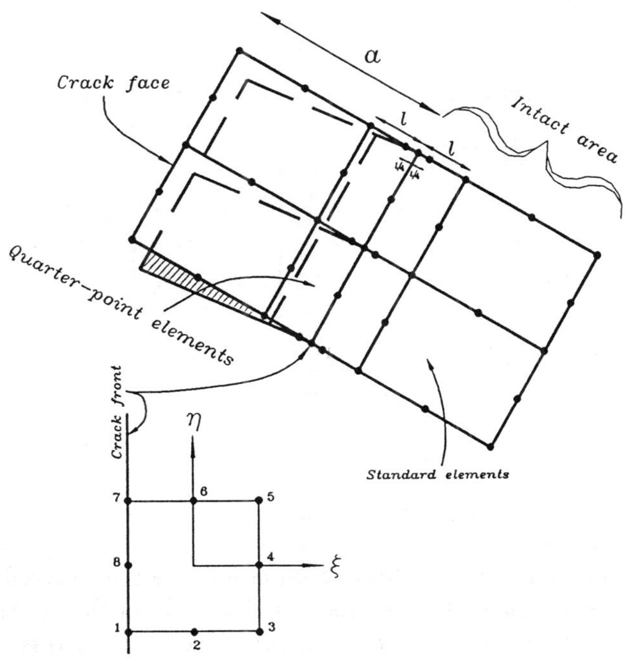

Figure 8: Quarter-point element implementation

singularity in the BEM formulation the traction fields of the quarter-point elements adjacent to the crack front on the intact area must be multiplied by $\sqrt{(l/\rho)}$, where l is the quarter-point element length, see Fig. 8.

It must be stressed at this stage that the quarter-point element has been used, in the past, see Cruse & Wilson,[77] to obtain stress intensity factors from extrapolation of either the displacements or stresses near the crack front, on the crack plane. But in this analysis it is used only to simulate a variation of $\rho^{-\frac{1}{2}}$ in the traction fields. The usual extrapolation techniques from the quarter-point elements have been shown[77] to depend on the quarter-point element length. However no extrapolation of the fields is required in the evaluation of the weight functions or the stress intensity factors.

6 Rectilinear crack front analysis

To demonstrate the validity of the regularisation procedure consider a rectangular bar with a planar rectilinear edge crack, Fig. 9. The crack is of length a in a bar

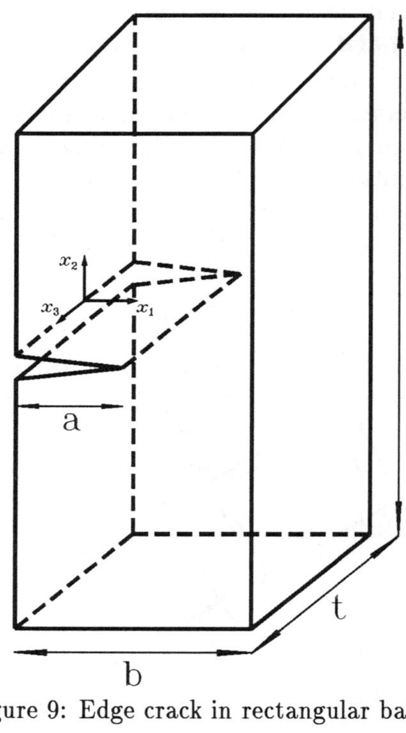

Figure 9: Edge crack in rectangular bar

of width b, thickness t and length $2h$. The bar consists of the following dimensions; $a/b = 0.5$, $t/b = 1.5$ and $h/b = 3.0$. Stress intensity factor solutions are required along the crack front for this configuration when it is subjected to various loadings on the remote boundaries that cause mode I deformation of the crack faces. Because of the symmetry of the geometry about the crack plane, only half the problem is discretized. Eight noded quadratic isoparametric elements together with eight Gaussian integration points have been used throughout this standard boundary element analysis.

This initial analysis only focuses attention at the center of the crack front and thus avoids the complications when the crack front intersects a boundary. Further, this allows direct comparison of the stress intensity factor with the two-dimensional plane strain value as obtained by Keer & Freedman.[78] A typical refined mesh, for half symmetry (i.e. $x_2 > 0$) consisted of 104 elements with 942 degrees of freedom. Details of the mesh employed can be found in Ref. 2. Stress intensity factor at the centre of the crack front are determined using three different techniques:

Technique 1: SFF only. Subtraction of fundamental fields (SFF) alone. No special crack front elements included adjacent to the rectilinear crack front to model the square root displacement variation or the square root singular traction variation.

Technique 2: SFF + DQP. SFF coupled with displacement quarter-point (DQP) elements. The midside nodes of element sides lying perpendicular to the crack front are repositioned to the quarter point relative to the front. Thus only the displacement variation on the crack face is modelled correctly. No account is taken of the singular traction variation ahead of the crack front.

Technique 3: SFF + TSQP. SFF coupled with displacement quarter-point elements and a traction-singular quarter-point (TSQP) elements. This is essentially the same as 2 above but with the addition that the interpolation functions of elements adjacent to the crack front, on the intact plane are modified to take into account the standard square root traction singularity.

For each technique, analysis was conducted to ascertain the dependence of the stress intensity factor on the quarter-point element length to crack length (l/a) ratio. The ratio (l/a) was varied from 0.05 to 0.3. The results obtained from all three techniques listed above are shown in Fig. 10, where they are compared with the plane strain value.[78] The results labelled SFF only, show that for ratios of (l/a) between 0.05-0.2, differences of less than 4.0% can be expected in the stress intensity factor. With increasing (l/a) the effect of the remaining standard stress singularity still present in the fields becomes dominant. The results show that the traction-singular quarter-point elements are better suited to model the remaining singularity in the traction fields than the displacement quarter-point elements. This is as expected since the remaining singularity in the traction fields is of order $\rho^{-\frac{1}{2}}$, where ρ is distance measured perpendicular to the crack front. Using the traction-singular quarter-point elements with subtraction of fundamental fields technique enable errors of less than 2.0% to be attained for all the ratios of (l/a) considered in this analysis.

From the analysis conducted thus far, SFF coupled was TSQP elements and (l/a) ratio of 0.2, was selected for the analysis of stress intensity factor variations along the crack front when the geometry depicted in Fig. 9 is subjected to the following loading cases:

1. Uniform tensile stress, $\pm\sigma$, acting on the planes $x_2 = \pm h$ of the bar; the applied traction $t_2 = \pm\sigma$, therefore the normalised stress intensity factor, obtained from eqn (15) is given by

$$\frac{K_I}{\sigma\sqrt{\pi a}} = \frac{E}{2\sqrt{2}(1-\nu^2)\sqrt{a}\pi^2}\int_{-(\frac{t}{2})}^{+(\frac{t}{2})}\int_0^b u_2^*(x_1, h, x_3)\mathrm{d}x_1\mathrm{d}x_3 \qquad (104)$$

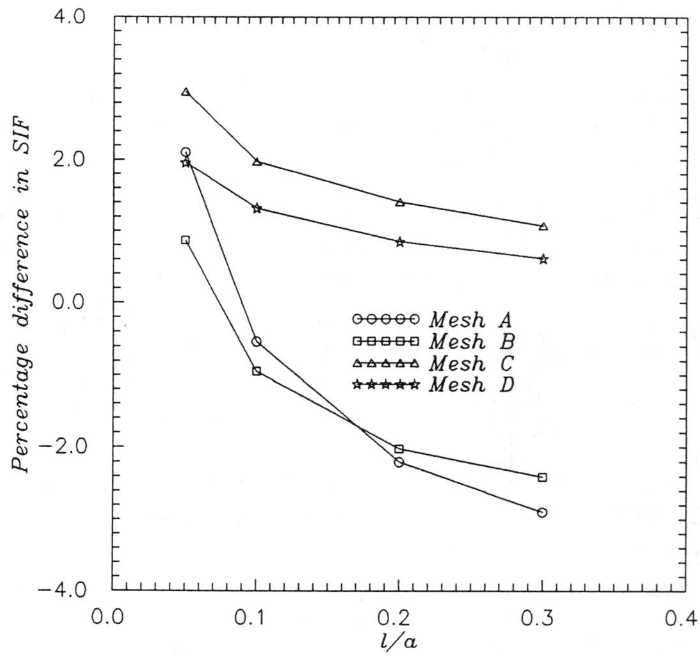

Figure 10: Edge crack:- SIF variation as a function of (l/a)

Equation (104) is obtained from eqn (15) by taking the following limit as the point force approaches the crack front:

$$\lim_{c \to 0} \left[\frac{P_2 \sqrt{c}}{\pi^2} \right] = 1$$

This limit is possible since the stress intensity factor K_I is independent of the product $P_2\sqrt{c}$.

2. Pure bending stress (x_1-variation): the applied traction on the planes $x_2 = \pm h$ is $t_2 = \pm\sigma(1 - 2x_1/b)$; thus the normalised stress intensity factor is given by

$$\frac{K_I}{\sigma\sqrt{\pi a}} = \frac{E}{2\sqrt{2}(1-\nu^2)\sqrt{a}\pi^2} \int_{-(\frac{t}{2})}^{+(\frac{t}{2})} \int_0^b \left(1 - \frac{2x_1}{b}\right) u_2^*(x_1, h, x_3) \mathrm{d}x_1 \mathrm{d}x_3 \qquad (105)$$

3. Bending stress (x_3-variation): the applied traction on the planes $x_2 = \pm h$ is $t_2 = \pm\sigma(1 - 2x_3/b)$; the normalised stress intensity factor is given by

$$\frac{K_I}{\sigma\sqrt{\pi a}} = \frac{E}{2\sqrt{2}(1-\nu^2)\sqrt{a}\pi^2} \int_{-(\frac{t}{2})}^{+(\frac{t}{2})} \int_0^b \left(1 - \frac{2x_3}{b}\right) u_2^*(x_1, h, x_3) \mathrm{d}x_1 \mathrm{d}x_3 \qquad (106)$$

4. Bending stress (x_1-variation): the applied traction on the planes $x_2 = \pm h$ is $t_2 = \pm\sigma(1 - x_1/b)$; the normalised stress intensity factor is given by

$$\frac{K_I}{\sigma\sqrt{\pi a}} = \frac{E}{2\sqrt{2}(1-\nu^2)\sqrt{a}\pi^2} \int_{-(\frac{t}{2})}^{+(\frac{t}{2})} \int_0^b \left(1 - \frac{x_1}{b}\right) u_2^*(x_1, h, x_3) \mathrm{d}x_1 \mathrm{d}x_3 \qquad (107)$$

5. Bending stress (x_1-variation): the applied traction on the planes $x_2 = \pm h$ is $t_2 = \pm\sigma(x_1/b)$; the normalised stress intensity factor is given by

$$\frac{K_I}{\sigma\sqrt{\pi a}} = \frac{E}{2\sqrt{2}(1-\nu^2)\sqrt{a}\pi^2} \int_{-(\frac{t}{2})}^{+(\frac{t}{2})} \int_0^b \left(\frac{x_1}{b}\right) u_2^*(x_1, h, x_3) \mathrm{d}x_1 \mathrm{d}x_3 \qquad (108)$$

The quantity $u_2^*(x_1, h, x_3)$ is a component of the weight function $\mathbf{u}^*$, acting in the x_2 coordinate direction, on the planes $x_2 = \pm h$. The loading cases represented above are defined in terms of the global co-ordinate system (x_1, x_2, x_3) with origin in the centre of the side bisected by the crack (see Fig. 9). The integrals shown above, for the stress intensity factors are evaluated numerically using Gaussian quadrature. A single computer run is only needed at each point on the crack front to generate weight functions for the geometry under consideration. Thus once the weight functions are obtained, stress intensity factors for all the load mentioned above can be obtained simultaneously for each point on the crack front. For uniform tensile loading, case 1, the variation of stress intensity factor along the crack front is shown in Fig. 11. Over

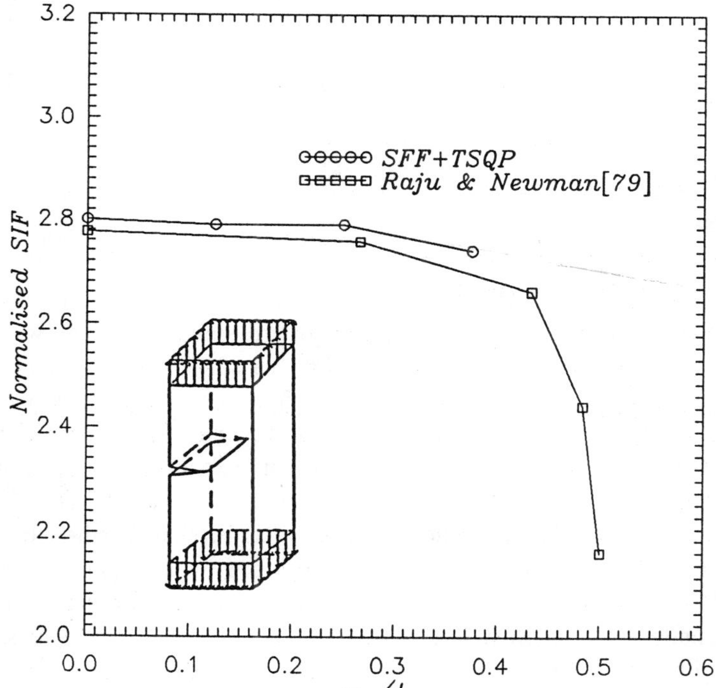

Figure 11: Edge crack:- Uniform stress

the range of (x_3/t) considered in Fig. 11, the stress intensity factor variation along the crack front is in agreement with the variation presented by Raju & Newman.[79] The maximum difference in the two analysis over the range of (x_3/t) considered is 1%. For the edge crack under uniform stress loading the stress intensity factor attains its maximum value at the centre of the crack front. In the present analysis, at $(x_3/t) = 0.375$ the stress intensity factor has decreased by 2% compared to its value at the centre of the crack front. Variation of the stress intensity factor along the crack front, for pure bending stress which varies linearly with x_1, case 2, and for bending

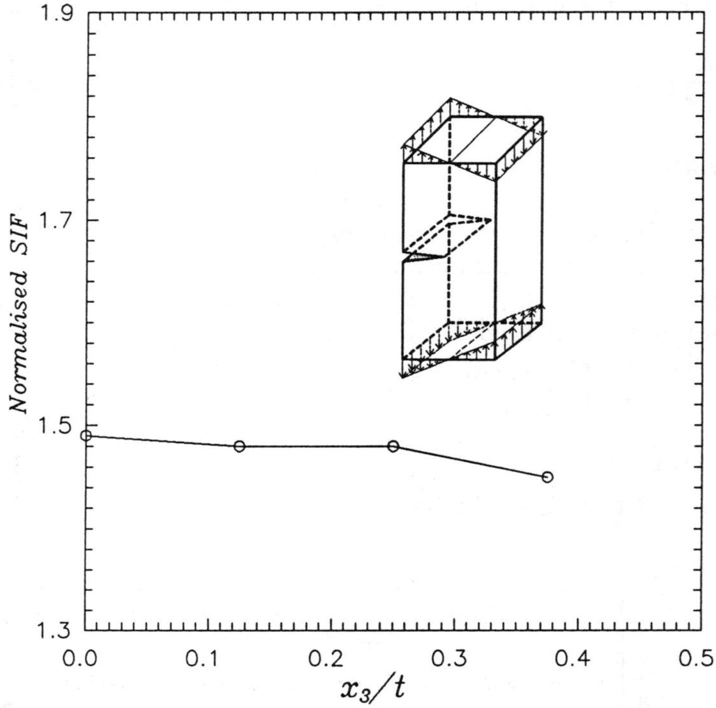

Figure 12: Edge crack:- Pure bending stress (x_1-variation)

stress which varies linearly with x_3, case 3, is shown in Figs 12 and 13 respectively. Results presented in Fig. 12 and Fig. 13 are in agreement with the work of Aliabadi *et. al.*[34] Results presented in Fig. 14 and Fig. 15 show the K_I variation along the crack front for loading case 4 and loading case 5, respectively. It can be shown that superposition of the load systems shown in Fig. 14 and Fig. 15, produces results for the uniform stress case, Fig. 11. A similar analysis can be conducted on the loading presented in Fig. 13.

The validity of the regularisation procedure is further verified by transforming the edge crack problem into a centre crack problem by a change of boundary conditions. Results for the variation of stress intensity factor along the crack front, for the case of tensile loading only, are shown in Fig. 16. All the results are for (l/a) ratio of 0.2. The stress intensity factor variation along the crack front follows the same pattern as those presented by Raju *et. al.*,[79] over the range of (x_3/t) covered in this analysis. The maximum difference in the two curves is 2%. For $(x_3/t) = 0.0$, the stress intensity factor value obtained by Raju *et. al.*,[79] and the present analysis differ by -0.8% and $+0.6\%$, respectively, relative to the two-dimensional collocation analysis of Newman.[80]

Stress intensity factors have not been evaluated in the vicinity of the crack front free surface intersection of the body. In this region the fields generated by the point force are not valid and the stress intensity factor is not defined. The work of Olaosebikan,[81] shows a change in the type of singularity as one approaches the

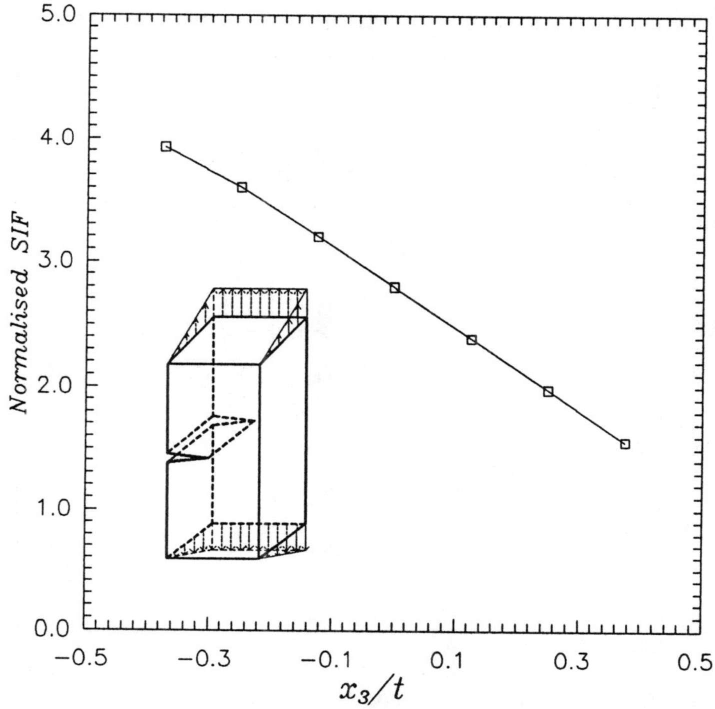

Figure 13: Edge crack:- Bending stress (x_3-variation)

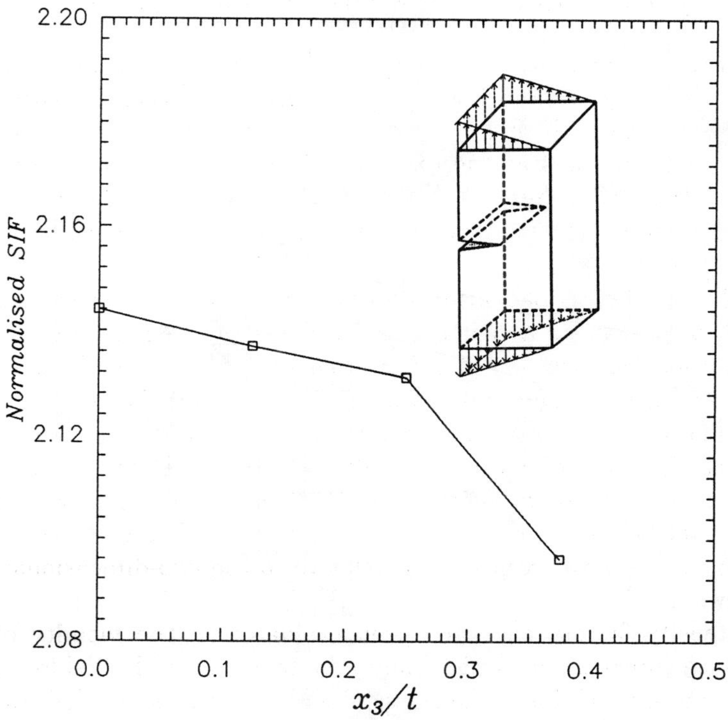

Figure 14: Edge crack:- Bending stress (x_1-variation)

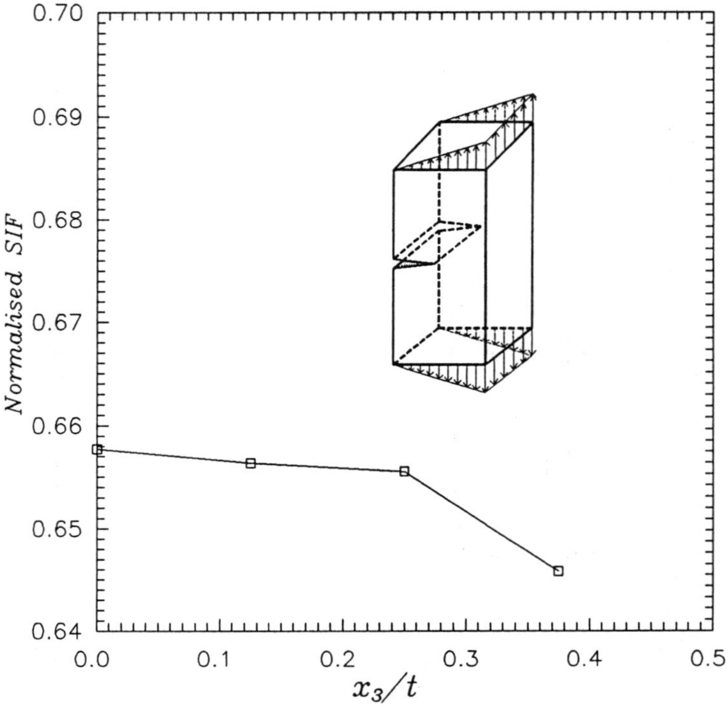

Figure 15: Edge crack:- Bending stress (x_1-variation)

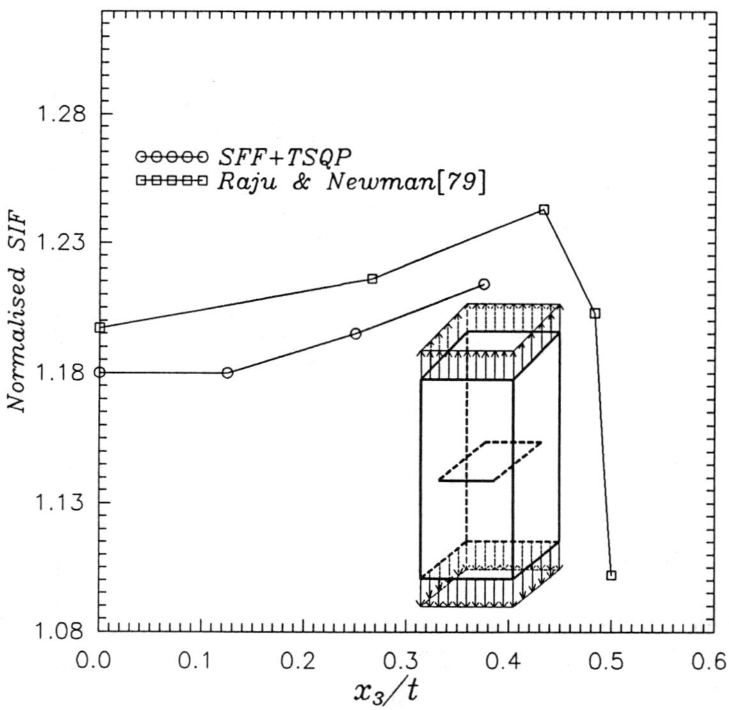

Figure 16: Centre crack:- Uniform stress

free surface of a finite-size body from its central interior; the singularity is no longer $O(\rho^{-\frac{1}{2}})$.

The analysis conducted thus far on the rectilinear edge crack, demonstrates that weight functions obtained by the use of SFF technique, for finite cracked structures, based on the notion of fundamental fields for infinity elastic bodies, can be used to generate stress intensity factors in an efficient and economical way. The present work will now be extended to cracks of penny-shaped and elliptical configurations in various geometries that are subjected to numerous loads that cause mode I deformation of the crack faces.

7 Cracks of penny-shaped configuration

Fundamental fields valid in the whole elastic domain containing an embedded penny-shaped crack were presented in Section 3.3. These can be employed in the regularisation procedure (SFF) given in Section 5 to generate weight functions for penny-shaped cracks in infinite or finite elastic bodies. Attention will also be focused on different crack front elements that can simulate the standard square root traction singularity. Stress intensity factor solutions are presented for embedded and corner penny-shaped cracks in various finite geometrical configurations.

7.1 Crack front modelling strategies

There are many special crack front elements that can be used to model square root traction singularity encountered at the crack front (away from the load point). A review of these elements can be found in Ref. 2. Analysis has not been conducted to date to ascertain if any one element is better suited than the others. In principle they should all give the same solutions but in reality they may show different dependencies on the mesh distribution employed. Further, for curved crack fronts it is generally difficult to find the correct position of certain nodes to simulate the stress singularity and for such classes of problems the Blackburn type or the inverse transformation of quarter-point elements, see Luchi & Rizzuti[82] and Bains,[2] may be better suited, these elements do not depend on the repositioning of nodes.

When employing the special crack front elements described in Ref. 2, it is further necessary to ascertain the sensitivity of the stress intensity factor solution with variation of the element length. This variation is important in selecting the optimum l/a ratio for a given crack problem; where l is the length of the special element adjacent to the crack front and measured perpendicular to that front, and a denotes the radius of the penny-shaped crack. These special elements are placed adjacent to the crack front only.

To this end, to clarify some of the above points the following special crack front elements were incorporated into the 3-D BEM library:-

Element type 1: Traction-singular quarter-point elements adjacent to the crack front on the intact plane and displacement quarter-point elements adjacent to the crack front on the crack faces. To ensure compatibility between the singular element and the standard element, the interpolation functions of nodes 5, 6 and 7 of the singular-elements were multiplied by 2, see Fig. 17. Certain midside

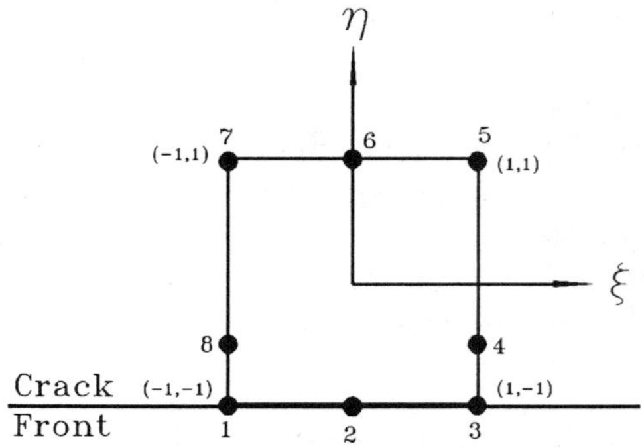

Figure 17: Local node numbering for crack front elements

nodes are shifted to the quarter-position, see Cruse *et. al.*,[77] and Bains.[2] This is the element used in the analysis of the rectilinear crack front.

Element type 2: Blackburn type traction elements adjacent to the crack front on the intact plane and Blackburn type displacement elements on the crack face adjacent to the crack front. Standard isoparametric interpolation functions modified to explicitly take into account the square root singularity, see Luchi *et. al.*,[82] and Bains.[2] All nodes of these elements are left in their usual positions.

Element type 3: Inverse transformation of quarter-point traction elements situated on the intact plane adjacent to the crack front and inverse transformation of quarter-point displacement elements next to the crack front on the crack faces. Standard isoparametric interpolation functions modified to take into account the square root singularity, see Luchi *et. al.*,[82] and Bains.[2] As in Element type 2 all nodes of these elements are left in their usual places.

Element type 4: Blackburn type traction elements adjacent to the crack front on the intact plane and the inverse transformation of quarter-point displacement elements adjacent to the front on the crack faces. All nodes are left in usual locations.

Element type 4 above is allowed in the BEM formulation since the tractions and displacements are independently represented from each other. The above four elements were implemented into the 3-D BEM code such that anyone of them could be coupled with the SFF technique to generate accurate weight functions and hence the stress intensity factors. When using the above special crack front elements care must be taken to ensure that the correct integration procedure is used for the particular special element employed, see Bains.[2]

7.2 Penny crack in infinite domain

To check the validity of the fundamental field formulation derived in Section 3.3 for the penny-shaped crack, the problem of determining the stress intensity factor for an embedded penny-shaped crack in homogeneous, isotropic, linearly elastic infinite domain under the action of a uniform tensile stress at infinity was considered. In this state of loading the penny-shaped crack will only experience mode I deformation.

The geometry under consideration is a bar of circular cross-section, with the center of the crack lying on the axis of the bar and with the plane of the crack perpendicular to that axis. The radius of the bar will be denoted by R, its height by $2H$. The crack is of radius a. To ensure that the penny-shaped crack is in an infinite domain (i.e. no interaction between the crack front and the boundaries of the domain), $a/R = 0.1$ and $H/R = 6$ was considered. The basic configuration is shown in Fig. 18. Due to the symmetry of the problem, only half the geometry, above the plane $x_3 = 0$,

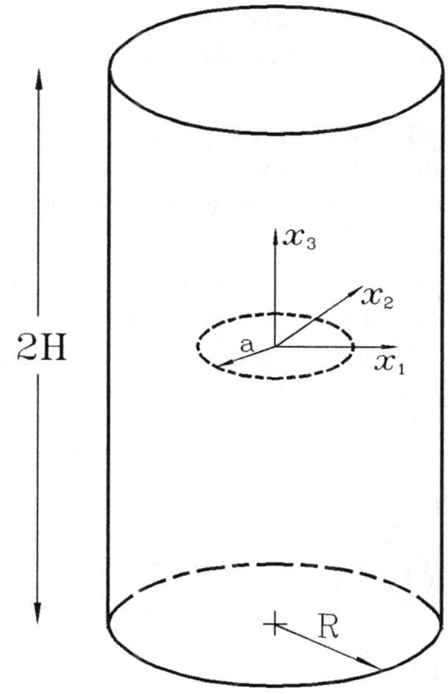

Figure 18: Geometry under analysis

is considered. For the numerical modelling eight noded isoparametric elements are used throughout this analysis. In the analysis that follows the subelement integration scheme proposed by Lachat & Watson[83] was implemented into the 3-D BEM code in accordance with the criteria described by Bains.[2] Each subelement contained only four Gaussian integration points. This scheme is used since it was envisaged that for arbitrary shaped crack fronts a large number of elements may be required for the modelling and consequently methods such as this are needed to reduce overall solution time while still retaining the solution accuracy. The stress intensity factor integral relationship for the penny-shaped crack configuration given by expression (88) can be normalised such that

$$\frac{K(\theta')}{K_\infty} = \frac{1}{2\pi\sigma\sqrt{2a}} \int_\Gamma \mathbf{t}^+ \cdot \mathbf{u}^*(\theta') d\Gamma \qquad (109)$$

where

$$K_\infty = 2\sigma\sqrt{\frac{a}{\pi}} \qquad (110)$$

for the case of uniform tensile loading σ, over the ends of the bar ($x_3 = \pm H$), and K_∞ is the stress intensity factor for an isolated ($R = \infty$) plane penny-shaped crack of radius a in a solid subjected to uniform stress σ at infinity. The percentage difference in the normalised stress intensity factor is defined as:

$$\%\text{Diff.} = \left(\frac{K(\theta')_{(\text{KS})} - K(\theta')}{K(\theta')_{(\text{KS})}} \right) \times 100 \tag{111}$$

where the subscript KS refers to the known solution.[85] For all the analysis presented in this report, σ equal to unity is considered. The integral presented in eqn (109) is evaluated numerically using Gaussian quadrature.

The boundary element mesh depicting the geometry and the fundamental field variation consisted of 120 elements and 1086 degrees of freedom. Furthermore, the mesh also contained a ring of special crack front elements as described in Section 7.1. These are placed on either side of the penny-shaped crack front. These elements are only used to simulate the traction singularity of order $(r - a)^{-\frac{1}{2}}$ in the crack plane, where r is distance measured from the centre of the penny-shaped crack. In Section 6, analysis was conducted to ascertain the dependence of the stress intensity factor on l/a the ratio of the quarter-point element length to the crack length for a straight crack front. The value of $l/a = 0.2$ obtained in the rectilinear edge crack analysis will be also used in this analysis. Later results will show the dependence of the stress intensity factor on the ratio l/a, when the crack front is circular.

Each of the four elements described in Section 7.1 were used to determine weight functions for the case when the point load were situated on the crack front at $\theta' = 0°$ for the geometry shown in Fig. 18. Stress intensity factor solutions at $\theta' = 0°$ obtained from integration of the weight functions are presented in Table 1.

Table 1: SIF variation for $a/R = 0.1$ with different crack front elements

Crack front elements	$\frac{K(\theta')}{K_\infty}$		
	$\theta' = 0.0°$	$\theta' = 45.0°$	$\theta' = 90.0°$
Element type 1	0.9959	0.9963	0.9959
Element type 2	0.9978	0.9982	0.9978
Element type 3	0.9966	0.9970	0.9966
Element type 4	0.9981	0.9984	0.9981

All the stress intensity factor solutions presented in Table 1 are in good agreement with the exact theoretical solution obtained by Ref. 84 as presented in Ref. 85 (i.e. $\frac{K(\theta')}{K_\infty} = 1$). These results show that the formulation as presented in Section 3.3 is correct and that any one of the special crack front elements can be employed to generate accurate stress intensity factors for the penny-shaped crack in infinite domains. All the special crack front elements generate results to within 0.5% of the exact solution for the penny-shaped crack configuration.

7.3 Penny-shaped crack in finite domain

To investigate the effect of a penny-shaped crack in finite size bar, the radius R was decreased, still maintaining $H/R = 6$ and the crack radius a kept constant. The initial analysis concentrated on $a/R = 0.53$. For this configuration the interaction between the crack front and remote boundaries of the bar becomes an important modelling criteria and as a direct consequence the mesh was designed so that the elements are concentrated in the region where the stress intensity factor solution is required. For this analysis $(l/a) = 0.2$ is still maintained. The typical mesh for this problem consisted of 230 elements and 2076 degrees of freedom. Stress intensity factor solutions are obtained for each of the four special crack front elements, see Section 7.1, when each is coupled in turn to the SFF technique.

The results are presented in Table 2 together with the percentage difference from the know solution for the same geometry as given in Ref. 85.

Table 2: SIF variation for $a/R = 0.53$ with different crack front elements

Crack elements	$\frac{K(\theta')}{K_\infty}$	% Diff.
Element type 1	1.067	2%
Element type 2	1.081	1%
Element type 3	1.122	-3%
Element type 4	1.087	<0.5%

These results show that for this particular mesh distribution all the special crack front elements give rise to stress intensity factor solutions that are within 3% of those presented in Ref. 85. The variation of the solution as shown in Table 2 for different crack front elements may be due to the different dependency on the (l/a) ratio for each element. In order to check this dependency stress intensity factor solutions were obtained for (l/a) ratios equal to 0.05, 0.1, 0.2 and 0.3, at $\theta' = 0°$ for a/R ratio of 0.53. The variation in the stress intensity factor solution for each of the special crack front elements as a function of (l/a) is shown in Fig. 19. To generate these solutions only the (l/a) ratio was varied, while the element distribution for $x_3 > 0$ is maintained constant.

The results show that any one of the special crack front elements can be employed to generate stress intensity factor solutions to within 3% of the solution as presented in Ref. 85. The traction-singular quarter-point elements in conjunction with displacement quarter-point elements (i.e. Element type 1, see Section 7.1) give consistent solutions within the range $0.05 \leq (l/a) \leq 0.3$ considered and are within 2% of the know solution. Element types 2 and 4 show similar behaviour and dependency on (l/a) to each other, while Element type 3 shows the greatest sensitivity to (l/a) variation covered. With the latter three special crack front elements the stress intensity factor solution depends on the (l/a) ratio used.

It could be argued that for the penny-shaped crack configuration an extremely large number of elements (i.e. 230) are required in the modelling of the fundamental fields. This is not necessarily true since the idea behind using such a large number

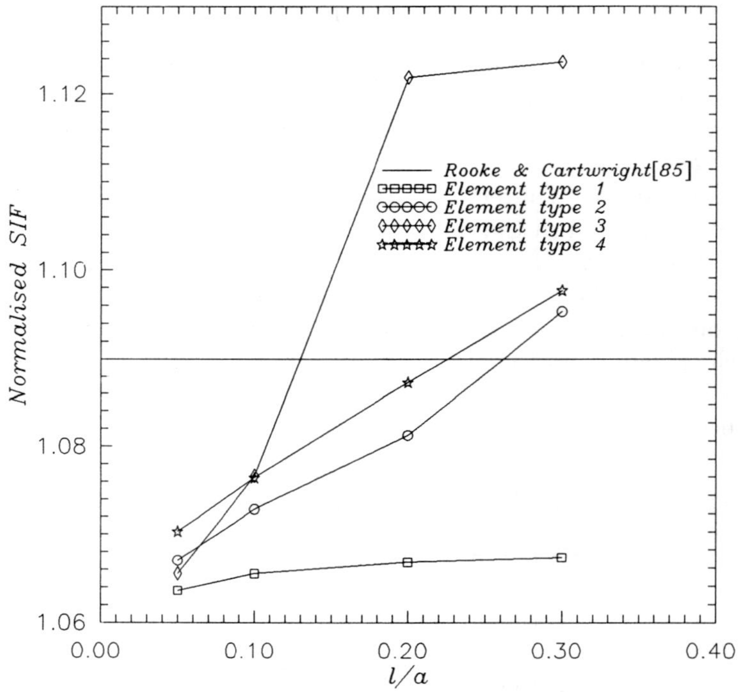

Figure 19: SIF variation with special crack front elements

of elements is to ensure that the solution required is no longer element distribution sensitive and further to accurately model the crack front interaction with the remote boundaries when a/R ratio is large. In Ref. 2 it was shown that accurate solutions can be obtained with fewer number of elements.

To investigate further the effect of a penny-shaped crack in a finite size bar, the following ratios of 0.1, 0.33, 0.53 and 0.75 for a/R were considered. Results have already been obtained for a/R equal to 0.1 and 0.53. The a/R ratio of 0.33 contained the same number of elements and degrees of freedom as for a/R equal to 0.53, while the a/R ratio of 0.75 contained 202 elements and 1824 degrees of freedom. The reason why the total number of elements is less for a/R ratio of 0.75 is that the ligament ahead of the crack front decreases as $R \rightarrow a$. Thus if the total number of elements on the intact area is maintained constant, then as $R \rightarrow a$ the element length to width ratio becomes extremely large. Furthermore, as $R \rightarrow a$, it was found that varying the discretization of the crack face produced virtually no change in the stress intensity factor. The stress intensity factor solution obtained from coupling Element type 4 (see Section 7.1) with the SFF technique are presented in Fig. 20. Results obtained from two independent techniques, namely integral transforms[85] and asymptotic approximation[86] are also plotted in Fig. 20. In Refs 85 and 86 the authors state the accuracies of their results to be better than 1%. At $a/R = 0.1$, all the results presented are in agreement with each other and the exact solution. For $a/R > 0.1$ all the stress intensity factor solutions show good agreement with Ref. 85 and are within 1%. At a/R ratio of 0.75 and 0.8 stress intensity factor results as presented by Refs 85 and 86 differ by 1.5% and 2.6% respectively. All the results have the same pattern;

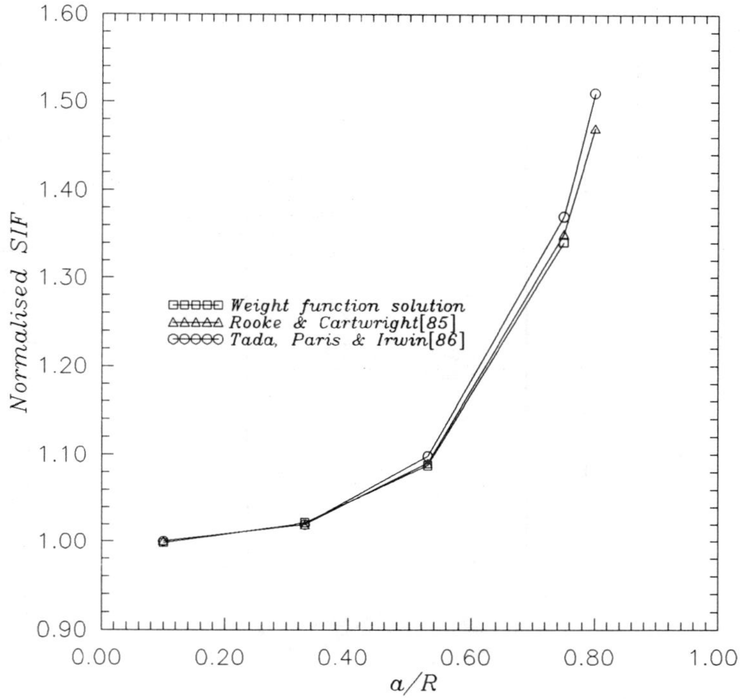

Figure 20: SIF variation as a function of a/R

with increasing a/R ratio the interaction between the crack front and the external surface boundaries becomes more dominant and the stress intensity factor increases.

7.4 Penny-shaped edge crack in cylindrical bar

Through knowledge of fundamental fields for the embedded penny-shaped crack that is situated in an infinite domain it is possible to demonstrate that with the SFF technique, stress intensity factors for edge cracks of penny-shaped configuration in finite elastic bodies can be solved. To this end consider a penny-shaped edge crack of radius a, in a cylindrical bar of radius R and height $2H$ as shown in Fig. 21. A cross-section through the crack plane is shown in Fig. 22. This figure also shows the location of the centre of coordinates of the penny-shaped edge crack on the boundary of the cylindrical bar. Stress intensity factors will be obtained for various points along the crack front when the geometry is subjected to numerous loads applied at the extremities of the bar, i.e. $(x_1, x_2, \pm H)$. These loads cause mode I deformation of the edge crack faces. Nisitani[87] solved the same configuration and obtained, for tensile loading only, stress intensity factors at the deepest point of the crack for various a/R ratios. In Ref. 87 the body force method was used to generate the solutions.

Since the geometry shown in Fig. 21 is symmetrical about the crack plane, only half the geometry needs to be discritised for the BEM analysis. Weight functions are obtained for the half configuration from the employment of the SFF technique with Element type 1, see Section 7.1. These elements are chosen since they are less sensitive to changes in the quarter-point element length, see Section 7.3. Since in the region of

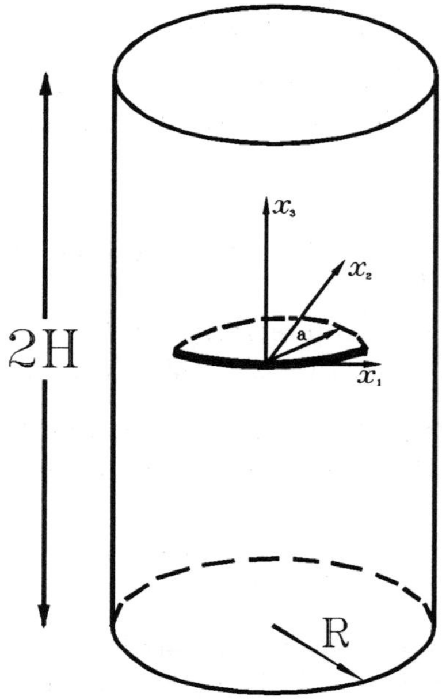

Figure 21: Penny-shaped edge crack in cylindrical bar

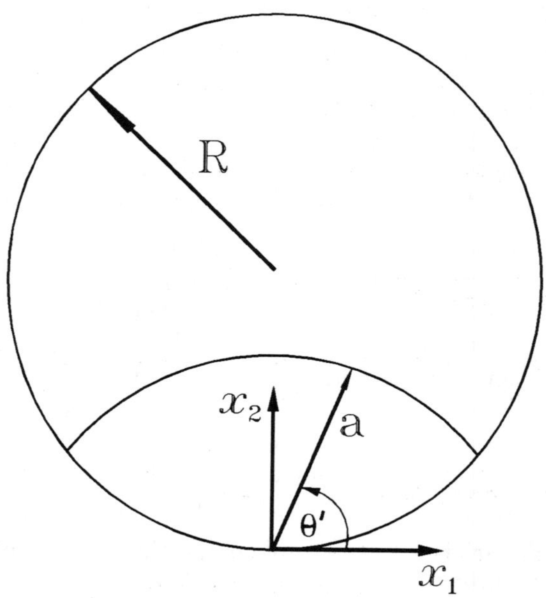

Figure 22: Definition of parameters for the penny-shaped edge crack

the crack front free boundary intersection, the point force solution is no longer valid, normalised stress intensity factors are obtained for points along the crack front in the range $115° \geq \theta' \geq 65°$, for the following load cases.

1. Uniform tensile stress, $\pm\sigma$, acting on the planes $x_3 = \pm H$ of the bar: the applied traction $t_3 = \pm\sigma$, therefore the normalised stress intensity factor as obtained from eqn (88) is given by

$$\frac{K(\theta')}{\sigma\sqrt{\pi a}} = \frac{1}{\sigma\pi^2\sqrt{2a}} \int\int_A u_3(\theta', x_1, x_2, H)\mathrm{d}x_1\mathrm{d}x_2 \tag{112}$$

   where the notation $\int\int_A$ refers to integration on the plane $x_3 = H$, where the applied load, for which the stress intensity factor is required, is acting.

2. Bending stress (x_1-variation): the applied traction on the planes $x_3 = \pm H$ is $t_3 = \pm\sigma(1 - x_1/R)/2$, thus the normalised stress intensity factor is given by

$$\frac{K(\theta')}{\sigma\sqrt{\pi a}} = \frac{1}{\sigma\pi^2\sqrt{2a}} \int\int_A \left(\frac{1}{2} - \frac{x_1}{2R}\right) u_3(\theta', x_1, x_2, H)\mathrm{d}x_1\mathrm{d}x_2 \tag{113}$$

3. Bending stress (x_1-variation): the applied traction on the planes $x_3 = \pm H$ is $t_3 = \pm\sigma(1 + x_1/R)/2$, thus the normalised stress intensity factor is given by

$$\frac{K(\theta')}{\sigma\sqrt{\pi a}} = \frac{1}{\sigma\pi^2\sqrt{2a}} \int\int_A \left(\frac{1}{2} + \frac{x_1}{2R}\right) u_3(\theta', x_1, x_2, H)\mathrm{d}x_1\mathrm{d}x_2 \tag{114}$$

4. Pure bending stress (x_1-variation): the applied traction on the planes $x_3 = \pm H$ is $t_3 = \pm\sigma(-x_1/R)$, thus the normalised stress intensity factor is given by

$$\frac{K(\theta')}{\sigma\sqrt{\pi a}} = \frac{1}{\sigma\pi^2\sqrt{2a}} \int\int_A \left(\frac{-x_1}{R}\right) u_3(\theta', x_1, x_2, H)\mathrm{d}x_1\mathrm{d}x_2 \tag{115}$$

5. Pure bending stress (x_2-variation): the applied traction on the planes $x_3 = \pm H$ is $t_3 = \pm\sigma(1 - x_2/R)$, thus the normalised stress intensity factor is given by

$$\frac{K(\theta')}{\sigma\sqrt{\pi a}} = \frac{1}{\sigma\pi^2\sqrt{2a}} \int\int_A \left(1 - \frac{x_2}{R}\right) u_3(\theta', x_1, x_2, H)\mathrm{d}x_1\mathrm{d}x_2 \tag{116}$$

6. Bending stress (x_2-variation): the applied traction on the planes $x_3 = \pm H$ is $t_3 = \pm\sigma(1 - x_2/(2R))$, thus the normalised stress intensity factor is given by

$$\frac{K(\theta')}{\sigma\sqrt{\pi a}} = \frac{1}{\sigma\pi^2\sqrt{2a}} \int\int_A \left(1 - \frac{x_2}{2R}\right) u_3(\theta', x_1, x_2, H)\mathrm{d}x_1\mathrm{d}x_2 \tag{117}$$

7. Bending stress (x_2-variation): the applied traction on the planes $x_3 = \pm H$ is $t_3 = \pm\sigma(x_2/(2R))$, thus the normalised stress intensity factor is given by

$$\frac{K(\theta')}{\sigma\sqrt{\pi a}} = \frac{1}{\sigma\pi^2\sqrt{2a}} \int\int_A \left(\frac{x_2}{2R}\right) u_3(\theta', x_1, x_2, H)\mathrm{d}x_1\mathrm{d}x_2 \tag{118}$$

The angle θ' denotes the position, relative to the coordinate axis x_1, of the location of the fundamental field point loads acting on the crack front. This position is also where the stress intensity factors are required. All the loads defined above are measured relative to coordinates centred at $(0, 0, \pm H)$. For the geometry depicted in Fig. 21 stress intensity factors are obtained for a/R ratios of 0.5 and 0.375 within the range of θ' specified above. The boundary element meshes for a/R ratios equal to 0.5 and 0.357 consisted of 210 elements with 1896 degrees of freedom and 240 elements with 2166 degrees of freedom, respectively. In these analysis the quarter-point element length to crack radius, (l/a), ratio is fixed at 0.2 and 0.1 for the larger and smaller a/R ratios respectively. Stress intensity factor solutions are normalised (i.e. $K(\theta')/\sigma\sqrt{\pi a}$ with respect to the an edge crack solution, of crack length equal to the radius of the penny-shaped crack a, in an infinite domain subjected to remote tensile loads σ at infinity. The normalised stress intensity factor results for a/R equal to 0.5 and 0.375 are plotted in Figs 23 and 24. Results were also generated for a/R equal to 0.375

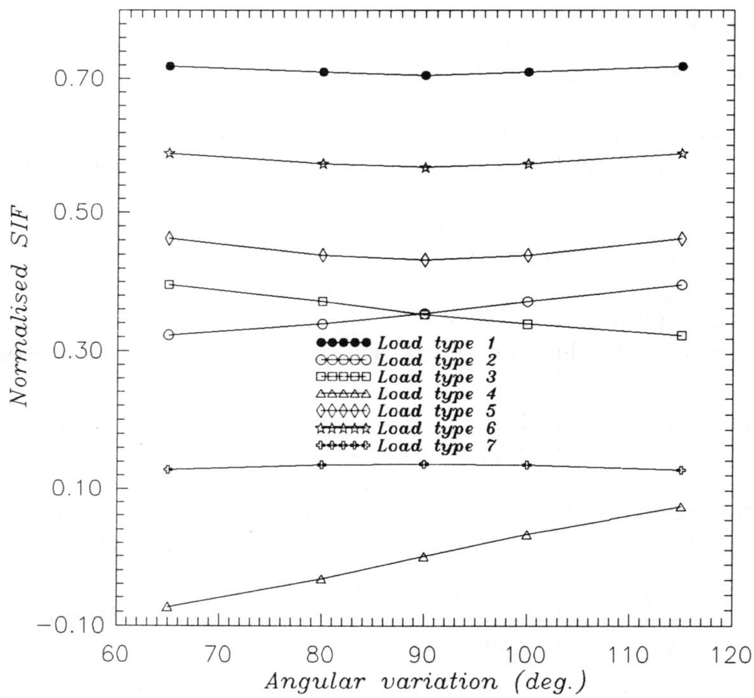

Figure 23: SIF variation with θ' for $a/R = 0.5$

at θ' equal to 90° using Element type 2 (see Section 7.1) rather than Element type 1. The two different solutions obtained by using Element type 1 and Element type 2 were within 0.5% of each other, see Bains.[2]

For load type 1 and at θ' equal to 90°, Nisitani[87] obtained, for a/R ratios of 0.5 and 0.375, normalised stress intensity factor values of 0.758 and 0.714, respectively. The results generated using the present analysis and those presented in Ref. 87 differ by 7% and 6% for a/R equal to 0.5 and 0.375 respectively.

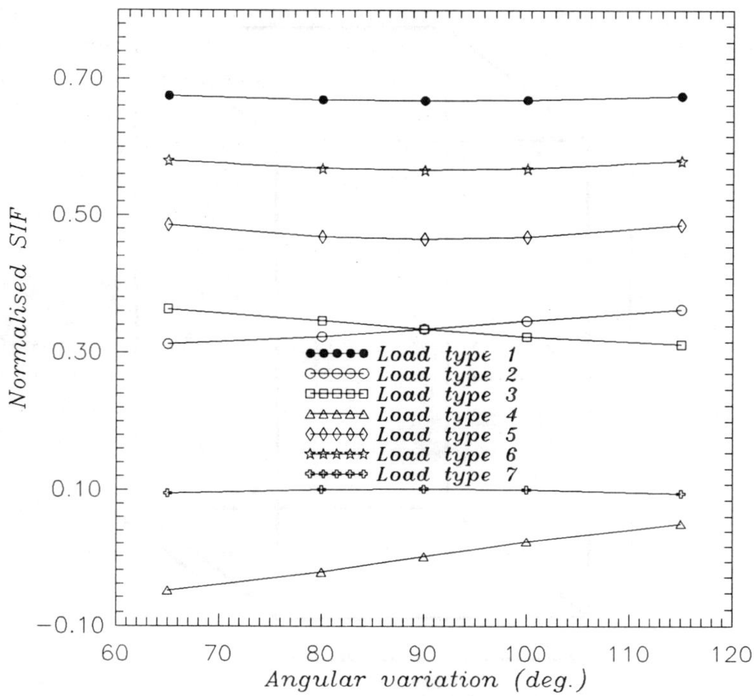

Figure 24: SIF variation with θ' for $a/R = 0.375$

7.5 Penny-shaped corner crack in square bar

The SFF technique will further be tested by considering a penny-shaped crack in a bar of square cross-section, see Fig. 25. The geometry consists of a corner crack of radius a in a bar of width w, thickness t and height $2H$. Stress intensity factors at different angular locations θ' on the crack front will be generated when the ends of the bar (i.e. $x_3 = \pm H$) are subjected to loads that produce mode I deformation of the crack faces. The angular location θ' denotes the position relative to the axis x_1 of the point loads acting on the crack front. The geometry depicted in Fig. 25 has half symmetry at the $x_3 = 0$ plane and this will be used in the BEM analysis. The geometry dimensions consist of w/a, H/w and t/w ratios of 10.0, 1.0 and 1.0 respectively and contains 253 elements with 2283 degrees of freedom. The mesh also contains traction-singular quarter-point elements situated adjacent to the crack front on the intact plane and displacement quarter-point elements situated adjacent to the front on the crack face, i.e. Element type 1, see Section 7.1, with (l/a) ratio of 0.1; where l is the special crack front element length measured perpendicular to the front. The material Poisson's ratio was assumed to be 0.3.

For the configuration of Fig. 25, normalised stress intensity factors were obtained at θ' equal to 30.0°, 37.5°, 45.0°, 52.5° and 60.0° for the following loading cases.

1. Uniform tensile stress, $\pm\sigma$, acting on the planes $x_3 = \pm H$ of the bar: the applied traction $t_3 = \pm\sigma$, therefore the normalised stress intensity factor (as defined by

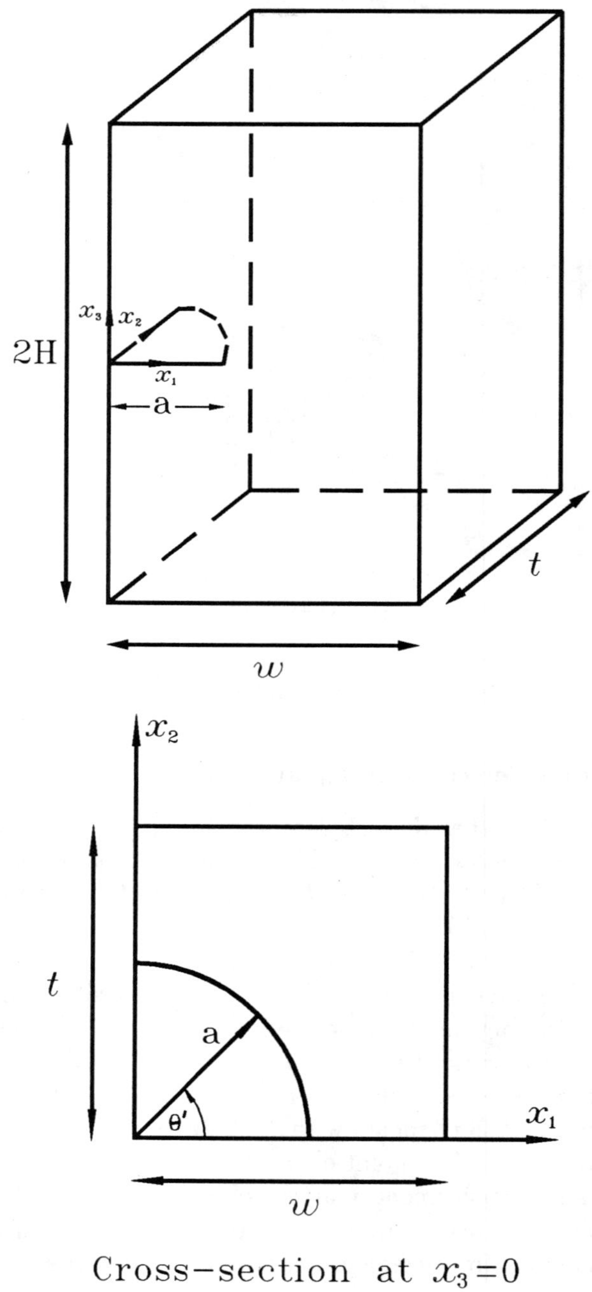

Figure 25: Penny-shaped corner crack in a bar of square cross-section

Newman & Raju[88]) as obtained from eqn (88) is given by

$$\frac{K(\theta')}{\sigma\sqrt{\pi a/Q}} = \frac{\sqrt{2.464}}{\sigma\pi^2\sqrt{2a}} \int_0^w \int_0^t u_3(\theta', x_1, x_2, H)\mathrm{d}x_1\mathrm{d}x_2 \tag{119}$$

2. Bending stress (x_1-variation): the applied traction on the planes $x_3 = \pm H$ is $t_3 = \pm\sigma(1 - x_1/w)$, thus the normalised stress intensity factor is given by

$$\frac{K(\theta')}{\sigma\sqrt{\pi a/Q}} = \frac{\sqrt{2.464}}{\sigma\pi^2\sqrt{2a}} \int_0^w \int_0^t \left(1 - \frac{x_1}{w}\right) u_3(\theta', x_1, x_2, H)\mathrm{d}x_1\mathrm{d}x_2 \tag{120}$$

3. Bending stress (x_1-variation): the applied traction on the planes $x_3 = \pm H$ is $t_3 = \pm\sigma(x_1/w)$, thus the normalised stress intensity factor is given by

$$\frac{K(\theta')}{\sigma\sqrt{\pi a/Q}} = \frac{\sqrt{2.464}}{\sigma\pi^2\sqrt{2a}} \int_0^w \int_0^t \left(\frac{x_1}{w}\right) u_3(\theta', x_1, x_2, H)\mathrm{d}x_1\mathrm{d}x_2 \tag{121}$$

4. Pure bending stress (x_1-variation): the applied traction on the planes $x_3 = \pm H$ is $t_3 = \pm\sigma(1 - 2x_1/w)$, thus the normalised stress intensity factor is given by

$$\frac{K(\theta')}{\sigma\sqrt{\pi a/Q}} = \frac{\sqrt{2.464}}{\sigma\pi^2\sqrt{2a}} \int_0^w \int_0^t \left(1 - \frac{2x_1}{w}\right) u_3(\theta', x_1, x_2, H)\mathrm{d}x_1\mathrm{d}x_2 \tag{122}$$

5. Pure bending stress (x_2-variation): the applied traction on the planes $x_3 = \pm H$ is $t_3 = \pm\sigma(1 - 2x_2/t)$, thus the normalised stress intensity factor is given by

$$\frac{K(\theta')}{\sigma\sqrt{\pi a/Q}} = \frac{\sqrt{2.464}}{\sigma\pi^2\sqrt{2a}} \int_0^w \int_0^t \left(1 - \frac{2x_2}{t}\right) u_3(\theta', x_1, x_2, H)\mathrm{d}x_1\mathrm{d}x_2 \tag{123}$$

6. Bending stress (x_2-variation): the applied traction on the planes $x_3 = \pm H$ is $t_3 = \pm\sigma(1 - x_2/t)$, thus the normalised stress intensity factor is given by

$$\frac{K(\theta')}{\sigma\sqrt{\pi a/Q}} = \frac{\sqrt{2.464}}{\sigma\pi^2\sqrt{2a}} \int_0^w \int_0^t \left(1 - \frac{x_2}{t}\right) u_3(\theta', x_1, x_2, H)\mathrm{d}x_1\mathrm{d}x_2 \tag{124}$$

7. Bending stress (x_2-variation): the applied traction on the planes $x_3 = \pm H$ is $t_3 = \pm\sigma(x_2/t)$, thus the normalised stress intensity factor is given by

$$\frac{K(\theta')}{\sigma\sqrt{\pi a/Q}} = \frac{\sqrt{2.464}}{\sigma\pi^2\sqrt{2a}} \int_0^w \int_0^t \left(\frac{x_2}{t}\right) u_3(\theta', x_1, x_2, H)\mathrm{d}x_1\mathrm{d}x_2 \tag{125}$$

In expressions (119)-(125) above, Q denotes the shape factor for an ellipse with major and minor axis equal to a_e and b_e, respectively. This shape factor is given by the square of the complete elliptic integral of the second kind. When the major and minor axis of the ellipse are equal (i.e. $a_e = b_e = a$) then Q takes the approximate value of 2.464.

Using finite element analysis Newman *et. al.*,[88] solved a similar geometry and crack configuration, but with a/t or a/w ratio of 0.2. The stress intensity factor solutions of the present weight function analysis can be compared with those of Newman *et. al.*,[88] by employing the empirical relationships presented in Ref. 88. These empirical equations are essentially curves fitted to the authors[88] finite element stress intensity factor solutions for different geometries and crack configurations subjected to remote uniform tensile load. At θ equal to 30.0°, 37.5°, 45.0°, 52.5° and 60.0° normalised stress intensity factor, $K(\theta')/(\sigma\sqrt{\pi a/Q})$ were obtained from the empirical equations of Newman *et. al.*,[88] for the cracked geometry configuration under analysis. These solutions together with the present weight function results for all the load configurations are plotted in Fig. 26. The stress intensity factor solutions for the

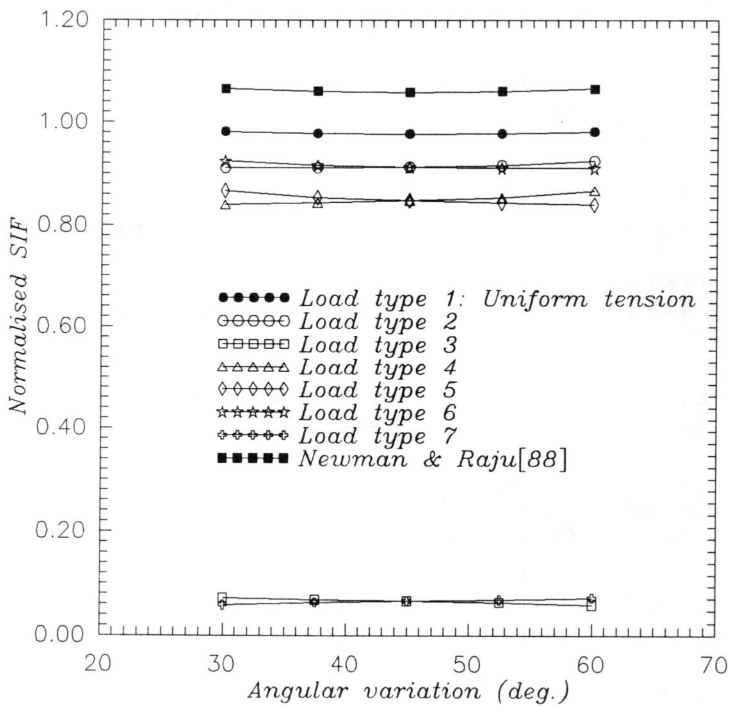

Figure 26: SIF solutions for a square bar with w/a ratio of 10.0

uniform tensile load case, as obtained from the present analysis are 8% lower (over all the range of θ' covered) than solutions obtained through the empirical equations.[88] As stated in Ref. 88 the empirical equations in general produce solutions to within 5% of the finite element results, if regions of the crack front intersection with the boundaries is neglected. It could be argued that the difference of 8% observed in the two solutions could be because the weight function solutions are not converged values. In order to check this the number of elements used to model the problem was increased to 421 elements with 3795 degrees of freedom. At θ' equal to 45.0° the seven load configurations described in eqns (119)-(125) were applied to the refined model that contained 421 elements ($a/w = 0.1$, $H/w = 1.0$, and $t/w = 1.0$) and normalised stress intensity factor generated. The solutions obtained for the two meshes, for load types 1, 2, 4, 5 and 6 were within 1% of each other and for load types 3 and 7 the solutions were within 3%. For load types 3 and 7 the stress intensity factor values are small relative to the values for the other load variations and rounding errors due to integration could explain the relatively large error observed after mesh refinement.

Normalised stress intensity factors are also presented for w/a ratios of 6, 5 and 4 for values of θ' specified above and for all the load configurations mentioned above. For w/a ratios of 6, 5 and 4 the solutions are plotted in Figs 27, 28 and 29 respectively. All these results are for H/w ratio of 2.0 and (l/a) ratio of 0.1. In all cases,

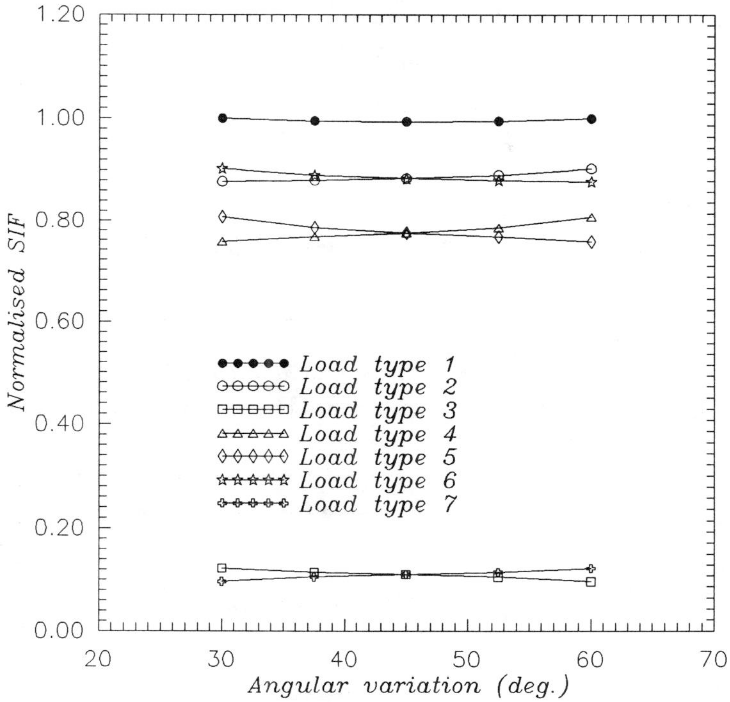

Figure 27: SIF solutions for a square bar with w/a ratio of 6.0

as expected, the uniform tensile loads produced higher stress intensity factors than the other loads. Also all the loads that are symmetric about the θ' equal to 45.0° and $x_3 > 0$ plane give rise to solutions that are symmetric about this plane. This is as expected because of the symmetric geometry configuration at θ' equal to 45.0°. In general all the solutions presented in Figs 27, 28 and 29 behave as expected along the crack front for each load variation. In regions above the crack front where the applied remote load variation increase, the stress intensity factor value along the crack front also increases as expected.

8 Cracks of elliptical configuration

8.1 Curvature matching

For this, more general crack configuration, exact complete solutions generated through

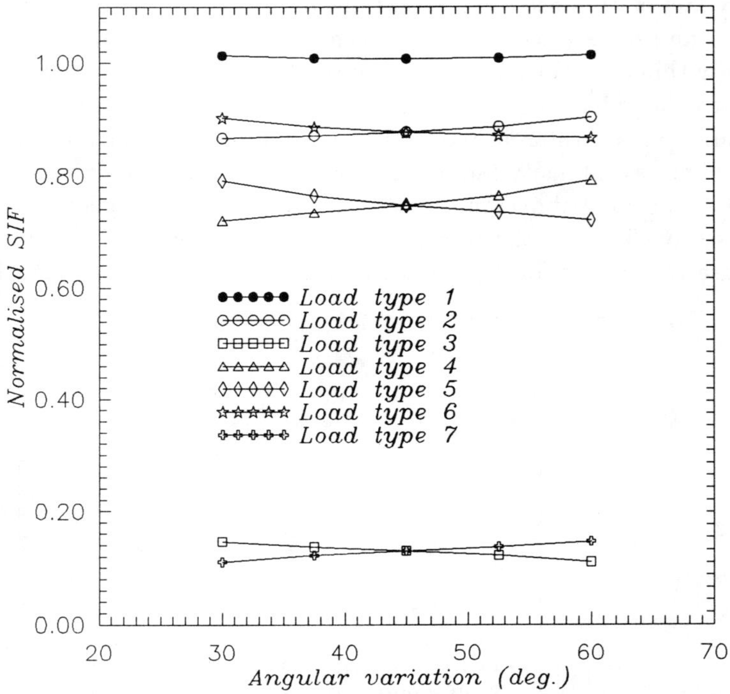

Figure 28: SIF solutions for a square bar with w/a ratio of 5.0

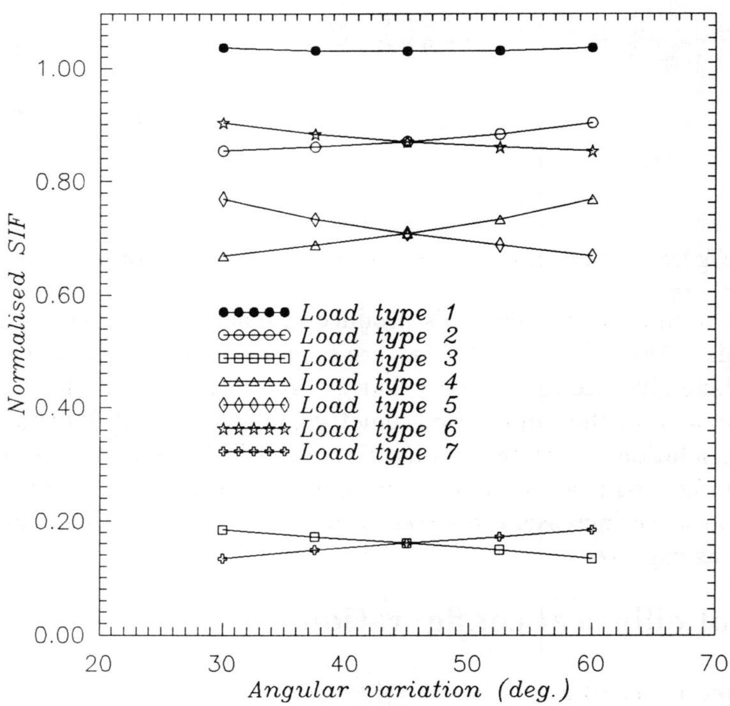

Figure 29: SIF solutions for a square bar with w/a ratio of 4.0

the notion of point loads acting at the crack front do not as yet exist. The penny-shaped crack can be considered as a special case of the elliptical crack, with the minor and major axis equal. A point force solution of the ellipse would have the same order singularities as the straight and penny-shaped crack fronts, since these crack geometries can be generated by a limiting process. This important observation was pointed out by Bueckner,[13] who considered the crack front to be composed of circular arcs of various curvatures. In the vicinity of the point of application of the point force, on one of these circular arcs the fundamental field variation would be governed by the field generated from $\Re(G)$, where $\Re$ denotes the real part of G as given by expression (73). So to determine the fundamental field associated with a more general crack front configuration, it suffices just to approximate portions of the general crack configuration by circular arcs.

Consider any point Ψ on an elliptical crack front. The weight functions resulting from the presence of a load at Ψ can be computed by matching the curvature variation at this point with a penny-shaped crack front. At Ψ the penny-shaped crack and the elliptical crack will have the same geometrical properties, (i.e. the same gradient and second derivative). As far as the point force is concerned at Ψ, it would not know whether it is situated on the elliptical crack front or the penny-shaped crack front, hence $\Re(G)$ can be used to obtain the weight functions and hence the stress intensity factor at this point. Thus stress intensity factors for any point on the elliptical crack front can be obtained by making use of the properties of fundamental fields at the point where the curvatures match. The matching of curvature at point Ψ will generate only one unique penny-shaped crack.

8.2 Modification of boundary conditions for elliptical cracks

Consider an elliptical crack subjected to point loads at angle β'' on the crack front relative to an arbitrary coordinate system, see Fig. 30. These point loads produce mode I deformation of the elliptical crack and generate displacement and traction fields that can be denoted by $u_j^e(\beta'')$ and $t_j^e(\beta'')$, respectively. It must be borne in mind that these fundamental fields are not available for this class of configuration. In order to approximate $u_j^e(\beta'')$ and $t_j^e(\beta'')$ consider a penny-shaped crack with point loads acting at an arbitrary position, β' on the front. The angle β' is measured relative to the arbitrary coordinate system defined in Fig. 30. If the load points on the elliptical crack front and the load points on the penny-shaped crack front are matched together with the curvature at this point of both crack configurations, then in the neighborhood of this coinciding point the properties of the fundamental fields for both configurations would be identical. Thus at the load points where the two curvatures are identical, fundamental fields $u_j^\infty(\beta')$ and $t_j^\infty(\beta')$ of the penny-shaped crack will be identical to the fundamental fields $u_j^e(\beta'')$ and $t_j^e(\beta'')$ of the elliptical crack. At this common point, β' equals β'' and if the coordinate system is moved to the centre of the penny-shaped crack then β' equals θ'.

In applying the SFF technique to this class of problem, eqn (99) is solved subject to the following new boundary conditions:

$$\overline{u}_j^R(\theta') = \overline{u}_j^e(\theta') - \overline{u}_j^\infty(\theta') \tag{126}$$

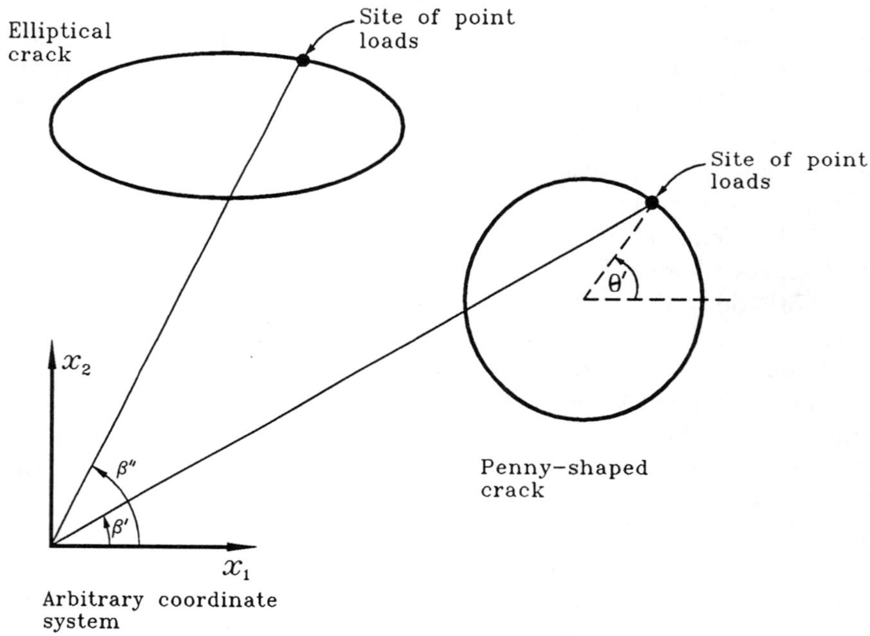

Figure 30: Fundamental fields for the elliptical crack

and

$$\bar{t}_j^R(\theta') = \bar{t}_j^e(\theta') - \bar{t}_j^{\infty}(\theta') \tag{127}$$

Where $\overline{u}_j^e(\theta')$ and $\bar{t}_j^e(\theta')$ are boundary values of the displacement and traction fields respectively, for an elliptical crack configuration subjected to point loads on the elliptical crack front at angle (θ'). The boundary values of $\overline{u}_j^{\infty}(\theta')$ and $\bar{t}_j^{\infty}(\theta')$ are generated by the point force at angle θ' on the penny-shaped crack front. Two interesting conditions arise as regards the application of these new boundary conditions to the elliptical crack plane, $x_3 = 0$, namely:

Condition 1: The elliptical crack front is not enclosed by the circular arc representing the penny-shaped crack front.

Condition 2: The elliptical crack front is enclosed by the circular arc representing the penny-shaped crack front.

These two conditions will play a central role in the determination of weight functions and thus the stress intensity factors using the SFF technique. In order to explain this, the general case of a point on an elliptical crack front, where the curvature is to be matched by a circular arc, will be investigated. Consider the elliptical crack front and the circular arc representing the penny-shaped crack front touching at point Ψ such that the elliptical crack is inscribed inside the penny-shaped crack. In order to explain the different regions where the new boundary conditions are applied it is necessary to define the elliptical crack face and the penny-shaped crack face. To this end, the elliptical crack face will be denoted by Ω_e^+ and the penny-shaped crack face by Ω_p^+. With respect to Fig. 31, on the $x_3 = 0$ plane, the following traction boundary condition could exist.

1. In $\Omega_e^+ \cap \Omega_p^+$ we have $\bar{t}_j^R(\theta') = 0$
2. In $\Omega_e^+ - \Omega_p^+$ we have $\bar{t}_j^R(\theta') = -\bar{t}_j^\infty(\theta')$
3. In $\Omega_p^+ - \Omega_e^+$ we have $\bar{t}_j^R(\theta') = \bar{t}_j^e(\theta')$
4. Outside $(\Omega_e^+ \cup \Omega_p^+)$ we have $\bar{t}_j^R(\theta') = \bar{t}_j^e(\theta') - \bar{t}_j^\infty(\theta')$

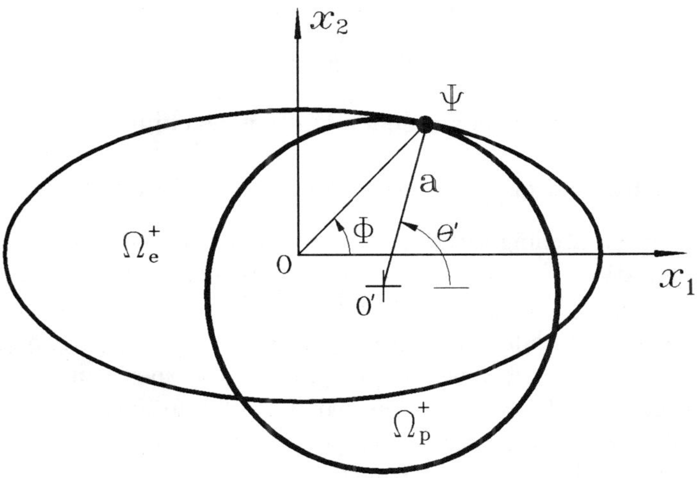

Figure 31: Modelling the elliptical crack configuration

Thus if any part of the periphery of Ω_p^+ is enclosed within Ω_e^+ then infinite traction boundary conditions are required on that part of the periphery. This implies that the penny-shaped crack obtained from the curvature match at Ψ may not be suitable for calculating the weight functions whenever it intersects Ω_e^+. Ideally, as far as the boundary conditions are concerned, the best situation is to have Ω_e^+ completely enclosed within Ω_p^+. In this situation the only traction boundary conditions required will be in the form of (1), (3) and (4) stated above, thus bypassing the need of specifying $\bar{t}_j^R(\theta')$ equal to $-\bar{t}_j^\infty(\theta')$ on part of Ω_e^+. Although there are an infinite number of penny-shaped cracks, which pass through the point Ψ and enclose Ω_e^+, interest will naturally lie in the smallest possible Ω_p^+ which encloses Ω_e^+ completely. The criteria used to determine the minimum radius and coordinates of the center O' of the circular arc, basically involves satisfying the requirement that every point on the elliptical crack front must simultaneously fall on or inside the penny-shaped crack forming the circular arc. With respect to the coordinate system denoted by x_1 and x_2, see Fig. 31, the point Ψ subtends an angle Φ. At Φ equal to 90°, the smallest penny-shaped crack which encloses the elliptical crack coincides with that obtained by matching the curvature at this point. At Φ equal to 0°, the smallest penny-shaped crack which encloses the elliptical crack will have its circumference passing through the two extreme points on the major axis of the elliptical crack. This penny-shaped crack is not the same as that which would be obtained by matching curvature at the point Φ equal to 0°; neither will it be the same for any Φ not equal to 90°.

In the determination of weight functions and thus the stress intensity factor, the procedure is to locate the minimum radius and coordinates of the center of the penny-shaped crack front passing through Ψ that completely encloses the elliptical crack. The global coordinate system is next shifted to the center of the penny-shaped crack O', and the weight functions determined for the unit point forces acting at $(a, \theta', 0)$ as shown in Fig. 31. With respect to the original and the shifted coordinate system, the stress intensity factor at the common point Ψ will have the same value, namely

$$K(\theta') = K(\Phi) \tag{128}$$

since both angles θ' and Φ define a unique point on the elliptical crack front.

8.3 Embedded elliptical crack in square cross-section bar

To test the curvature matching technique of generating weight functions, an elliptical crack front configuration, with major and minor axis denoted by $2a_e$ and $2b_e$ respectively, will be considered. The elliptical crack is embedded with its centre coincident with the centre of a rectangular bar of square cross-section, see Fig. 32, of width $2w$, height $6w$ and $a/w = 0.5$. Because of the symmetry of the geometry only half the problem domain need be discretized. The bar is loaded uniformly on the ends, such that the elliptical crack experiences mode I deformation only. At Ψ the weight functions have been obtained using the SFF technique coupled in turn to the four special crack front elements, see Section 7.1. These special elements ($l/a_e = 0.2$), were placed on either side of the elliptical crack front to model the standard stress singularity.

For aspect ratios of 0.9 and 0.5 stress intensity factors are obtained at three locations on the elliptical crack front. These locations are at Φ equal to 0°, 43.517° and 90°, for aspect ratios of 0.9, and Φ equal to 0°, 35.078° and 90°, for aspect ratios of 0.5. All the stress intensity factor results $K(\Phi)$, have been normalised with respect to K_∞, where K_∞ is evaluated for a penny-shaped crack of radius equal to the major axis of the elliptical crack in an infinite domain. The percentage difference in the normalised stress intensity factor is obtained relative to the normalised solutions presented in Ref. 85. It must be noted that to compare solutions with those presented in Ref. 85, the angle Φ must be converted to the parametric angle defined in Ref. 85.

For both aspect ratios the meshes consisted of 172 elements and 1554 degrees of freedom. Stress intensity factor solutions obtained at three points on the elliptical crack front, using all the special crack front elements, see Section 7.1, when each one is coupled to the SFF technique are shown in Fig. 33 and Fig. 34 for aspect ratios of 0.9 and 0.5 respectively. These results for both aspect ratios show that the

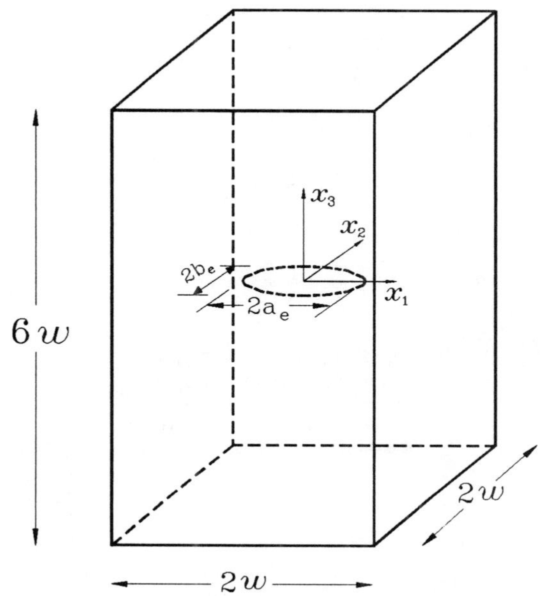

Figure 32: Elliptical crack in square cross-section bar

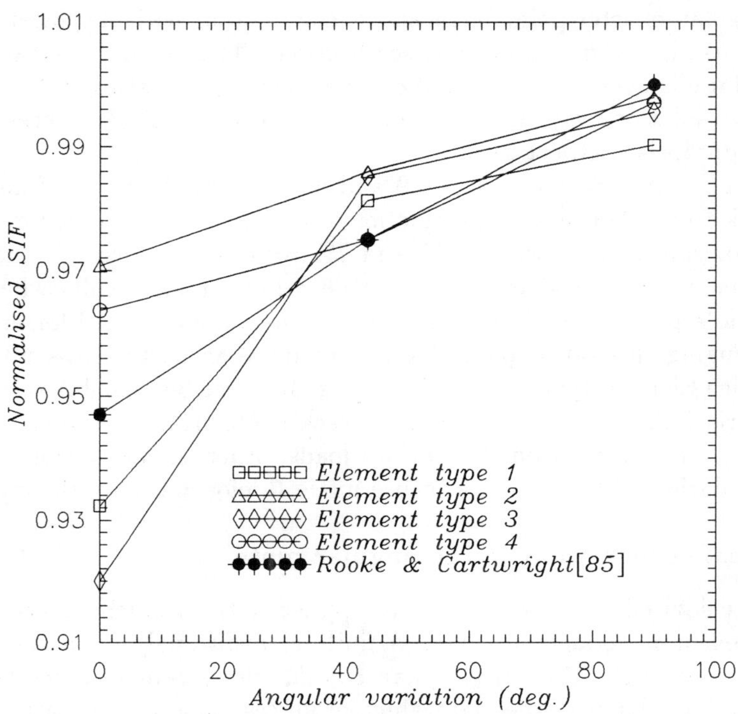

Figure 33: SIF variation for $b_e/a_e = 0.9$

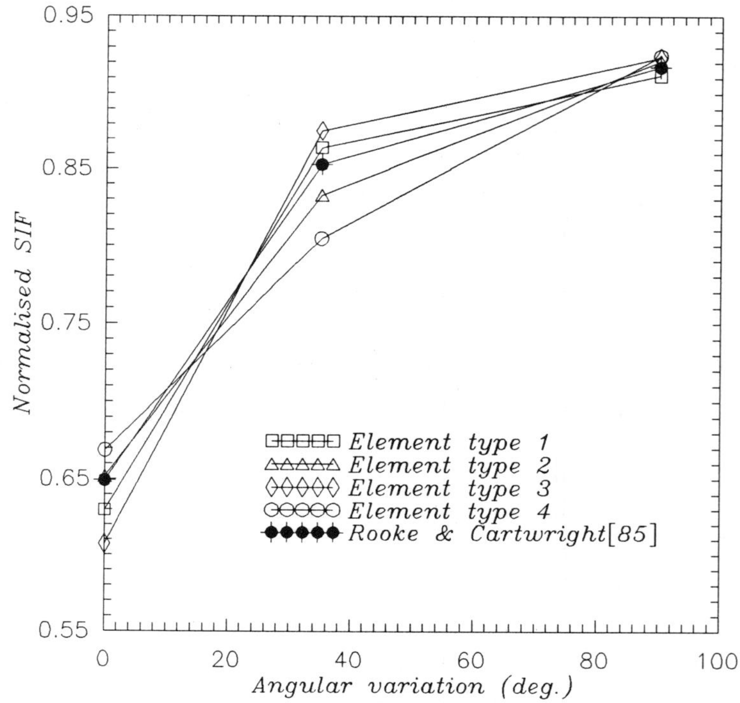

Figure 34: SIF variation for $b_e/a_e = 0.5$

traction-singular quarter-point elements in conjunction with displacement quarter-point elements (i.e. Element type 1, see Section 7.1) give the most accurate stress intensity solutions over the range of Φ covered. For aspect ratios of 0.9 and 0.5 these elements give solutions within of 1% and 3% respectively of those presented in Ref. 85 for the appropriate aspect ratio.

In general, for aspect ratios of 0.9 and 0.5, the employment of any one of the special crack front elements in conjunction with the SFF technique gives stress intensity factor solutions to within 3% and 7% respectively, over all the range of Φ covered. Element type 3 gives the largest deviation in the results when Φ is equal to 0° and the aspect ratio is 0.5. All these solutions are obtained for an (l/a_e) ratio of 0.2. Changing this ratio could give rise to more accurate stress intensity factor solutions since Element types 2, 3 and 4 were shown (see Section 4.2.3) to be sensitive to (l/a_e) variation, for the penny-shaped crack configuration. Bains[2] also obtained stress intensity factor solutions for various loads, other than uniform tensile, on the remote boundaries of the body that cause mode I deformation of the crack faces.

8.4 Elliptical corner crack in rectangular plate

The principle devised in Section 8.1 will be applied to the problem of a corner elliptical crack situated in a rectangular plate. A finite element analysis has been conducted on this type of configuration by Newman *et. al.*,[88] for a remote uniform tensile load acting on the boundaries of the plate which produces mode I deformation of the crack faces.

For the analysis, a rectangular plate of height $2H$, thickness t and width w, with a corner elliptical crack will be considered. The major and minor axis of the elliptical crack will be denoted by a_e and b_e, respectively. A typical plate and crack geometry is presented in Fig. 35. The cross-section through the crack plane at $x_3 = 0$ shows the angular location θ, with respect to axis x_1 on the elliptical crack front where the stress intensity factor solutions will be required, when various mode I loads are applied on the $x_3 = \pm H$ planes.

Since the geometry of Fig. 35 is symmetrical about the $x_3 = 0$ plane, only half the problem was discritised for the BEM analysis. A typical mesh of this problem is shown in Fig. 36. The mesh of Fig. 36 consists of 253 elements with 2280 degrees of freedom. Stress intensity factors at the appropriate angle θ on the elliptical crack front are obtained with the aid of the curvature matching technique described in Section 8.1. When applying the boundary conditions needed for the SFF technique care must be taken to ensure that no part of the elliptical crack front is cut by the circular arc representing the penny-shaped crack front, as generated in the process of matching the curvature at the appropriate angle θ, see Section 8.2. The material Poisson's ratio was assumed to be 0.3.

Stress intensity factors at θ equal to 20.71°, 34.07° and 50.93° on the elliptical crack front have been generated when the ends of the plates, $x_3 = \pm H$ are subjected to the following load cases;

1. Uniform tensile stress: the applied traction on the planes $x_3 = \pm H$ is

$$t_3 = \pm\sigma \tag{129}$$

2. Bending stress (x_1-variation): the applied traction on the planes $x_3 = \pm H$ is

$$t_3 = \pm\left(1 - \frac{x_1}{w}\right)\sigma \tag{130}$$

3. Bending stress (x_1-variation): the applied traction on the planes $x_3 = \pm H$ is

$$t_3 = \pm\left(\frac{x_1}{w}\right)\sigma \tag{131}$$

4. Pure bending stress (x_1-variation): the applied traction on the planes $x_3 = \pm H$ is

$$t_3 = \pm\left(1 - \frac{2x_1}{w}\right)\sigma \tag{132}$$

5. Bending stress (x_2-variation): the applied traction on the planes $x_3 = \pm H$ is

$$t_3 = \pm\left(1 - \frac{x_2}{t}\right)\sigma \tag{133}$$

6. Bending stress (x_2-variation): the applied traction on the planes $x_3 = \pm H$ is

$$t_3 = \pm\left(\frac{x_2}{t}\right)\sigma \tag{134}$$

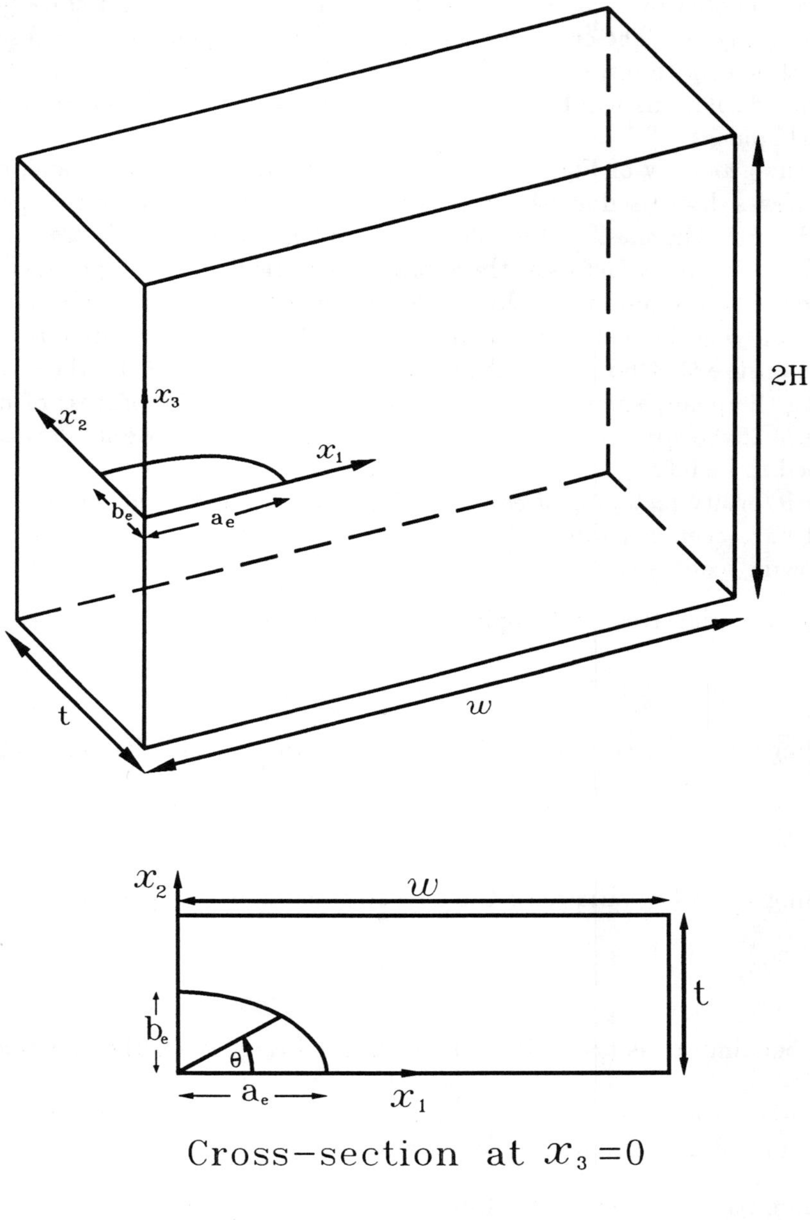

Figure 35: Plate geometry with corner elliptical crack

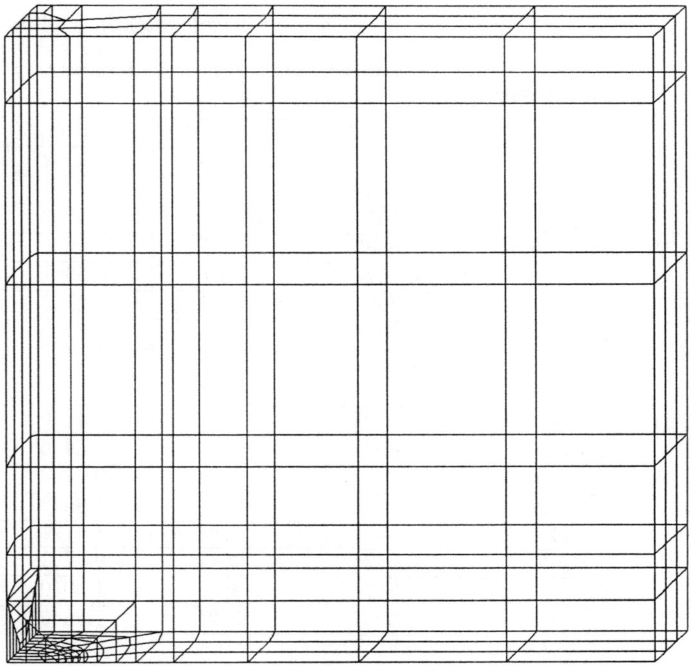

Figure 36: A typical mesh for corner elliptical crack in a plate

All the above load variations are measured relative to the origin of coordinates centred at $(0, 0, \pm H)$. Stress intensity factors as defined in eqn (88) are obtained from

$$\frac{K(\theta)}{\sigma\sqrt{\pi b_e/Q}} = \frac{\sqrt{Q}}{\sigma\pi^2\sqrt{2b_e}} \int_0^w \int_0^t t_3\, u_3(\theta, x_1, x_2, H) \mathrm{d}x_1 \mathrm{d}x_2 \tag{135}$$

where Q is the shape factor for the ellipse, see Section 7.5, and t_3 denotes the load configurations as described by load types 1-6 above. All the stress intensity factors presented hereafter have been generated with the SFF technique coupled with Element type 1, see Section 7.1. It must be stressed that a single computer run is all that is needed to obtain the weight functions, from which one derived all the stress intensity factors at the angular location θ for all the loads described above; the boundary element analysis need not be repeated for each different load configuration.

Three a/w ratios of 0.1, 0.125 and 0.25 were studied and stress intensity factor solutions obtained at different angular locations θ on the crack front. All solutions generated are for constant b_e/a_e and w/H ratios of 0.5 and 1.0 respectively, with an (l/a_e) ratio of 0.1, where l denotes the length of the quarter-point element. Results along the crack front for all the different plate geometries and θ variation are presented in Table 3.

The present uniform tensile solutions, see Table 3, cannot be directly compared with the finite element solutions as obtained by Newman *et. al.*,[88] since they did not solve for the elliptical crack configuration, with b_e/a_e ratio of 0.5, as shown in Fig. 36. A comparison for this configuration can be made by employing the empirical equations for the stress intensity factors as presented in Ref. 88. The authors[88]

Table 3: SIF solutions for the corner elliptical crack in a plate

Angle θ	Load type t_3	$\frac{K(\theta)}{\sigma\sqrt{\pi b_e/Q}}$ $a/w = 0.100$	$a/w = 0.125$	$a/w = 0.250$
20.71°	$\pm\sigma$	1.1084	1.1163	1.2504
	$\pm\sigma\left(1-\frac{x_1}{w}\right)$	1.0033	0.9905	1.0221
	$\pm\sigma\left(\frac{x_1}{w}\right)$	0.1051	0.1258	0.2283
	$\pm\sigma\left(1-\frac{2x_1}{w}\right)$	0.8982	0.8647	0.7938
	$\pm\sigma\left(1-\frac{x_2}{t}\right)$	0.7024	0.7375	0.9961
	$\pm\sigma\left(\frac{x_2}{t}\right)$	0.4059	0.3788	0.2543
34.07°	$\pm\sigma$	1.2246	1.2376	1.6269
	$\pm\sigma\left(1-\frac{x_1}{w}\right)$	1.1327	1.1268	1.3932
	$\pm\sigma\left(\frac{x_1}{w}\right)$	0.0919	0.1108	0.2336
	$\pm\sigma\left(1-\frac{2x_1}{w}\right)$	1.0408	1.0160	1.1596
	$\pm\sigma\left(1-\frac{x_2}{t}\right)$	0.4029	0.4850	1.3476
	$\pm\sigma\left(\frac{x_2}{t}\right)$	0.8217	0.7526	0.2793
50.93°	$\pm\sigma$	1.3975	1.4140	1.8532
	$\pm\sigma\left(1-\frac{x_1}{w}\right)$	1.3551	1.3563	1.6771
	$\pm\sigma\left(\frac{x_1}{w}\right)$	0.0424	0.0577	0.1761
	$\pm\sigma\left(1-\frac{2x_1}{w}\right)$	1.3127	1.2986	1.5010
	$\pm\sigma\left(1-\frac{x_2}{t}\right)$	0.5217	0.6124	1.5540
	$\pm\sigma\left(\frac{x_2}{t}\right)$	0.8757	0.8016	0.2993

fitted empirical relationships to their finite element solutions and obtained empirical equations for the stress intensity factors for a wide variety of three-dimensional crack configurations subjected to remote uniform tensile load. These empirical equations related the stress intensity factor at a particular point on the crack front to the crack depth, crack length and plate thickness for numerous embedded and edge/corner crack configurations. In Ref. 88, the authors state that the empirical stress intensity factor relations are generally within 5% of their finite element results, except within a thin 'boundary layer' where the crack front intersects a free boundary.

Using the empirical equations in Ref. 88 normalised stress intensity factor solutions, $K(\theta)/(\sigma\sqrt{\pi b_e/Q})$ have been obtained for b_e/a_e ratio of 0.5 and geometry configuration depicted in Fig. 36. In using the empirical equations of Ref. 88 the angle θ as defined in this work cannot be used directly in the empirical equations and must be converted to the appropriate parametric angle for the ellipse as defined by Newman *et. al.*[88] For θ equal to 20.71°, 34.07° and 50.93° the empirical equations produced normalised stress intensity factor values of 1.0870, 1.1800 and 1.2631 respectively. Thus with increasing θ, the results of the present weight function technique and those obtained from the empirical relationships depart from each other. In both cases the stress intensity factor solutions behave as expected; that is with increasing θ the ligament ahead of the crack front decreases resulting in greater interaction of the remote boundaries with the crack front and hence the stress intensity factor values increase. The percentage difference between the present solutions and those of Ref. 88 is 2%, 4% and 10% for θ equal to 20.71°, 34.07° and 50.93° respectively. Thus the present analysis shows a greater sensitivity to the crack front free surface interaction than the empirical results.

For a/w ratio of 0.1, the normalised stress intensity factor solutions obtained along the crack front for all the other remote load configurations defined above (i.e. 2-6) are also presented in Table 3. All the solution variations along the elliptical crack front are in agreement with the intuitive solution variation expected for each load configuration; that is in regions of high remote load variations the stress intensity factor must have a higher value relative to regions of lower remote load variation (all other parameters maintained constant), this is what is observed. With the aid of linear superposition it can be shown that for all θ covered, addition of the stress intensity factor solution for load case 4 with twice the stress intensity factor solution value as obtained from load case 3 generates the stress intensity factor solutions for the uniform tensile remote case.

In Table 3 normalised stress intensity factor solutions for plates with a/w ratios of 0.125 and 0.25 are also presented. For both of these a/w ratios the mesh depicted in Fig. 36 was slightly modified and more elements added across the elliptical crack front to improve the modelling since for large a/w ratios the crack front becomes relatively large compared to the dimensions of the plate. Meshes for a/w ratios of 0.125 and 0.25 both contained 289 elements and 2607 degrees of freedom with similar element distributions. The solutions presented in Table 3 cannot be compared since no solutions for this geometry and load configuration are available in the literature. All the solutions, for both a/w ratios, show the variation as would be expected from each of the load variations described above.

8.5 Corner crack in single-edge-notched specimen

Finally the curvature matching technique is applied to a single-edge-notched specimen containing a quarter-elliptical corner crack as depicted in Fig. 37. The quarter-elliptical corner crack is assumed to emanate from the notch root. The specimen geometry consists of a plate with width w, thickness t and height $2H$. The edge notch is semi-circular with radius denoted by R, see Fig. 37. The crack is situated such that its minor and major axis b_e and a_e are coincident with coordinate axis x_1 and x_2, respectively. The specimen configuration of Fig. 37 has half symmetry at $x_3 = 0$ plane and thus only one-half of the crack specimen will be modelled in this analysis. The material Poisson's ratio was assumed to be 0.3.

The only known solutions for this specimen configuration are due to Shivakumar & Newman.[89] In Ref. 89 the authors produced, using the equivalent domain integral (EDI) method[90] in conjunction with the three-dimensional finite element analysis, stress intensity factors for various ratio's of a_e/t and R/t while the ratio a_e/b_e was maintained constant at a value equal to 2. A typical finite element model of one-half of the crack specimen in Ref. 89 consisted of 4375 elements and 5380 nodes with 16,140 degrees of freedom. Stress intensity factor solutions were presented in Ref. 89 along the crack when uniform tensile loads are applied at the ends ($x_3 = \pm H$) of the specimen that cause Mode I deformation of the crack faces. The authors[89] presented their stress intensity factors in terms of a parametric angle θ_s, as shown in Fig. 38. The angle θ_s was measured relative to the minor axis of the elliptical crack and defines a unique point on the crack front which subtends an angle θ with respect to the x_1

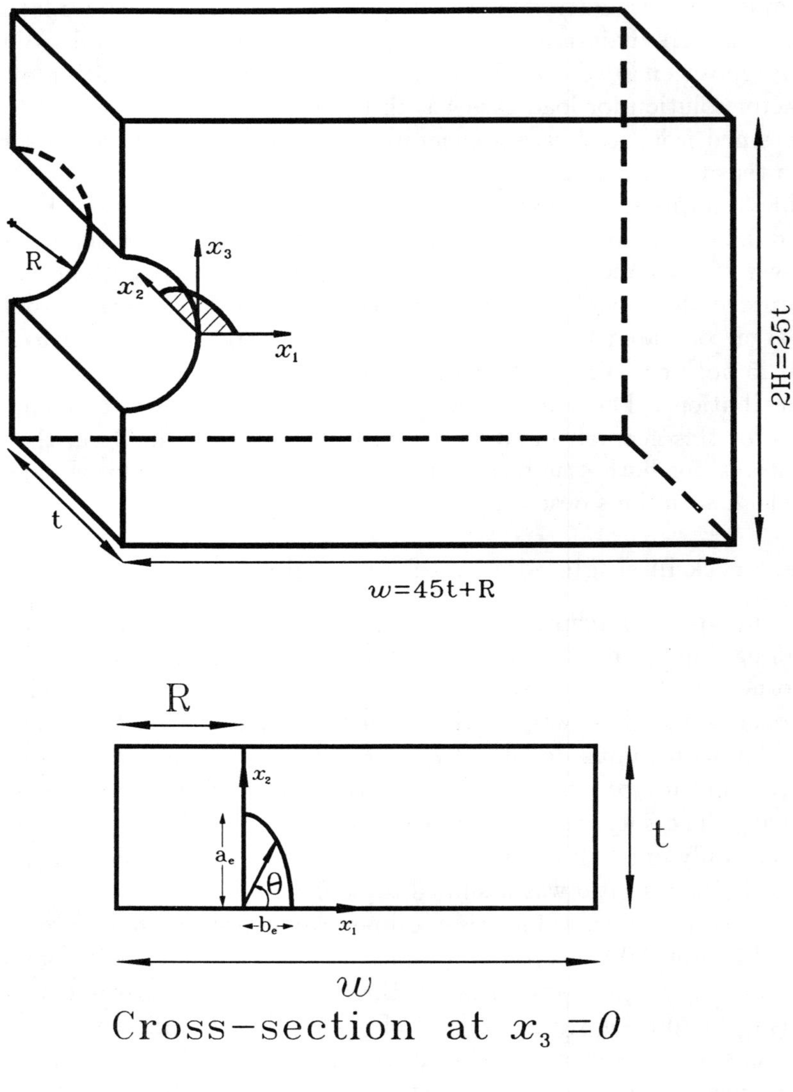

Figure 37: Quarter-elliptical corner crack in single-edge-notched specimen

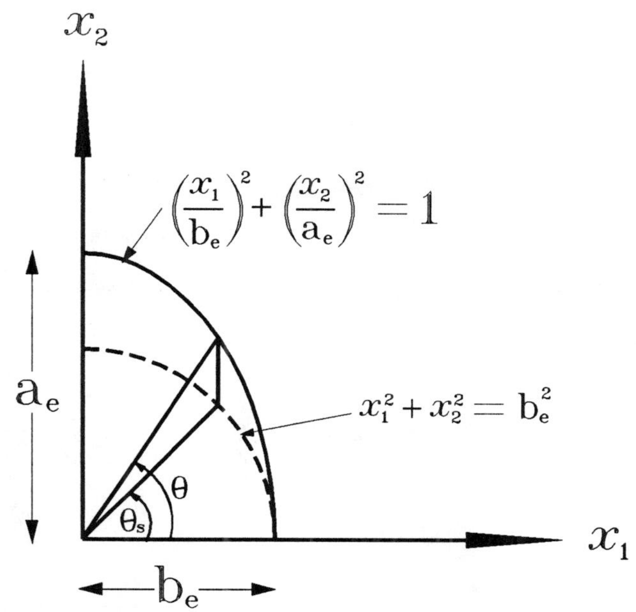

Figure 38: Definition of parametric angle θ_s

coordinate direction, see Fig. 38.

To compare the stress intensity factor solutions of the present weight function technique with those of Shivakumar *et. al.*,[89] the specimen width w was taken as equal to $(45t + R)$ and half height H as equal to $12.5t$. Further, as in Ref. 89, a_e/t, a_e/b_e and R/t ratios were taken as 0.5, 2.0 and 1.5, respectively. A typical boundary element mesh for the half configuration of Fig. 37 contained 624 quadratic elements and 1874 nodes with 5622 degrees of freedom. The mesh also contained traction-singular quarter-point elements on the intact plane adjacent to the elliptical crack front and displacement quarter-point elements on the crack face adjacent to the front, with (l/a_e) ratio equal to 0.2; where l denotes the singular element length measured perpendicular to the crack front.

Normalised stress intensity factor solutions, $K(\theta)/(\sigma\sqrt{\pi a_e/Q})$ are obtained by employing the curvature matching technique Section 8.1, along the crack front at angular location θ on the elliptical crack front for various values of the parametric angle θ_s, see Fig. 38. These solutions are obtained for $64.62° \geq \theta_s \geq 27.0°$ which corresponds to $76.66° \geq \theta \geq 45.54°$ for the case of uniform tensile stress, $\pm\sigma$, acting on the planes $x_3 = \pm H$ of the single-edge-notched specimen: the applied traction $t_3 = \pm\sigma$, therefore the stress intensity factor as obtained from eqn (88) is given by

$$\frac{K(\theta)}{\sigma\sqrt{\pi a_e/Q}} = \frac{\sqrt{Q}}{\sigma\pi^2\sqrt{2a_e}} \int_{-R}^{+45t} \int_0^t u_3(\theta, x_1, x_2, H)\mathrm{d}x_1\mathrm{d}x_2 \tag{136}$$

Normalised stress intensity factor obtained in Ref. 89 at parametric angle θ_s equal to 18.0°, 27.0°, 36.0°, 45.0°, 54.0° and 63.0° together with the present weight function solutions are plotted in Fig. 39. In general the solutions from the present

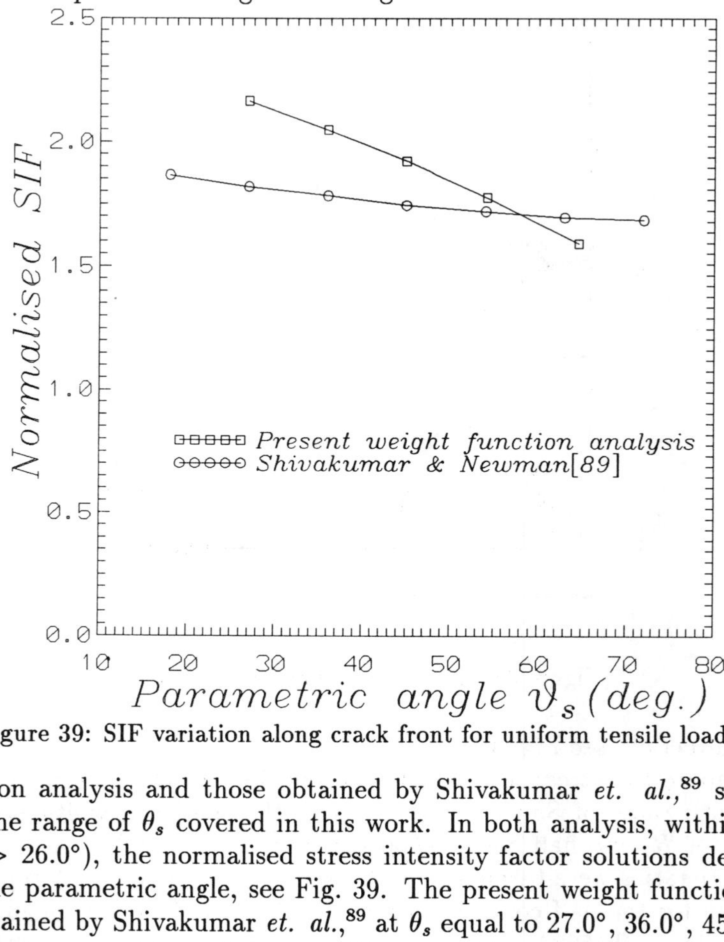

Figure 39: SIF variation along crack front for uniform tensile load

weight function analysis and those obtained by Shivakumar *et. al.*,[89] show similar trends over the range of θ_s covered in this work. In both analysis, within the range ($65.0° > \theta_s > 26.0°$), the normalised stress intensity factor solutions decrease with increase in the parametric angle, see Fig. 39. The present weight function solutions and those obtained by Shivakumar *et. al.*,[89] at θ_s equal to 27.0°, 36.0°, 45.0°, 54.169° and 64.616° differ by -19%, -15%, -10%, -3% and +6% respectively.

9 Conclusions

A regularisation procedure incorporating the fundamental fields generated by the action of a pair of unit collinear loads acting at the crack front that is situated in an infinite elastic body has been used in conjunction with standard BEM code to produce weight functions and thus stress intensity factors for 3-D cracks in finite geometrical configurations. The subtraction of fundamental fields (SFF) technique means that the resulting numerical problem is no longer highly singular. The remaining usual crack singularity can easily be modelled by standard quarter-point boundary elements to determine accurate weight functions. Subsequent computation of the stress intensity factor is reduced to a simple integration over the boundaries of the elastic body. This novel combination is more efficient than existing methods, since the stress intensity factor for a given three-dimensional geometrical configuration under any arbitrary loading can be obtained from a single computer solution of the equations of elasticity.

Acknowledgement

This work has been carried out with the support of Procurement Executive, Ministry of Defence, UK.

References

1. Aliabadi, M.H. & Rooke, D.P. *Numerical Fracture Mechanics*, Computational Mechanics Publications and Kluwer Academic Publishers, 1991.

2. Bains, R.S. Boundary element analysis of 3-D cracks problems: weight function techniques, PhD Thesis, CNAA, UK, 1992.

3. Bueckner, H.F. A novel principle for the computation of stress intensity factors, *Z. Angew. Math. Mech.*, 1970, **50**, 529-546.

4. Muskhelishvili, N.I. *Some Basic Problems of the Mathematical Theory of Elasticity*, Noordhoff, Leyden, 1953.

5. Betti, E. Teori dell elasticita, Il Nuovo Cimento, 1872, pp. 7-10.

6. Bueckner, H.F. Field singularities and related integral representations, *Methods of Fracture I: Methods of Analysis and Solution of Crack Problems*, G.C. Sih (ed), Noordhoff, Leyden, 1973, pp. 239-314.

7. Bueckner, H.F. Weight functions for the notched bar, *Z. Angew. Math. Mech.*, 1971, **51**, 97-109.

8. Bueckner, H.F. The weight functions of the configuration of collinear cracks, *Int. J. Fract.*, 1975, **11**, 71-83.

9. Bueckner, H.F. The edge crack in a half space, *Proc. of an Int. Conf. on Fracture Mechanics and Technology*, Hong Kong, March 21-25, Vol. II, G.C Sih & C.L. Chow (ed), Sijthoff and Noordhoff International Publishers, 1977, pp. 1091-1107.

10. Bueckner, H.F. The weight functions of mode I of the penny-shaped and of the elliptic crack, *Proc. of an Int. Conf. on Fracture Mechanics and Technology*, Hong Kong, March 21-25, Vol. II, G.C. Sih & C.L. Chow (ed), Sijthoff and Noordhoff International Publishers, 1977, pp. 1069-1089.

11. Dyson, F.W. The potentials of ellipsoids of variable densities, *Quart. J. Math.*, Oxford, 1891, Ser. 25, 259-288.

12. Hobson, E.W. On some general formulae for the potentials of ellipsoids, shells and disks, *Proc. London Math. Soc.*, 1896, **27**, 519-544.

13. Bueckner, H.F. Weight functions and fundamental fields for the penny-shaped and half-plane crack in three-space, *Int. J. Solids & Struct.*, 1987, **23**(1), 59-93.

14. Gao, H. Weight functions for the external circular cracks, *Int. J. Solids & Struct.*, 1989, **25**(2), 107-127.

15. Bueckner, H.F. Observations on weight functions, *Engng. Analysis with Boundary Elements*, 1989, **6**(1), 1-18.

16. Rice, J.R. Some remarks on elastic crack tip stress fields, *Int. J. of Solids and Struct.*, 1982, **8**, 751-758.

17. Rice, J.R. First order variation in elastic fields due to variations in location of a planar crack front, ASME, *J. Appl. Mech.*, 1985, **52**, 571-579.

18. Gao, H. & Rice, J.R. Shear stress intensity factors for a planar crack with slightly curved crack-front, ASME, *J. Appl. Mech.*, 1989, **53**, 774-778.

19. Gao, H. & Rice, J.R. Nearly circular connections of elastic half spaces, ASME, *J. Appl. Mech.*, 1987, **54**, 627-634.

20. Gao, H. & Rice, J.R. Somewhat circular tensile crack, *Int. J. of Fract.*, 1987, **33**, 155-174.

21. Gao, H. Nearly circular shear mode cracks, *Int. J. Solids & Struct.*, 1988, **24**, 177-193.

22. Vainshtok, V.A. & Varfolomeyev, I.V. Application of the weight function method for determining stress intensity factors of semi-elliptical cracks, *Int. J. of Fract.*, 1987, **35**, 175-186.

23. Rice, J.R. Three-dimensional elastic tip interaction with transformation strains and dislocations, *Int. J. Solids & Struct.*, 1985, **21**, 781-791.

24. Rice, J.R. Weight function theory for three-dimensional elastic crack analysis, *Fracture Mechanics: Perspectives and Directions* (Twentieth Symposium), ASTM STP 1020, R.P. Wei & R.P. Gangloff (ed), American Society for Testing and Materials, Philadelphia, 1989, pp. 29-57.

25. Anderson, P.M. & Rice, J.R. The stress field and energy of a three-dimensional dislocation loop at a crack tip, *J. Mech. & Physics of Solids*, 1987, **35**, 743-769.

26. Kirchner, H.O.K. Description of loading and screening of cracks with the aid of universal weight functions, *Int. J. of Fract.*, 1986, **31**, 173-181.

27. Kirchner, H.O.K. & Michot, G. Cracks screened by plastic zones, *Mater. Sci. & Engng.*, 1986, **79**, 169-173.

28. Paris, P.C. & McMeeking, R.M. Efficient finite element methods for stress intensity factors using weight functions, *Int. J. of Fract.*, 1975, **11**, 354-358.

29. Paris, P.C., McMeeking, R.M. & Tada, H. The weight function method for determining stress intensity factors, cracks and fracture, ASTM STP 601, American Society for Testing and Materials, 1976, 471-189.

30. Cartwright, D.J. & Rooke, D.P. An efficient boundary element method for calculating Green's functions in fracture mechanics, *Int. J. of Fract.*, 1985, **27**, R43-R50.

31. Aliabadi, M.H., Rooke, D.P. & Cartwright, D.J. Mixed-mode Bueckner weight functions using boundary element analysis, *Int. J. of Fract.*, 1987, **38**, 131-147.

32. Aliabadi, M.H. & Rooke, D.P. Improved modelling of crack-tip fields in weight function analysis, *Int. J. of Fract.*, 1989, **40**, R73-R75.

33. Rooke, D.P., Cartwright, D.J. & Aliabadi, M.H. Boundary elements combined with singular fields for three-dimensional cracked solids, *Proc. 4th Int. Conf. on Numerical Methods in Fracture Mechanics*, Pineridge Press, Swansea, 1987, pp. 15-26.

34. Aliabadi, M.H., Rooke, D.P. & Cartwright, D.J. Three-dimensional weight functions for a cracked bar, *Boundary Elements X*, C.A. Brebbia (ed), Vol. 3, Computational Mechanics Publications co-published with Springer-Verlag, September 1988, pp. 195-210.

35. Bains, R., Aliabadi, M.H. & Rooke, D.P. Fracture mechanics weight functions in three dimensions: subtraction of fundamental fields, *Int. J. Num. Meth. Engng.*, 1992, to appear.

36. Aliabadi, M.H., Rooke, D.P. & Cartwright, D.J. An improved boundary element formulation for calculating stress intensity factors: application to aerospace structures, *J. Strain Anal.*, 1987, **22**(4), 203-207.

37. Aliabadi, M.H. An enhanced boundary element method for determining fracture parameters, *Proc. 4th Int. Conf. on Numerical Methods in Fracture Mechanics*, San Antonio, Texas, Pineridge Press, March 1987, pp. 27-39.

38. Portela, A. & Aliabadi, M.H. On the accuracy of the boundary and finite element techniques for crack problems in fracture mechanics, *Advances in Boundary Elements*, Vol. 1: Computations and Fundamentals, C.A. Brebbia & J.J. Connor (ed), Computational Mechanics Publications, 1989, pp. 123-137.

39. Portela, A., Aliabadi, M.H. & Rooke, D.P. Boundary element analysis of v-notch plates, *Boundary Element Techniques: Applications in Engineering*, C.A. Brebbia & N.G. Zamani (ed), Proc. 4th Int. Conf. on Boundary Element Technology, Windsor, Canada, Computational Mechanics Publications, June 1989, pp. 147-155.

40. Bains, R., Aliabadi, M.H., & Rooke, D.P. Boundary element analysis of three- dimensional crack problems - weight function technique, *Localized Damage – Computer Aided Assessment and Control*, Vol. 3: Advanced Computational Methods, M.H. Aliabadi, C.A. Brebbia & D.J. Cartwright (ed), Computational Mechanics Publications co-published with Springer-Verlag, 26-28 June 1990, pp. 365-379.

41. Bains, R., Aliabadi, M.H. & Rooke, D.P. Three-dimensional weight functions for circular cracks in cylindrical bars, *Boundary Element Technology VI*, C.A. Brebbia (ed), Computational Mechanics Publications co-published with Elsevier Applied Science, 1991, pp. 393-405.

42. Bains, R.S. Weight functions for the penny-shaped and elliptical cracks using SFF technique, Report No. 3, Research Agreement No. 2235/01 XR/ST, Wessex Institute of Technology, Computational Mechanics Institute, March 1991.

43. Parks, D.M. & Kamenetzky, E.M. Weight functions from virtual crack extension, *Int. J. Num. Meth. Engng.*, 1979, **14**, 1693-1706.

44. Vanderglas, M.L. A stiffness derivative finite element technique for determination of influence functions, *Int. J. of Fract.*, 1978, **14**, R291-R294.

45. Parks, D.M. A stiffness derivative finite element technique for determination of crack tip stress intensity factors, *Int. J. of Fract.*, 1974, **10**, 487-502.

46. Hellen, T.H. On the method of virtual crack extension *Int. J. Num. Meth. Engng.*, 1975, **9**, 187-207.

47. Sham, T.-L. A unified finite element method for determining weight functions in two and three dimensions, *Int. J. Solids & Struct.*, 1987, **23**(10), 1357-1372.

48. Sham, T.-L. & Zhou, Y. Computation of three-dimensional weight functions for circular and elliptic cracks, *Int. J. Fract.*, 1989, **41**, 51-75.

49. Sham, T.-L. & Zhou, Y. Weight functions and fundamental fields in layered media, *Engng. Analysis with Boundary Elements*, 1989, **6**(1), 38-47.

50. Nikishkov, G.P. & Piminov, V.A. Computation of weight functions for three-dimensional cracks by the boundary element method, *Computational Mechanics*, 1991, **9**, 211-220.

51. Cruse, T.A. & Besuner, P.M. Residual life prediction for surface cracks in complex structural details, *J. of Aircraft*, 1975, **12**(4), 369-375.

52. Besuner, P.M. The influence function method for fracture mechanics and residual fatigue life analysis of cracked components under complex stress fields, *Nuclear Engng. and Design*, 1977, **43**, 115-154.

53. Labbens, R.C., Heliot, J. & Pellissier-Tannon, A. Weight functions for three-dimensional symmetrical crack problems, *Cracks and Fracture*, ASTM, STP 601, 1976, 448-470.

54. Grandt, A.F. Stress intensity factors for some through-cracked fastener holes, *Int. J. Fract.*, 1975, **11**, 283-294.

55. Heliot, J., Labbens, R.C. & Pellissier-Tannon, A. Semi-elliptical cracks in a cylinder subjected to stress gradients, *Fracture Mechanics*, ASTM, STP 677, 1979, 341-364.

56. Morawietz, P., Mattheck, C. & Munz, D. Calculation of approximate weight functions in fracture mechanics by FEM, *Int. J. Num. Meth. Engng.*, 1985, **21**, 1487-1497.

57. Fett, T., Mattheck, C. & Munz, D. Approximate weight functions for 2D and 3D problems, *Engng. Analysis with Boundary Elements*, 1989, **6**(1), 38-47.

58. Saff, C.R. & Sanger, K.B. Part-through flaw stress intensity factors developed by a slice synthesis technique, *Fracture Mechanics*, R.J. Stanford (ed), ASTM, STP 833, 1984, 24-43.

59. Fujimoto, W.T. Determination of crack growth and fracture toughness parameters for surface flaws emanating from fastener holes, *Proc. AIAA/ASME/SEA, 17th Structures, Structural Dynamics and Materials*, 1976.

60. Zhao, W., Wu, X.R. & Yan, M.G. Weight function method for three-dimensional crack problems-I. Basic formulation and application to an embedded elliptical crack in finite plates, *Int. J. Fract.*, 1989, **34**(3), 593-607.

61. Zhao, W., Wu, X.R. & Yan, M.G. Weight function method for three-dimensional crack problems-II. Basic formulation and application to an embedded elliptical crack in finite plates, *Int. J. Fract.*, 1989, **34**(3), 609-624.

62. Nishioka, T. & Atluri, S.N. The first-order variation of the displacement field due to geometrical changes in an elliptical crack, *J. Appl. Mech.*, 1990, **57**, 639-646.

63. Niu, X. & Glinka, G. Weight functions for edge and surface semi-elliptical cracks in flat plates and plates with corners, *Int. J. Fract.*, 1990, **36**(3), 459-475.

64. Oore, M. & Burns, D.J. Estimation of stress intensity factors for embedded irregular cracks subjected to arbitrary normal stress fields, *J. Pres. Ves. Tech.*, Trans. of the ASME, May 1990, **102**, 202-211.

65. Khattab, M.A.A., Burns, D.J., Pick, R.J. & Thompson, J.C. Opening mode stress intensity factors for embedded rectangular and irregular planar defects, *J. Pres. Ves. Tech.*, Trans. of the ASME, May 1986, **108**, 41-49.

66. Rooke, D.P., Cartwright, D.J & Aliabadi, M.H. Boundary elements combined with singular fields for three-dimensional cracked solids, *Proc. 4th Int. Conf. on Numerical Methods in Fracture Mechanics*, San Antonio, Texas, Pineridge Press, March 1987, pp. 15-26.

67. Sih, G.C. & Liebowitz, H. Mathematical theories of brittle fracture, *Treatise on Fracture*, H. Liebowitz (ed), Vol. II, 67, Academic Press, New York, 1968.

68. Kassir, M.K. & Sih, G.C. Semi-infinite plane cracks, *Three-Dimensional Crack Problems, Mechanics of Fracture 2*, Chapter 5, Academic Press, New York, 1975, pp. 134-176.

69. Neuber, H. *Theory of Notch Stresses*, Chapter 3, Edwards, Michigan, 1946, pp. 9-30.

70. Rooke, D.P & Aliabadi, M.H. Weight functions for crack problems using boundary element analysis, *Engng. Analysis with Boundary Elements*, March 1989, **6**(1), 19-29.

71. Scheid, F. *Schaum's Outline Series, Theory and Problems of Numerical Analysis*, McGraw-Hill Book Company, 1968.

72. Grobner, W., Hofreiter, M., Hofreiter, N., Laub, J. & Peschl, E. *Table of Integrals, Part I, Indefinite Integrals*, Luftfahrt-Forschungsanstalt Braunschweig, March 1944, pp. 30.

73. Meade, K.P. & Keer, L.M. On the problem of a pair of point forces applied to the faces of a semi-infinite plane crack, *J. of Elasticity*, 1984, **14**, 3-14.

74. Aliabadi, M.H., Cartwright, D.J. & Rooke, D.P. Fracture-mechanics weight-functions by the removal of singular fields using boundary element analysis, *Int. J. of Fract.*, 1989, **40**, 271-284.

75. Barsoum, R.S. On the use of isoparametric finite elements in linear fracture mechanics, *Int. J. Num. Meth. Engng.* 1976, **10**(1), 25-37.

76. Henshell, R.D. & Shaw, K.G. Crack tip elements are unnecessary, *Int. J. for Num. Meth. Engng.*, 1975, **9**, 495-509.

77. Cruse, T.A & Wilson, R.B. Boundary-integral equation method for elastic fracture mechanics analysis, Report No. AFOSR-TR-78-0355, Pratt & Whitney Aircraft Group, United Technologies Corporation, East Hartford, Conn. 06108, Nov. 1977.

78. Keer, L.M. & Freedman, J.M. Tensile strip with edge cracks, *Int. J. Engng. Sci.*, 1973, **11**, 1265-1275.

79. Raju, I.S. & Newman, J.C., Jr. Three-dimensional finite-element analysis of finite-thickness fracture specimens, NASA, Report TN D-8414, May 1977, 1-40.

80. Newman, J.C., Jr. An improved method of collocation for the stress analysis of cracked plates with various shaped boundaries, NASA Report TN D-6376, 1971.

81. Olaosebikan, L. On the variation of the stress along the front of cracks and surface flaws, *Engng. Frac. Mech.*, 1990, **37**(1), 221-235.

82. Luchi, M.L. & Rizzuti, S. Boundary elements for three-dimensional elastic crack analysis, *Int. J. Num. Meth. Engng.*, 1987, **24**, 2253-2271.

83. Lachat, J.C. & Watson, J.O. Effective treatment of boundary integral equations: a formulation for three-dimensional elastostatics, *Int. J. Num. Meth. Engng.*, 1976, **10**, 991-1005.

84. Sneddon, I.N & Welch, J.T. A note on the distribution of stress in a cylinder containing a penny-shaped crack, *Int. J. Engng. Sci.*, 1963, **1**, 411-419.

85. Rooke, D.P. & Cartwright, D.J. Compendium of stress intensity factors, Procurement Executive, Ministry of Defence, HMSO, 1976.

86. Tada, H., Paris, P.C. & Irwin, G.C. *The Stress Analysis of Cracks Handbook*, Second Edition, Paris Production Incorporation and Del Research Corporation, 1985.

87. Nisitani, H. Body force method for determination of stress intensity factors, *J. Aeronautical Society of India*, February 1985, 21-41.

88. Newman, J.C., Jr. & Raju, I.S. Stress-intensity factors for cracks in three-dimensional finite bodies, NASA-TM-83200, August 1981, 1-49.

89. Shivakumar, K.N. & Newman, J.C., Jr. Stress intensity factors for large aspect ratio surface and corner cracks at a semi-circular notch in a tension specimen, *Engng. Frac. Mech.*, 1991, **38**(6), 467-473.

90. Nikishkov, G.P. & Atluri, S.N. Calculation of fracture mechanics parameters for an arbitrary three-dimensional crack, by the equivalent domain integral method, *Int. J. Num. Meth. Engng.*, 1987, **9**, 1801-1821.

Chapter 7

Application of the boundary element method to anisotropic crack problems

H. Ishikawa

Department of Mechanical and Control Engineering, University of Electro-Communications, 1-5-1, Chofugaoka, Chofu, Tokyo 182, Japan

Abstract

Two- and three-dimensional boundary element analysis of some cracks in anisotropic bodies are carried out. The bodies used are the orthotropic plate, transversely isotropic three-dimensional body, and anisotropic three-dimensional body. To obtain the stress intensity factor of the crack, the elastic stress field near the crack tip in the anisotropic body is considered. Based on this consideration, the stress intensity factors of some cracks are obtained by using the numerical results of the boundary element calculation. The accuracy of the results is very good. This gives a numerical assurance of the usefulness of the boundary element method.

1 Introduction

Recently, various kinds of composite materials have been developed and used for structural components. In the progress of the development, the analysis of the mechanical behaviour of the composite is essential to the design of the composite structures. Some composite materials, such as fiber reinforced plastics and unidirectionally solidified eutectic alloys, show the anisotropic elasticity. The analytical analyses of the elastostatic problem for the anisotropic plate have been carried out by Ikeda,[1] Hayashi,[2,3] Lekhnitskii,[4] and Benjumea & Sikarskie.[5] However, for the actual structural components, such as a body with a crack, numerical techniques of stress analysis are more useful. One of the techniques is the boundary element method. The analysis of the anisotropic plates by the boundary element method have been carried out by Rizzo & Shippy,[6] Cruse & Swedlow,[7] Wilson & Cruse,[8] Snyder *et al.*[9] and Nishimura & Kobayashi.[10] In the present study, the author's attention is given to the crack problems of the anisotropic bodies. The anisotropic bodies used are the anisotropic plate, the transversely isotropic three-dimensional body, and anisotropic three-dimensional body. To obtain the stress intensity factor of the crack, the elastic stress field near the crack tip in the anisotropic body is considered. In the case of the three-dimensional anisotropic body, the effect of the accuracy of the fundamental solutions obtained by the numerical integration on the boundary element analysis is also discussed. Based on these considerations, the stress intensity factors for some cracks, such as a center crack, a surface crack and a penny-shaped crack, are obtained by using the numerical results of the boundary element calculations.

2 Procedure of analysis

One begins by stating the boundary integral equation for the linear elastostatics in its general form. The integral equation for the boundary point of the domain under consideration is

$$cu_j = \int_s (U^*_{ji}t_i - T^*_{ji}u_i)\mathrm{d}s \tag{1}$$

where u_i and t_i are the displacement and traction force, respectively, and c is the value obtained from the condition of the rigid displacement. U^*_{ji} and T^*_{ji} are the fundamental solutions for the displacement and traction force, respectively.

By using the boundary element, eqn (1) is discretized. In the present study, quadratic isoparametric elements with three nodes for two-dimensional problems and eight nodes for three-dimensional problems are used. The outline of the discretization will be explained for the three-dimensional case. By using the interpolation function, Φ_j, the coordinate, displacement, and traction force are given as,

$$x_i(\xi) = \sum_{j=1}^{\infty} \Phi_j(\xi_1, \xi_2)x_i^j$$

$$u_i(\xi) = \sum_{j=1}^{\infty} \Phi_j(\xi_1, \xi_2)u_i^j \tag{2}$$

$$t_i(\xi) = \sum_{j=1}^{\infty} \Phi_j(\xi_1, \xi_2)t_i^j$$

In eqn (2), the subscript i (= 1,2,3) corresponds to the coordinate direction, x_i^j, u_i^j and t_i^j mean the ith component of the coordinate, displacement and traction force at the nodal point, j, respectively. The interpolation function, $\Phi_j(\xi_1, \xi_2)$ with the variables of the homogeneous coordinates, $\xi_1$and$\xi_2$$(-1 \leq \xi_1, \xi_2 \leq 1)$, is given by the following equations.

$$\Phi_1 = (1 - \xi_1)(1 - \xi_2)(-\xi_1 - \xi_2 - 1)/4$$

$$\Phi_2 = (1 - \xi_1^2)(1 - \xi_2)/2$$

$$\Phi_3 = (1 + \xi_1)(1 - \xi_2)(\xi_1 - \xi_2 - 1)/4$$

$$\Phi_4 = (1 - \xi_1^2)(1 + \xi_2)/2$$

$$\Phi_5 = (1 + \xi_1)(1 + \xi_2)(\xi_1 + \xi_2 - 1)/4$$

$$\Phi_6 = (1 - \xi_1^2)(1 + \xi_2)/2$$

$$\Phi_7 = (1 - \xi_1)(1 + \xi_2)(-\xi_1 + \xi_2 - 1)/4$$

$$\Phi_8 = (1 - \xi_1)(1 - \xi_2^2)/2 \tag{3}$$

By using eqns (2) and (3), eqn (1) can be discretized as

$$\begin{aligned} c_{ij}u_j(x^{(n)}) \quad + \quad & \sum_{k=1}^{M}\sum_{m=1}^{8} u_j^m(x^k) \int_{\Gamma_k} T_{ij}(x^{(n)}, x^k)\Phi_m(\xi)|J|\,\mathrm{d}\xi_1\mathrm{d}\xi_2 \\ = \quad & \sum_{k=1}^{M}\sum_{m=1}^{8} t_j^m(x^k) \int_{\Gamma_k} U_{ij}(x^{(n)}, x^k)\Phi_m(\xi)|J|\,\mathrm{d}\xi_1\mathrm{d}\xi_2 \end{aligned} \tag{4}$$

Here, $x^{(n)}$ is the coordinates of a node n, and x^k is, also, the coordinate of the point on an element k. $u_j^m(\xi^k)$ and $t_j^m(\xi^k)$ are the displacement and traction force at a node m on an element k, respectively. Γ_k is the area of an element k. J is Jacobian of the coordinate transformation.

On the other hand, the fundamental solutions in eqn (1) are given for the anisotropic plate by Cruse & Swedlow,[7] for the transversely isotropic body by Pan & Chou[11] and for the anisotropic three-dimensional body by Wilson & Cruse.[8] These are shown in the subsequent sections. Substituting the fundamental solutions into the integral equation, we can get the following equation in the matrix form by eqn (4).

$$[H_{ij}]\{U_j\} \; = \; [G_{ij}]\{t_j\} \tag{5}$$

The condition of the rigid displacement is given by the result that the summation of each row vector in matrix $[H_{ij}]$ is equal to zero. Since c in eqn (1) is in the diagonal elements of the matrix, it is obtained from the summation.

The numerical integral on an each element Γ_k in eqn (4) is done by Gaussian integration formula with two special techniques. The integration for the diagonal element in matrix $[H_{ij}]$ is singular. Then, one of the techniques improves the numerical accuracy of the singular integration by a method of Lachat & Watson.[12] In the method, the region of line (two-dimensional problem) or area (three-dimensional problem) integral on each element is divided into small sub-regions, each of which involves the singular point P. Examples of the sub-regions are shown in Fig. 1. For the case where the corner point of an element is the singular one, two sub-regions, each of which involves the singular point P, are considered, shown in Fig. 1(a). For the case where the middle point along the element boundary is the singular one, the element is divided into three sub-regions, shown in Fig. 1(b).

Another is the effective determination technique of the number of the numerical integration points, that is, Gaussian points. This technique is based on the error estimate method for the Gaussian integral and is also presented by Lachat & Watson.[12]

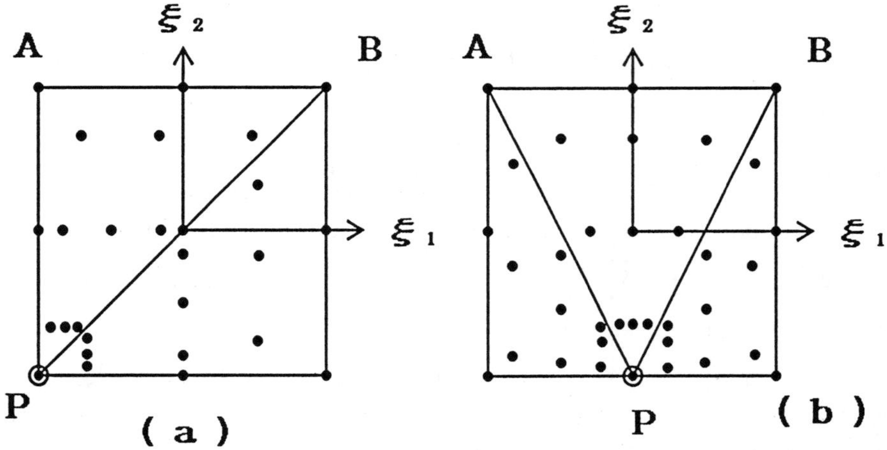

Figure 1: Improvement of the numerical accuracy of the singular integration. (a) A case where the corner point of an element is the singular one. (b) A case where the middle point along the element boundary is the singular one.

The boundary element program in the present study automatically determines the number of 2 to 15 by this technique.

Taking the prescribed boundary values of the displacement and traction force for eqn (5), the equation can be solved. Then, the unknown values of the displacement and traction force on the boundary are obtained.

3 Fundamental solutions

3.1 Orthotropic plate

For the orthotropic plate, such as a unidirectional fiber reinforced plate, the fundamental solutions given by Cruse & Swedlow[7] are used, and those are shown as

$$U_{ji}^{*} = 2\mathrm{Re}[P_{i1}A_{j1}\log Z_1 + P_{i2}A_{j2}\log Z_2] \tag{6}$$

$$T_{ji}^{*} = 2\mathrm{Re}[Q_{i1}(\mu_1 n_x - n_y)A_{j1}/Z_1 + Q_{i2}(\mu_2 n_x - n_y)A_{j2}/Z_2] \tag{7}$$

where Re denotes the real part of the complex number, n_x and n_y are the direction cosines of the unit vector normal to the boundary. P and Q are material constants, defined by

$$\begin{aligned} P_{1k} &= B_{11}\mu_k^2 + B_{12} - B_{13}\mu_k\ , \quad P_{2k} = B_{12}\mu_k + B_{23}/\mu_k - B_{23} \\ Q_{1k} &= \mu_k, \qquad Q_{2k} \;=\; -1 \qquad (k = 1, 2) \end{aligned} \tag{8}$$

In eqn (8), B_{ij} $(i, j = 1, 2, 3)$ is the compliance coefficient (strain-stress relation) of the orthotropic materials, and μ_k is the root of the characteristic equation (Lekhnitskii,[4] Sih, Paris & Irwin[13]) of the material, which is given by

$$B_{11}\mu^4 - 2B_{13}\mu^3 + (2B_{12} + B_{33})\mu^2 - 2B_{23}\mu + B_{22} = 0 \qquad (9)$$

The complex number Z_k ($k = 1, 2$) in eqns (6) and (7) is related to μ_k by the coordinate of x and y, as

$$Z_k = x + \mu_k y \qquad (10)$$

A_{jk} in eqns (6) and (7) is the complex number that is obtained from the multi-valuedness of complex function, log Z_k, and the condition of the single-valuedness of the displacement.

3.2 Transversely isotropic body

In the case of the transversely isotropic three-dimensional body, the fundamental solutions of the displacement and the traction force are obtained from the point force solutions[11] of the stress and displacement at an arbitrary point in the infinite medium. The medium, that is, the transversely isotropic material, has the anisotropy in the x_3 direction that is normal to the $x_1 - x_2$ plane of isotropy.

When the unit point force is applied at the origin in the transversely isotropic body, the displacement, u, and stress, σ_{ij}, at an arbitrary point in the body are as follows:

For the case of unit force applied in x_3 direction,

$$U_j = \sum [v_i A_i X_j/(R_i R_i^*)(\delta_{1j} + \delta_{2j})$$

$$+ (C_{11} - C_{44}v_i^2)A_i/\{(C_{13} + C_{44})R_i\}\,\delta_{3j}] \qquad (j = 1, 3) \qquad (11)$$

$$\sigma_{jj} = \sum [(C_{11} - C_{13}v_i^2 k_i)A_i v_i Z_i/R_i^3 -$$

$$2C_{66}v_i A_i\,(1/(R_i R_i^*) - x_{3-j}^2\{1/(R_i^3 R_i^*) + 1/(R_i^2 R_i^{*2})\})]$$

$$\sigma_{33} = \sum [(C_{13} - C_{33}v_i^2 k_i)A_i v_i Z_i/R_i^3] \qquad (12)$$

$$\sigma_{12} = -\sum [2C_{66}v_i A_i X_1 X_2\{1/(R_i^3 R_i^*) + 1/(R_i^2 R_i^{*2})\}]$$

$$\sigma_{3j} = -\sum [C_{44}(C_{11} + C_{13}v_i^2)A_i/\{(C_{13} + C_{44})R_i^3\}\,X_j] \qquad (j = 1, 2)$$

For the case of unit force applied in x_1 direction,

$$U_j = \sum [2v_i A_i'(\delta_{1j}/R_i^* - X_1 X_j/(R_i R_i^{*2})]$$

$$-(-1)^j \mathrm{D}\{\delta_{1j}/R_3^* - X_2 X_{3-j}/(R_3 R_3^{*2})\} \qquad (\varphi = 1, 2)$$

$$U_3 = -\sum [2(C_{11} - C_{44}v_i^2)A_i' X_1/\{R_i R_i^*(C_{13} + C_{44})\}] \qquad (13)$$

$$\sigma_{jj} = \sum \left[-2C_{44}v_i^3(1+k_i)A_i'X_1/R_i^3 - 4C_{66}v_iA_i'X_1\left(-(1+2\delta_{2j})/(R_iR_i^{*2}) + X_{3-j}^2\ \{1/(R_i^3R_i^{*2}+2/(R_i^2R_i^{*3})\}\right)\right]$$

$$-2\mathrm{D}(-1)^iC_{66}X_1\left(-1/(R_3R_3^{*2}) + X_2^2\{1/R_3^3R_3^{*2}) + 2/(R_3^2R_3^{*3})\}\right] \qquad (j=1,2)$$

$$\sigma_{33} = -\sum\left[2(C_{13}-C_{33}v_i^2k_i)A_i'v_iX_i/R_i^3\right]$$

$$\sigma_{3j} = \sum\left[2C_{44}(C_{11}+C_{13}v_i^2)/(C_{13}+C_{44})A_i'\left[-\delta_{1j}/R_iR_i^*\right) + X_1X_j\{1/(R_i^3R_i^*) + 1/(R_i^2R_i^{*2})\}\right]\right] - (-1)^jC_{44}v_3\mathrm{D}\left[-\delta_{1j}/(R_3R_3^*) + X_2X_{3-j}\left\{1/(R_3^3R_3^*) + 1/(R_3^2R_3^{*3})\right\}\right] \qquad (j=1,2) \qquad (14)$$

$$\sigma_{12} = \sum\left[4C_{66}v_iA_i'X_2\left(-1/(R_iR_i^{*2}) + X_1^2\{1/(R_i^3R_i^{*2}) + 2/(R_i^2R_i^{*3})\}\right)\right] + C_{66}\mathrm{D}X_2\left(-2/(R_3R_3^{*2}) + (X_2^2-X_1^2)\{1/(R_3^3R_3^{*2}) + 2/(R_3^2R_3^{*3})\}\right)$$

Solutions for the case of unit force applied in x_2 direction are the same as those for the case of x_1 direction.

Meanings of the symbols in equations mentioned above are as follows:

C_{ij}: Young's modulus, δ_{ij}: Kronecker's delta, $P_{13} = (C_{11}C_{33})^{0.5}$

$$v_j = -(-1)^j\left[(P_{13}+C_{13})(P_{13}-C_{13}-2C_{44})/(4C_{33}C_{44})\right]^{0.5} + \left[(P_{13}-C_{13})(P_{13}+C_{13}+2C_{44})/(4C_{33}C_{44}\right]^{0.5} \qquad (j=1,2)$$

$$v_3 = (C_{66}/C_{44})^{0.5}, \quad Z_i = v_iX_3 \ (i=1,3), \quad R_i = (X_1^2+X_2^2+Z_i^2)^{0.5} \quad (i=1,3)$$

$$R_i^* = R_i + Z_i \ (i=1,3), \quad k_i = (C_{11}v_i^{-2} - C_{44})/(C_{13}+C_{44}) \quad (i=1,3),$$

$$\mathrm{D} = 1/(4\pi C_{44}v_3)$$

$$v_1A_1 = -v_2A_2 = (C_{13}+C_{44})/\left\{4\pi C_{33}C_{44}(v_2^2-c_1^2)\right\}$$

$$A_j' = -(-1)^j(C_{44}-C_{33}v_j^2)/\left\{8\pi C_{33}C_{44}(v_2^2-v_1^2)v_j^2\right\} \qquad (j=1,2)$$

3.3 Three-dimensional anisotropic body

The fundamental solutions for the three-dimensional anisotropic case is obtained, as follows.[8] By using the summation convention, the equilibrium equation of the three-dimensional elastic body can be expressed as

$$C_{ijkm}U_{lk,jm} = \delta_{il}\delta(X,Y) \qquad (15)$$

where C_{ijkm} $(i,j,k,m = 1,2,3)$ is the elastic constant, U is the displacement, the subscripts, l and k, mean the directions of the unit force and the displacement, respectively, and $U_{lk,jm}$ is the partial derivative of the displacement, U_{lk}, with respect to $x_j$andx_m. X and Y mean the observation point and the source point of the unit force, respectively. δ_{il} is the Kronecker delta, and $\delta(X,Y)$ is the Dirac's delta function.

Taking the Fourier transform, $U(\xi)$, of eqn (15) and its inverse Fourier transform, $U(X)$, and by using the characteristics of the delta function, eqn (15) can be reduced to

$$U_{ij}(X) = \int_c K_{ij}^{-1} d\phi / 8\pi^2 R = G_{ij}(\theta_1, \theta_2)/8\pi^2 R \tag{16}$$

where

$$K_{ij} = C_{iljm}\xi_l \xi_m \tag{17}$$

The line integration path C is the unit circle which lies on the plane perpendicular to the vector $\mathbf{X}$ from the origin to the observation point. R is the absolute value of $\mathbf{X}$. Equation (16) is the fundamental solution of the displacement of the anisotropic body which has the elastic constant, C_{iljm}. Equation (16) is a function of the directions, θ_1 and θ_2, of the vector $\mathbf{X}$ and the angle ϕ as shown in Figs 2 and 3. ϕ is defined on the plane perpendicular to $\mathbf{X}$. The starting line for measuring ϕ can be arbitrary. The coordinate at the arbitrary point on the integration path is related to the angles, θ_1, θ_2andϕ, as

$$\left\{ \begin{array}{c} \xi_1 \\ \xi_2 \\ \xi_3 \end{array} \right\} = \left\{ \begin{array}{cc} \cos\theta_2 \cos\theta_1 & -\sin\theta_2 \\ \sin\theta_2 \cos\theta_1 & \cos\theta_2 \\ -\sin\theta_1 & 0 \end{array} \right] \left\{ \begin{array}{c} \cos\phi \\ \sin\phi \end{array} \right\} \tag{18}$$

The representation, eqn (16), cannot generally be evaluated in closed form. For an isotropic material, it reduces analytically to the well-known Kelvin's solution. For an anisotropic material, one of the methods to evaluate the function, U_{ij}, is a direct numerical method. K_{ij} can be obtained from eqn (17). Then, taking the inverse matrix, K_{ij}^{-1}, by the Cramer's formula, the fundamental solution can be evaluated by the numerical integration and the distance, R, from the source point of the point load to the observation point. The representation for the traction force solution, T_{ij}, can be derived from the derivative of the fundamental solution of the displacement, U_{ij}, and the elastic constant. Taking the partial derivative of eqn (16) with respect to x_k,

$$\frac{\partial U_{ij}}{\partial x_k} = -\frac{x_k G_{ij}}{8\pi^2 R^3} + \frac{1}{8\pi^2 R}\left\{ \frac{\partial G_{ij}}{\partial \theta_1}\frac{\partial \theta_1}{\partial x_k} + \frac{\partial G_{ij}}{\partial \theta_2}\frac{\partial \theta_2}{\partial x_k} \right\} \tag{19}$$

On the other hand, from the characteristics of the inverse matrix, K_{ij}^{-1}, we can get

$$K_{ij}K_{ij}^{-1} = \delta \tag{20}$$

The partial derivative of eqn (20) with respect to x_k gives

$$\frac{\partial K_{ij}^{-1}}{\partial \theta_a} = -K_{ij}^{-1}\frac{\partial K_{ij}}{\partial \theta_a}K_{ij}^{-1} \tag{21}$$

From eqns (19) and (20), $\partial G_{ij}/\partial \theta_a$, is obtained. Substituting the result into eqn (19), the strain component can be obtained. Then, from the strain and the elastic constant, we can finally get the representation, T_{ij}, for the fundamental solution of traction force.

As mentioned above, the fundamental solutions for the three-dimensional anisotropic body are not directly computable, and can be numerically obtained. In the present

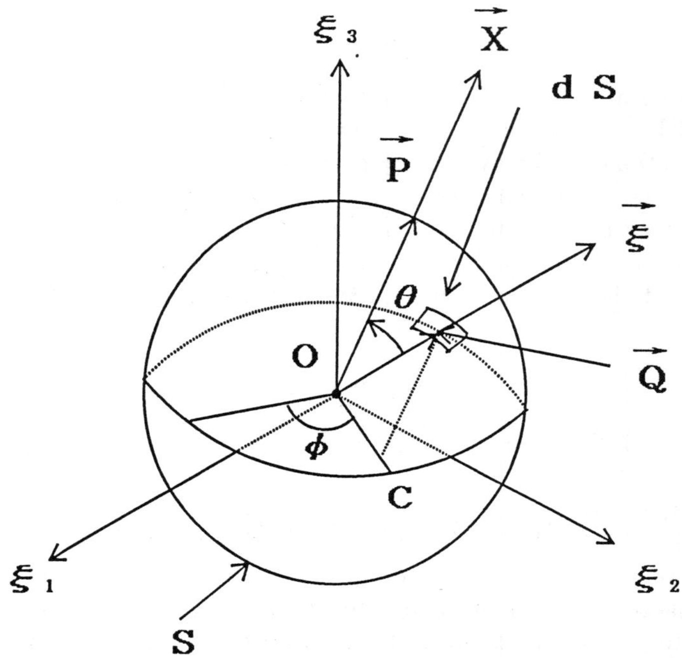

Figure 2: Transform to the sphere coordinate system

Table 1: Effect of the number of integration points on the accuracy of the fundamental solution

	Isotropic body		Anisotropic body	
n	Displacement	Traction	Displacement	Traction
2	2.3913552D-6	5.2714083D-3	4.3564570D-5	3.8788446D-3
3	1.7599300D-6	2.9176217D-3	3.4361388D-5	3.5095851D-3
5	1.7599300D-6	2.9176218D-3	3.4518163D-5	2.6253455D-3
7	1.7599300D-6	2.9176218D-3	3.4510469D-5	3.0861472D-3
10	1.7599300D-6	2.9176218D-3	3.4518163D-5	3.0827040D-3
15	1.7599300D-6	2.9176218D-3	3.4510159D-5	3.0851408D-3
20			3.4510157D-5	3.0851284D-3
30			3.4510159D-5	3.0851408D-3
40			3.4510159D-5	3.0851408D-3
50			3.4510159D-5	3.0851408D-3

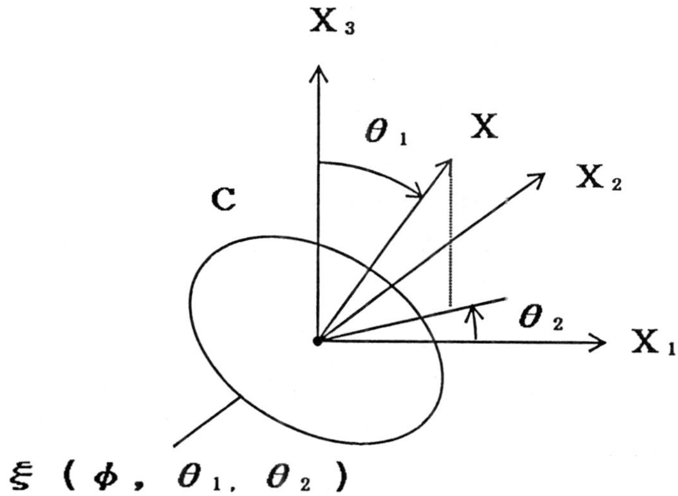

Figure 3: Position of the observation point and the integration point

study, the trapezoidal rule for the numerical integration is used. To discuss the accuracy of the integration, an example of the observation point is adopted. The example point has the coordinates of x_1 = 1mm, x_2 =2 mm, x_3 = 3mm, and the normal directions of $n_1 = n_2 = n_3 = 1/\sqrt{3}$. The effects of the numbers, n, of the integration points on the accuracy are shown in Table 1 for the isotropic and anisotropic bodies. The isotropic case has Young's modulus of 206 GPa and Poisson's ratio of 0.3. The anisotropic case has the different Young's modulus, E_1, E_2, E_3, in each direction of the coordinate axes, x_1, x_2, x_3, that is, E_1 of 9.8 GPa, E_2 of $2E_1$, and E_3 of $3E_1$.

From Table 1, for the isotropic case, the lesser number, n, brings good convergence of the fundamental solutions of the displacement and traction force. However, for the anisotropic case, the larger number, n, is necessary. The fundamental solutions are fairly converged with the number of 30. In the present research, the numerical integrations are carried out with the number of 40.

On the other hand, the numerical value of the fundamental solutions depends on the geometry of the given problem and the element mesh pattern for the problem. Then, if we carry out the numerical integration for all the calculations to get the value of the fundamental solutions, even the case of one mesh pattern of the given problem requires a huge amount of the calculation. So, in the present study, the author uses an interpolation technique.[8] Since the numerical integration is carried out for the angles, θ_1 and θ_2, in Fig. 3, determining the position of the observation point, the two-dimensional table for the discrete values of θ_1 and θ_2 with the equal interval of $\Delta\theta_1 = \Delta\theta_2 = \Delta\theta$ is made. The numerical values of the fundamental solutions for the actual observation points are obtained by the interpolation from the values

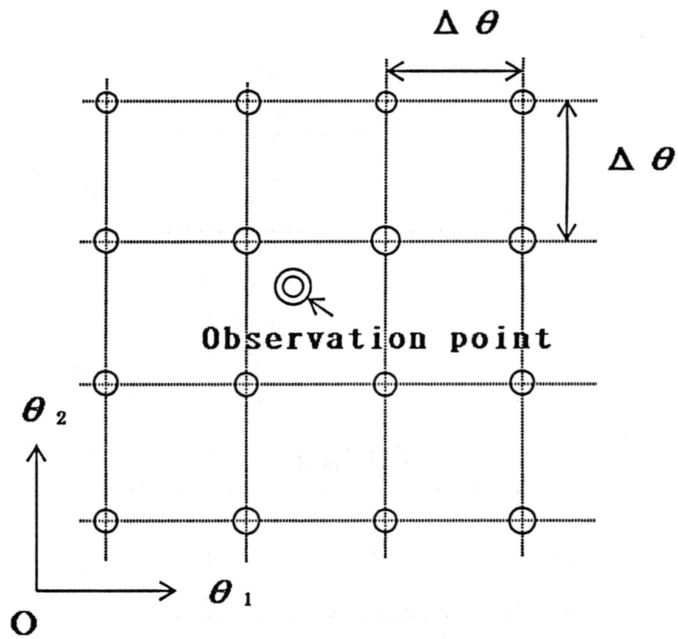

Figure 4: Interpolation of the fundamental solution

in the table. The Lagrange's interpolation technique is used to obtained the value of the fundamental solutions. Namely, the value at the actual observation point is interpolated from the values at 16 points around the observation point in the table, as shown in Fig. 4.

To discuss the effect of the $\Delta\theta$ on the accuracy of the boundary element calculation, the five kinds of $\Delta\theta$ (10°, 15°, 22.5°, 30°, 45°) are used. By using the $\Delta\theta$, the following example problems are solved:

(a) an uniform tension problem of a square rod, shown in Fig. 5;
(b) an uniform tension problem of a square rod with a center through crack, shown in Fig. 6.

The materials of the example problems are the isotropic materials that have Young's modulus of 206 GPa and Poisson's ratio of 0.3. These problems have the geometrical symmetries. So, one eighth region of the body is analysed. In the problem (a), the displacement at the loading point is calculated, and its error, E_{r1}, is discussed by comparing with the analytical value. In the problem (b), the stress intensity factor at a point on the crack front is calculated by the method mentioned in the following section, and its error, E_{r2}, is discussed by comparing with the numerical results (Yagawa *et al.*[14]) of the finite element method. These results are shown in Table 2, it is found that the results are almost converged with $\Delta\theta$ of 15°. In the present study, the $\Delta\theta$ of 10° is used for the boundary element calculation of the anisotropic body.

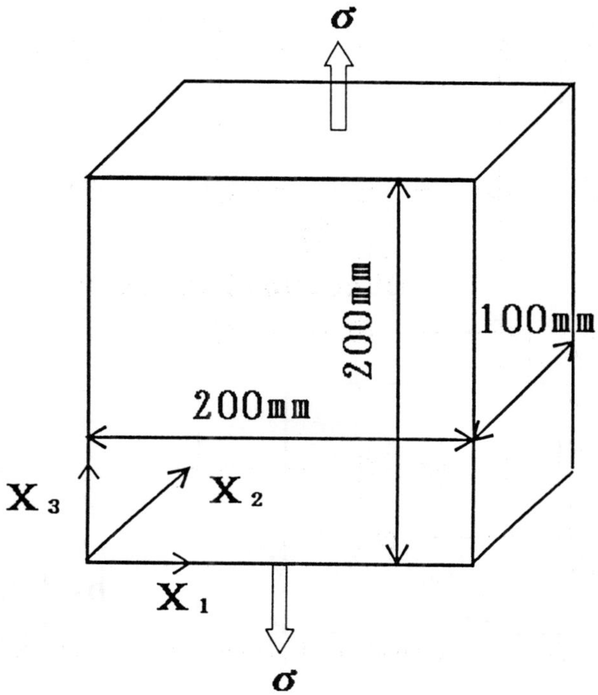

Figure 5: Uniform tension problem of a square rod

Table 2: Effect of the interval of the interpolation of the fundamental solutions on the boundary element solutions

$\Delta\ \theta$	Er_1 (%)	Er_2 (%)
10°	0.04	1.63
15°	0.03	1.70
22.5°		2.37
30°	0.68	3.86
45 °		10.61

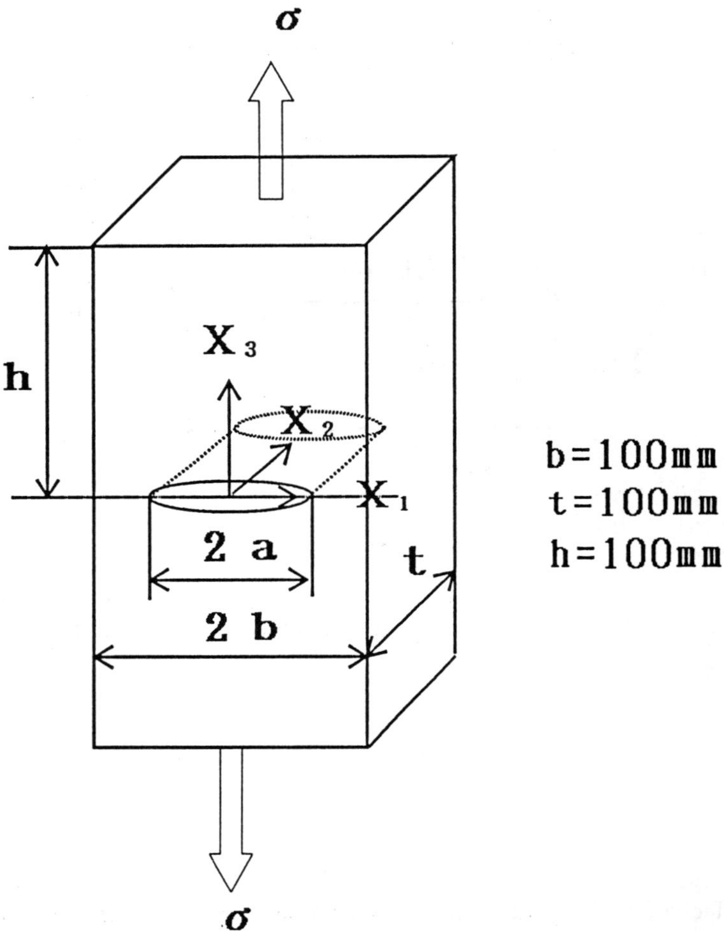

Figure 6: Uniform tension problem of a square rod with a center through-crack

4 A method to obtain the stress intensity factor

To obtain the stress intensity factor of a crack in the anisotropic body, the stress field near the crack tip in the anisotropic body is discussed. The author considers the stress field in a plane which is normal to the crack front. The displacement near the crack tip in an anisotropic plate is expressed in terms of the stress functions, $\Psi_k(z_k), (k = 1, 2)$ as

$$u = \mathrm{Re}[P_{11}\Phi_1(z_1) + P_{12}\Phi_2(z_2)], \quad v = 2\mathrm{Re}[P_{21}\Phi_1(z_1) + P_{22}\Phi_2(z_2)] \qquad (22)$$

where P_{ij} is a material constant which is given by eqn (8). The singular terms of the stress functions are given by

$$\Phi_j(z_j) = 2\lambda_{j0}(z_j - z_o)^{1/2} \quad (j = 1, 2) \qquad (23)$$

where z_j is the complex coordinate system which has the origin at the center of the crack, z_0 is the complex coordinate of the crack tip. The coefficient, λ_{j0}, is defined as

$$\lambda_{10} = -\left[\left\{\mu_2/(2\sqrt{2(\mu_1 - \mu_2)}\right\}(K_I + K_{II}/\mu_2)\right]$$

$$\lambda_{20} = \left[\left\{\mu_2/2\sqrt{2(\mu_1 - \mu_2)})\right\}(K_I + K_{II}/\mu_1)\right] \qquad (24)$$

In eqn (24), K_I and K_{II} are the stress intensity factors of mode I and mode II, respectively. By using the polar coordinate system which have the origin at the crack tip,

$$z_j - z_0 = \mathrm{r}e^{i\theta} \qquad (25)$$

Substituting eqns (23), (24) and (25) into eqn (22), we can obtain the crack opening displacement, v.

$$v = \sqrt{(2r/\pi)}K_I B_{22}\mathrm{Im}[-(\mu_1 + \mu_2)/\mu_1\mu_2] \qquad (26)$$

where Im means the imaginary part of the complex number. Equation (26) shows that the relationship between the crack opening displacement, v, and the square root of the distance, r, from the crack tip is linear, and we can obtain the value of K_I by the slop of the relationship. The crack opening displacement can be obtained from the displacement at nodal points of the elements near the crack tip. In the present study, to improve the accuracy of the value of K_I, the two types of crack element, A and B (shown in Fig. 7) which show the stress singularity of the crack are used at the crack tip. One (element A) is the quarter element, where a middle point node is moved to a quarter point on the element side. Another (element B) that is a neighbour of the crack tip element, A, can also gives the stress singularity by moving a middle node of the element to the position, r_2, defined by eqn (27) (Lynn & Ingraffea[16]).

$$r_2 = (\sqrt{r_1} + \sqrt{r_3})^2/4 \qquad (27)$$

where r_1 and r_3 are the distance from the crack tip to either side node of the element, shown in Fig. 7. These crack elements are used for every crack problems analysed in the present study.

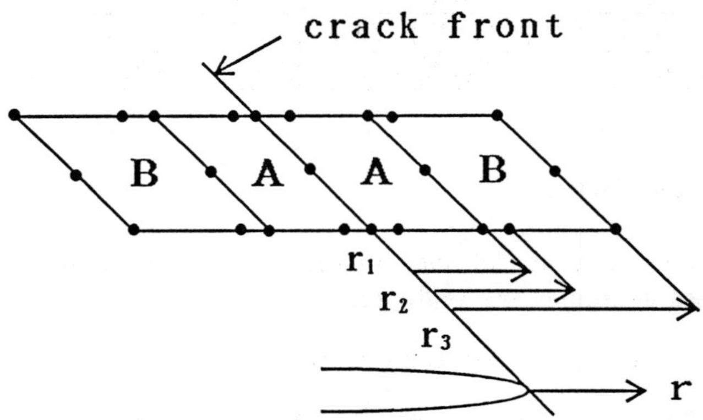

Figure 7: Crack tip element

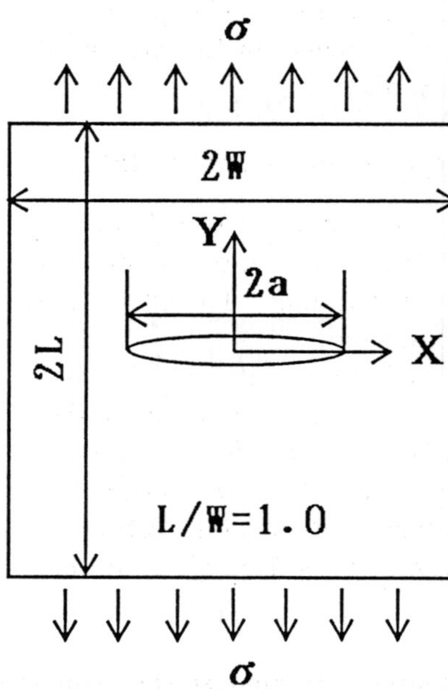

Figure 8: Uniform tension problem of a square plate with a center crack

Table 3: Anisotropic plate materials

	E_x (GPa)	E_y (GPa)	G_{xy} (GPa)	ν_x
(a) CFRP	9.02	142	4.81	0.021
(b) GFRP	11.99	33.8	4.33	0.117
(c)	$E_x/E_y = 2.5$		7.07	0.33

5 Numerical examples

In order to discuss the accuracy of the numerical results obtained by the present method, the following examples are analysed.

1. A centre crack in a square anisotropic plate;
2. A centre through-crack in a transversely isotropic square rod;
3. A semi-circular surface crack in a transversely isotropic square rod;
4. A centre through-crack in a three-dimensional anisotropic square rod;
5. A penny-shaped crack in a three-dimensional anisotropic body.

In the calculation, .there is a problem that the traction discontinuities over the boundary cannot be properly simulated. The problem is caused by the interpolation function of isoparametric elements. In the present study, the concept of double nodes[15] is used to solve the problem.

5.1 A center crack in a square anisotropic plate

An uniform tension problem of a square plate with a center crack (Fig. 8) is analysed by the present method. From the numerical results of the crack opening displacement, the stress intensity factor, K_I, is obtained. The effects of the material anisotropy and the crack length on K_I are discussed.

The number of elements is 38. In the present calculation, three kinds of anisotropic materials are used and their mechanical properties are shown in Table 3. In this table, CFRP and GFRP are the carbon fiber reinforced epoxy and the glass fiber reinforced epoxy, respectively. And, also, E_m, G_{xy}andν_{xy} mean Young's modulus in the m direction ($m = x, y$), shear modulus and Poisson's ratio, respectively. In the case of the third material in the table, the ratio of Young's modulus in the x and y directions is given.

Bowie & Freeze[17] used this material to analyse the same crack problem by the collocation method. Thus, compared with their results, the accuracy of the present results can be discussed. Crack length ratio, a/w, used are 0.1, 0.3, 0.5, 0.7.

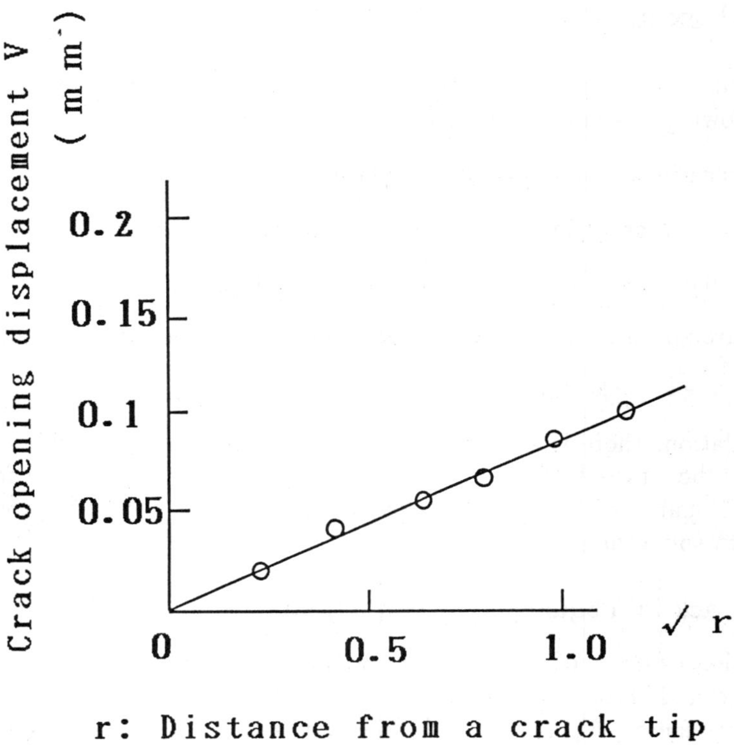

Figure 9: Relationship between the crack opening displacement and the square root of the distance r from the crack tip

An example of the results of the relationship between the crack opening displacement and the square root of the distance from the crack tip is shown in Fig. 9 (material (a) in Table 3, $a/w = 0.5$). From the slope of the relationship, and eqn (26), the stress intensity factor, K_I is obtained. The numerical results of K_I are shown in Table 4. In this table, K_I is normalized as $F_I = K_I/\sigma\sqrt{\pi a}$. ($\sigma$: applied stress, a: half crack length). Compared with the results of material (c) by Bowie & Freeze, we can find that the accuracy of the present results are fairly high.

5.2 A center through-crack in a transversely isotropic square rod

A transversely isotropic three-dimensional rod with a through-crack under uniform tension, as shown in Fig. 6, is analysed. From the numerical results of the crack opening displacement, the stress intensity factor, K_I, of the crack is obtained, and the effects of the material anisotropy and crack length on the stress intensity factor are discussed. The values of K_I are obtained at a quarter point ($x_2 = 25$ mm in Fig. 6) of the crack front width (100 mm). Because of the symmetry of the problem, one eighth region of the cracked body is analysed. Crack length ratio, a/b (2a: crack length, 2b: square rod width), is 0.4 and 0.5. The number of elements is 58 for the case of a/b of 0.5, and 62 for the case of a/b of 0.4. In Fig. 6, the $x_1 - x_2$ plane is the isotropic one and the x_3 is the anisotropic axis. Five kinds of transversely isotropy are analysed, and their material elastic constants are given in Table 5. In Table 5, (a-2) is a transversely isotropic material that has material constants close to the completely isotropic material, (a-1), that consists of epoxy. Namely, the difference in elastic constants between (a-1) and (a-2) is taken in shear modulus, and it is very small. Because the fundamental solutions in the present study are not available for the completely isotropic materials, and there seem to be no references on the stress intensity factor, K_I, of the transversely isotropic body, the material, (a-2), is used. The value of K_I of a crack for the material (a-2) may be compared with that for the material (a-1), obtained by the finite element method. The material, (e), is the glass fiber reinforced epoxy mentioned above. The materials, (b), (c) and (d), are the transversely isotropic materials that have the elastic constants making linear change of E_{x1}, E_{x3}, and G_{x1x3} between (a-1) and (e).

The problem given in Fig. 6 has the geometrical symmetries. So, one eighth region of the body of the problem is analysed. An example of the boundary element mesh for the problem is shown in Fig. 10. The stress intensity factor, K_I, is obtained from the relationship between the crack opening displacement, v, and the square root of r. The numerical results is given in Table 6. The values in this table are the normalized stress intensity factor, F_I. For a case of completely isotropic material, (a-1), with crack length ratio, a/b, of 0.5, the result of the finite element computation (Yagawa *et al.*[14]) is given in Table 6. From the comparison of the materials (a-1) and (a-2) in Table 6, it is found that the numerical results in the present study are obtained with fairly good accuracy. Table 6, also, shows that the stronger the anisotropy in the x_3 direction is, the greater the K_I value is. These results are similar to those of anisotropic plates, mentioned above.

Table 4: Numerical results of the anisotropic crack plate

	a/w	F_1 Present results	F_1 Results by Bowie *et al.*
(a) CFRP	0.1	1.095	
	0.3	1.501	
	0.5	2.082	
	0.7	2.955	
(b) GFRP	0.1	1.043	
	0.3	1.219	
	0.5	1.591	
	0.7	2.021	
(c)	0.1	1.016	1.01
	0.3	1.084	1.08
	0.5	1.235	1.24
	0.7	1.569	1.53

Table 5: Elastic constants of the transversely isotropic materials

	(a-1)	(a-2)	(b)	(c)	(d)	(e)
E_{x1} (MPa)	2452	2452	4835	7228	9611	11994
E_{x2} (MPa)	2452	2452	10297	18143	25989	33834
G_{x1x2} (MPa)	922	902	1765	2618	3481	4335
ν_{x1x2}	0.33	0.33	0.33	0.33	0.33	0.33
ν_{x1x2}	0.33	0.33	0.33	0.33	0.33	0.33

Table 6: Non-dimensional stress intensity factor of a center crack of the square rod of transversely isotropic material (analytical fundamental solution)

a/b	(a-1)	(a-2)	(b)	(c)	(d)	(e)
0.4		1.211	1.213	1.243	1.256	1.263
0.5	1.347	1.337	1.360	1.404	1.422	1432

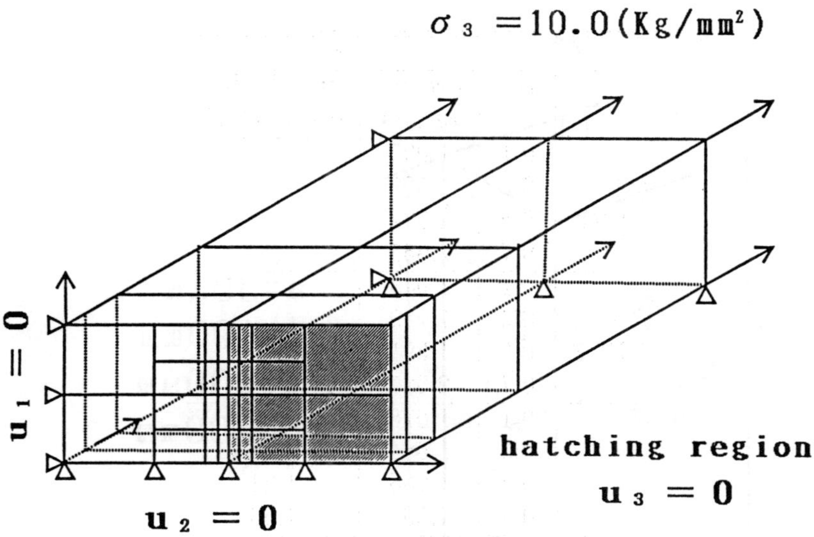

Figure 10: Element mesh pattern for the uniform tension problem of a square rod with a center through-crack

5.3 A semi-circular surface crack in a transversely isotropic square rod

The normalized stress intensity factors, F_I, of a semi-circular surface crack in a three-dimensional square rod under uniform tension, shown in Fig. 11(a), are obtained at five positions, θ, on the crack front. The θ, shown in Fig. 11(b), is 0°, 22.5°, 45°, 67.5°, 90°. Because of the symmetry of the problem, a quarter region of the cracked body is analysed. An example of the boundary element mesh is shown in Fig. 12. The number of elements is 61. The material of the body is the glass fiber reinforced epoxy, mentioned above, and has the transversely isotropic property, that is, anisotropic direction is similar to that in the previous case. Crack size, a/b, is 0.3. The numerical results of the normalized stress intensity factors, F_I, are shown in Table 7. The value of F_I has the least value at the bottom of surface crack, and takes greater value along the crack front to the surface point of the body.

5.4 A center through-crack in a three-dimensional anisotropic square rod

A center through-crack in a square rod under uniform tension, σ, shown in Fig. 6, is solved. The material is the transversely isotropic material being one of the general anisotropic three-dimensional materials. The material used is the unidirectional glass-fiber reinforced epoxy, GFRP, the elastic properties of which are the same as that in the previous Section 5.3. Because of the symmetry of the problem, one eighth region of the cracked body is analysed. An example of the number of elements is 58

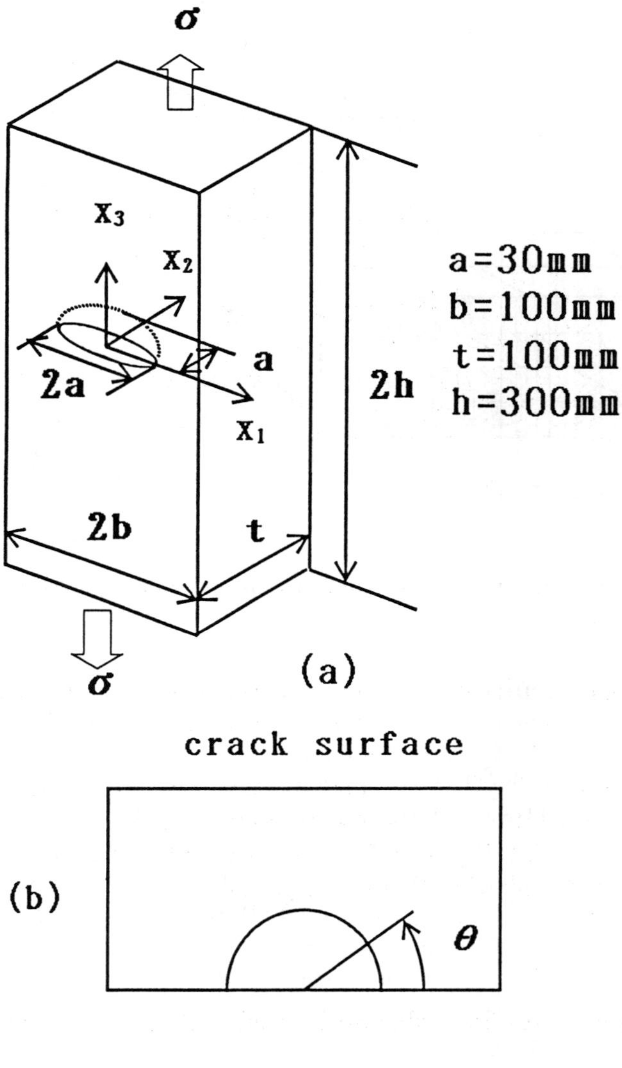

Figure 11: (a) A semi-circular surface crack in a square rod under uniform tension; (b) A semi-circular surface crack.

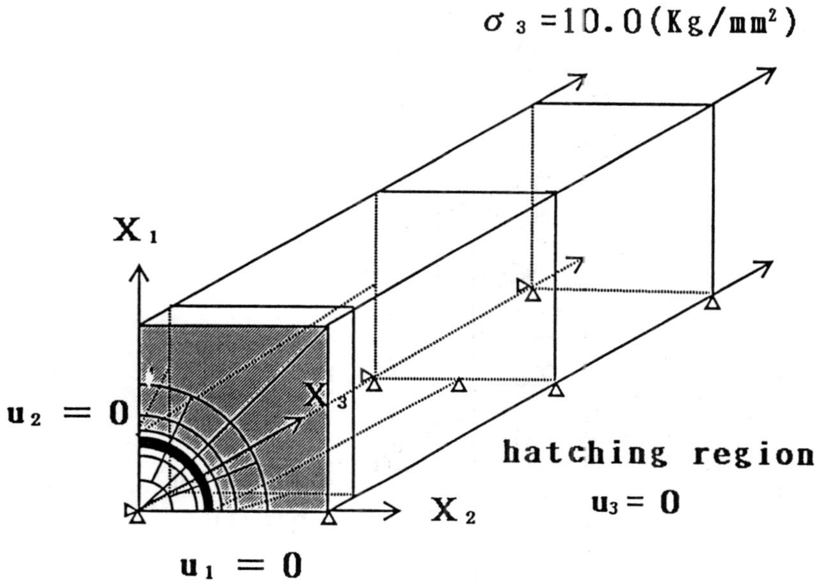

Figure 12: Element mesh pattern for the uniform tension problem of a square rod with a semi-circular surface crack

for a case of crack length ratio, a/b (2a: crack length, 2w: square rod width) of 0.5. The ratio used is 0.4, 0.5 and 0.6. The stress intensity factor, K_I, is obtained by the same method as the cases of the previous examples.

The results are shown in Table 8. The present problem was analysed with the analytical fundamental solutions of the transversely isotropic material presented in the previous Section 5.2. F_I^* in Table 8 is the value given by using the analytical fundamental solutions. Comparing these results, the difference between them is less than 0.6%. Then, it is said that the present calculation is carried out with fairly good accuracy.

5.5 A penny-shaped crack in a three-dimensional anisotropic body

A uniform tension problem of a cylindrical rod with a penny-shaped crack, shown in Fig. 13, is analysed. The crack radius, a/b (a: radius of the penny-shaped crack, b: radius of the cylinder), is 0.5. The distribution of the normalized stress intensity factors, F_I, along the crack front is obtained. The position to determine the value of F_I is defined by the angle, θ, measured from the x_1-axis, and shown in Table 9. Because of the symmetry of the problem, one eighth of the region of the body is analysed. An example of the boundary element mesh for the problem is shown in Fig. 14. The mechanical properties of the material used are shown in Table 10. The numerical results are given in Table 9. The non-dimensional stress intensity factor, F_I, is increased in the counter-clockwise directions from the x_1-axis.

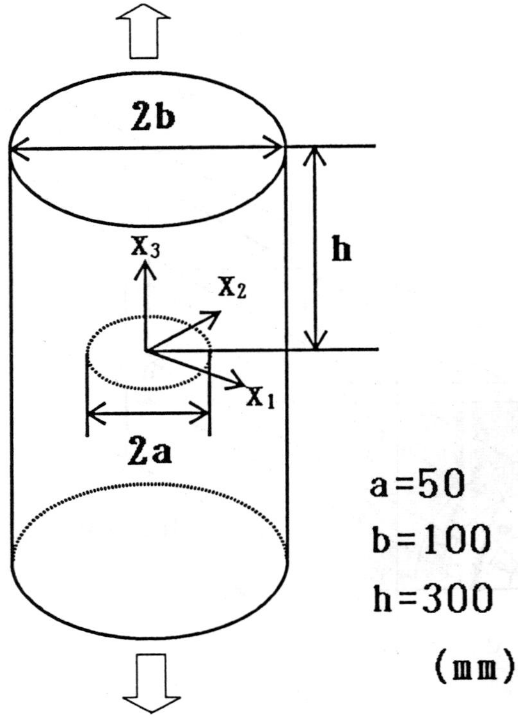

Figure 13: Uniform tension problem of a cylindrical rod with a penny-shaped crack

Table 7: Non-dimensional stress intensity factor of a semi-circular surface crack

a/b	0°	22.5°	45°	67.5°	90°
0.3	0.6594	0.5811	0.5766	0.5686	0.5671

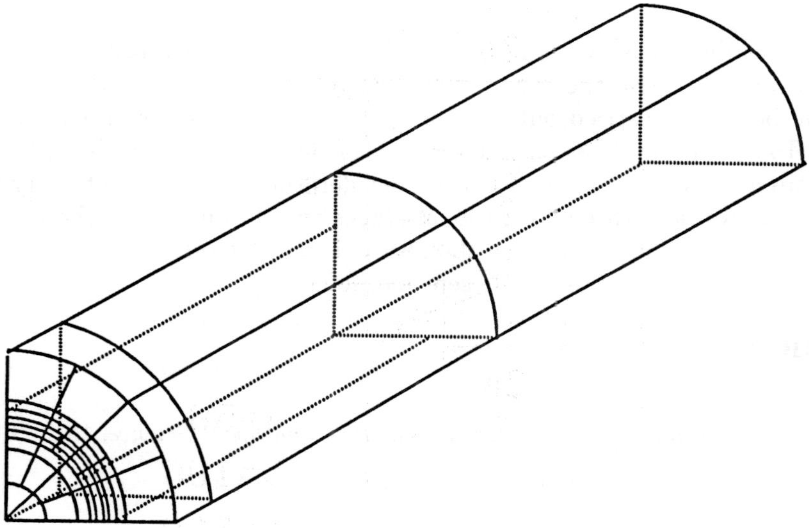

Figure 14: Element mesh pattern for the cylindrical rod with a penny-shaped crack

Table 8: Non-dimensional stress intensity factor of a center crack of the square rod of transversely isotropic material (numerical fundamental solution)

a/b	F_1	F_1^*
0.4	1.270	1.263
0.5	1.434	1.432

Table 9: Non-dimensional stress intensity factor of a penny-shaped crack of the cylindrical rod

$\theta(°)$	0.0	22.5	33.75	45.0	56.25	67.5	90.0
F_1	0.598	0.608	0.612	0.617	0.624	0.631	0.641

Table 10: Three-dimensional anisotropic material

$Ex1$	$Ex2$	$Ex3$	Gij	νij
9.8 GPa	$3Ex1$	$5Ex1$	3.92 GPa	0.33

6 Concluding remarks

Two- and three-dimensional boundary element analyses of the anisotropic bodies, such as the anisotropic plates, the transversely isotropic body and the three-dimensional anisotropic body, are carried out. Numerical examples of some crack problems are analysed. To obtain the stress intensity factors of the crack, the elastic stress field near the crack tip is considered. Based on this consideration, the stress intensity factors of some kinds of two- and three-dimensional cracks are obtained from the boundary element calculations. The accuracy of the results is fairly good. This gives a numerical assurance of the usefulness of the present method.

References

1. Ikeda, K. *J. of the Japan Soc. for Aeronautical and Space Science*, 1942, **9**, 1209.

2. Hayashi, Y. *Trans. JSME*, 1959, **25**, 1125.

3. Hayashi, Y. *Trans. JSME*, 1960, **26**, 516.

4. Lekhnitskii, S.G. *Theory of Elasticity on An Anisotropic Elastic Body*, Holden-Day, Inc., San Francisco, 1963.

5. Benjumea, R. & Sikarskie, D.J. *J. Appl. Mech.*, 1972, 801.

6. Rizzo, F.J. & Shippy, D.J. *J. Composite Materials*, 1970, **4**, 36.

7. Cruse, T.A. & Swedlow, J.L. Air Force Rep., No. AFML-TR-71-268, 1971.

8. Wilson, R.B. & Cruse, T.A. *J. Numer. Meth. Engng*, 1978, **12**, 1383.

9. Snyder, M.D. & Cruse, T.A. *Int. J. Fract.*, 1975, **11**, 315.

10. Nishimura, N. & Kobayashi, S. *Proc. 5th Int. Conf. Boundary Element*, 1983, pp. 345.

11. Pan, P.C. & Chou, T.W. *J. Appl. Mech.*, 1976, **43**, 608.

12. Lachat, J.C. & Watson, J.O. *Int. J. Numer. Meth. Engng*, 1976, **10**, 991.

13. Sih, G.C., Paris, P.C. & Irwin, G.R.*Int. J. Fract. Mech.*, 1975, **1**, 189.

14. Yagawa, G., Ichinomiya, M. & Ando, Y. *Preprint Japan. Soc. Mech. Engineers*, (in Japanese), 1977, **770**(1), 29.

15. Brebbia, C.A., Telles, J.C.F. & Wrobel, L.C. *Boundary Element Techniques*, Springer-Verlag, 1975, pp. 204.

16. Lynn, P.P. & Ingraffea, A.R. *Int. J. Numer. Methods Eng*, 1978, **12**, 1031.

17. Bowie, O.L. & Freeze, C.E. *Int. J. Fract. Mech.*, 1972, **8**, 49.